The Composite Materials Han

VOLUME 2

Polymer Matrix Composites: Materials Properties

CRC Press
Taylor & Francis Group
Boca Raton London New York

CRC Press is an imprint of the
Taylor & Francis Group, an **informa** business

CRC Press
Taylor & Francis Group
6000 Broken Sound Parkway NW, Suite 300
Boca Raton, FL 33487-2742

First issued in paperback 2019

ISBN 13: 978-0-367-44738-0 (pbk)
ISBN 13: 978-1-56676-970-9 (hbk)

Visit the Taylor & Francis Web site at
http://www.taylorandfrancis.com

and the CRC Press Web site at
http://www.crcpress.com

ACKNOWLEDGEMENT

The services necessary for the development and maintenance of the Composite Materials Handbook (MIL-HDBK-17) are provided by the handbook Secretariat, Materials Sciences Corporation. This work is performed under contract with the US Army Research Laboratory (Contract Number DAAL01-97-C-0140).

The primary source of funding for the current contract is the Federal Aviation Administration. Other sources include NASA, Army, Department of Energy, and Air Force. Volunteer committee members from government, industry, and academia coordinate and review all the information provided in this handbook. The time and effort of the volunteers and the support of their respective departments, companies, and universities make it possible to insure completeness, accuracy, and state-of-the-art composite technology.

FOREWORD

1. This handbook is approved for use by all Departments and Agencies of the Department of Defense.

2. This handbook is for guidance only. This handbook cannot be cited as a requirement. If it is, the contractor does not have to comply. This mandate is a DoD requirement only; it is not applicable to the Federal Aviation Administration (FAA) or other government agencies.

3. Every effort has been made to reflect the latest information on polymeric composites. The handbook is continually reviewed and revised to ensure its completeness and correctness. Documentation for the secretariat should be directed to: Materials Sciences Corporation, MIL-HDBK-17 Secretariat, 500 Office Center Drive, Suite 250, Fort Washington, PA 19034.

4. MIL-HDBK-17 provides guidelines and material properties for polymer (organic) and metal matrix composite materials. The first three volumes of this handbook currently focus on, but are not limited to, polymeric composites intended for aircraft and aerospace vehicles. The fourth volume will focus on metal matrix composites (MMC). Ceramic matrix composites (CMC) and carbon/carbon composites (C/C) will be covered in separate volumes as developments occur.

5. This standardization handbook has been developed and is being maintained as a joint effort of the Department of Defense and the Federal Aviation Administration.

6. The information contained in this handbook was obtained from materials producers, industry, reports on Government sponsored research, the open literature, and by contact with research laboratories and those who participate in the MIL-HDBK-17 coordination activity.

7. All information and data contained in this handbook have been coordinated with industry and the U.S. Army, Navy, and Air Force prior to publication.

8. Copies of this document and revisions thereto may be obtained from the Defense Automated Printing Service (DAPS), 700 Robbins Avenue, Building 4D, Philadelphia, PA 19111-5094.

9. Beneficial comments (recommendations, additions, deletions) and any pertinent data which may be of use in improving this document should be addressed to: Director, U.S. Army Research Laboratory, Weapons and Materials Research Directorate, ATTN: AMSRL-WM-M, Aberdeen Proving Ground, MD 21005-5069, by using the Standardization Document Improvement Proposal (DD Form 1426) appearing at the end of this document or by letter.

CONTENTS

CONTENTS

CONTENTS

CONTENTS

CONTENTS

CONTENTS

FIGURE PAGE

CONTENTS

SUMMARY OF CHANGES IN REVISION MIL-HDBK-17-2E

Self Cover

The title of the handbook has been changed to:
COMPOSITE MATERIALS HANDBOOK
VOLUME 2. POLYMER MATRIX COMPOSITES
MATERIALS PROPERTIES

Foreword

Item number 4 has been revised in order to reflect the upcoming metal matrix composites volume. Item number 9 has been revised to show the correct address for sending comments to improve the handbook.

Chapter 1

The index that only applies to this chapter has been moved to page 1-54.

Chapters 1, 4 and the Appendix

ASTM cancellation information has been added to reference 1.7(c) and 1.7.2(a) on page 1-53, at the bottom of the table on pages 4-71, 4-149, 4-150, and to reference A1.3.4.4 on page A1-72.

Chapters 4, 6 and 10

A data set description has been added preceding each data set presented in Chapters 4, 6, and 10. This addition is reflected in a revision of Chapter 1, Section 1.2.

Chapters 4 and 6

New data sets were added in Sections 4.2.26, 6.2.1, 6.2.2, and 6.2.3.

Chapter 6

Current military specification information has been added at the bottom of the table on pages 6-11, 6-12, 6-25, and 6-26.

CHAPTER 1 GENERAL INFORMATION

1.1 INTRODUCTION

The standardization of a statistically-based mechanical property data base, procedures used, and overall material guidelines for characterization of composite material systems is recognized as being beneficial to both manufacturers and governmental agencies. It is also recognized that a complete characterization of the capabilities of any engineering material system is primarily dependent on the inherent material physical and chemical composition which precede, and are independent of, specific applications. Therefore, at the material system characterization level, the data and guidelines contained in this handbook are applicable to military and commercial products and provide the technical basis for establishing statistically valid design values acceptable to certificating or procuring agencies.

This standardization handbook has been developed and is maintained as a joint effort of the Department of Defense and the Federal Aviation Administration. It is oriented toward the standardization of methods used to develop and analyze mechanical property data on current and emerging composite materials.

1.2 PURPOSE AND SCOPE OF VOLUME 2

This handbook is for guidance only. This handbook cannot be cited as a requirement. If it is, the contractor does not have to comply. This mandate is a DoD requirement only; it is not applicable to the Federal Aviation Administration (FAA) or other government agencies.

The purpose of this handbook is to provide a standard source of statistically-based mechanical property data for current and emerging composite materials. Strength and strain-to-failure properties will be reported either in terms of B-values or S-values. A B-basis value is the mechanical property value above which at least 90 percent of the population is expected to fall, with a confidence of 95 percent. An S-basis value is the mechanical property value which is usually the specified minimum value of the appropriate government specification for the material. Physical, chemical, and mechanical values of the composite constituents - the fibers, matrix material, and prepreg - will be reported where applicable. Later chapters will include data summaries for the various composite systems. Individual chapters focus on particular reinforcement fibers.

Statistically-based strength properties will be defined for each composite material system over a range of potential usage conditions. The intent will be to provide data at the upper and lower limits of the potential environmental conditions for a particular material, so that application issues do not govern the mechanical property characterizations. If data are also available at intermediate environmental conditions, they will also be used to more exactly define the relationship between the material properties and the effect of the environment on those properties. The statistically-based strength data which are available are tabulated in Volume 2. These data are useful as a starting point for establishing structural design allowables when stress and strength analysis capabilities permit lamina level margin of safety checks. Depending on the application, some structural design allowables will have to be determined empirically at the laminate and composite level, since MIL-HDBK-17 does not provide these data. Additional information and properties will be added as they become available and are demonstrated to meet the guideline criteria.

All data included herein are based on test specimens only. Test specimen dimensions conform with those specified for the particular test method which is used. Standard test methods are recommended where possible (ASTM standards are the primary source). The designer and all users must be responsible for any translation of the data contained herein to other coupon dimensions, temperature, humidity, and other environmental conditions not covered in this document. Problems such as scale-up effects and the influence of the test method selected on properties are also not addressed in this document. The manner in which S-basis values are used is also up to the discretion of the designer. In general, decisions concerning which properties to use for a specific application or design are the responsibility of the designer and are outside the scope of this handbook.

Information about each data set has been included at the beginning of each data set. The first part is a description of the material. The material form section describes the material tested in the data set. The information found under the heading of "General Supplier Information" is based on vendor supplied data. The information supplied by the vendor may vary slightly from that of the actual material tested in the data set. The purpose of the description is to provide a quick summary and guideline to the material and its usage. Specific requirements and each material's suitability must be determined by the end user. A more detailed description of fibers and/or matrix materials may be found in Volume 3, Chapter 2, Materials and Processes. The second part of the data set contains pertinent information on the statistical analysis of the data. It alerts the users to items such as pooling of data sets, outliers and batch variations.

1.3 ORGANIZATION OF DATA IN HANDBOOK

The data in Volume 2 is divided into chapters of fiber properties, resin properties, and composite properties organized by fiber and then resin.

1.3.1 Fiber properties

Chapter 2 in Volume 2 will provide data for fiber properties. Sections are to be included for different types of fiber, e.g., glass fibers and carbon fibers. In each section, the general characteristics of the type of fiber will be given, as well as an index of suppliers, designations, and abbreviations. For each specific fiber, data will be organized in the following manner. The X's in the subsection number will be determined by the type of fiber and the specific fiber described.

2.X.X.1 Supplier and product data
2.X.X.2 Chemical and physical properties
2.X.X.2.1 Typical range of chemical constituents
2.X.X.2.2 Expected bound in physical properties
2.X.X.3 Electrical properties
2.X.X.4 Thermal-mechanical properties
2.X.X.4.1 Stress-strain curves
2.X.X.4.2 Environmental effects

1.3.2 Matrix properties

Matrix or resin properties will be included in Chapter 3 which will be divided into sections according to the type of resin. For example, Section 3.2 will give data for epoxies and Section 3.3 will provide data for polyester resins. The subsections for each resin will be the same as those in Chapter 2 given above.

1.3.3 Composite properties

The remaining chapters of Volume 2 will provide data for prepreg, lamina, laminate, and joint properties. There will be individual chapters for each family of composites based on fiber type. For example, Chapter 4 describes carbon fiber composites. Within each chapter, there is expected to be an index of suppliers, designations, and abbreviations. Sections will be included based on the resin type used with the fiber described in the chapter, e.g., Section 4.3 will provide properties for carbon/epoxy composites.

Properties will be organized in the following manner for each specific composite:

X.X.X.1 Supplier and product data
X.X.X.2 Prepreg chemical and physical properties
X.X.X.2.1 Physical description
X.X.X.2.2 Resin content

X.X.X.2.3	Fiber content
X.X.X.2.4	Volatiles content
X.X.X.2.5	Moisture content
X.X.X.2.6	Inorganic fillers and additives content
X.X.X.2.7	Areal weight
X.X.X.2.8	Tack and drape
X.X.X.2.9	Resin flow
X.X.X.2.10	Gel time
X.X.X.3	Lamina chemical properties
X.X.X.4	Lamina physical properties
X.X.X.5	Lamina mechanical properties
X.X.X.5.1	Data Summaries
X.X.X.5.2	Typical stress-strain curves
X.X.X.6	Thermal properties
X.X.X.7	Electrical properties
X.X.X.8	Laminate thermal-mechanical properties
X.X.X.8.1	Index of properties by lay-up
X.X.X.8.2	Strength properties
	a. Lay-up No. 1
	b. Lay-up No. 2
X.X.X.8.3	Thermal properties
X.X.X.8.4	Electrical properties
X.X.X.9	Joint thermal-mechanical properties
X.X.X.9.1	Index of properties by joint and composite system
X.X.X.9.2	Bearing strength
	a. System No. 1
	b. System No. 2

1.4 PRESENTATION OF DATA

This section describes how the data are presented and organized in this volume (MIL-HDBK-17-2).

1.4.1 Properties and definitions

The properties and their definitions are found in the appropriate chapters of Volume 1. Fiber properties and methods for obtaining them are discussed in Chapter 3. Resin properties are presented in Chapter 4. Methods for characterizing prepreg materials are discussed in Chapter 5 and properties and definitions for laminae and laminates are presented in Chapter 6. Properties for structural elements are presented in Chapter 7. The statistical methods used in determining these properties are discussed in Chapter 8. Material system codes and laminate orientation codes are defined in this chapter.

1.4.1.1 Sign convention

All compressive values, represented by a superscript c, are reported as positive numbers. Thus, a positive compression strength indicates failure due to a load applied in the opposite direction of a positive tensile failure.

1.4.2 Table formats

The table formats for mechanical property data presentation are given in Tables 1.4.2(a), and 1.4.2(c) through (g). Table 1.4.2(a) shows the summary pages giving information about the material system and the properties for which data are available. The following notes apply to this table:

❶ Handbook section title and number. Sections are titles using the following information:

{Fiber} {Filament-Count}/{Matrix} {Tape/Weave Type} {Critical processing information}

The tape/weave type includes unidirectional tape (UT), plain weave (PW), or five-harness satin (5HS). Critical processing information, such as bleed or no-bleed, is included when it is necessary to discriminate between data sets. If a warning regarding data documentation is included for the data set, an asterisk follows the section title.

❷ The first set of information in a data section is a summary table containing information on the materials, processing, etc. The box with a heavy border in the upper right-hand corner identifies the first summary table.

{Fiber Class}/{Matrix Class} {Nominal FAW} - {Tape/Weave Type} **{Fiber}/{Matrix}** **Summary**

This box contains the fiber/matrix class of the material, such as carbon/epoxy, identified using the material system codes in Section 1.5.1 With the fiber and matrix class is the nominal fiber areal weight and the abbreviated tape/weave type. The material identification is summarized by the fiber and matrix names.

❸ Material information is presented for the composite, the prepreg, the fiber, and the matrix. Composite material identification is presented as:

{Fiber} {Filament-Count}/{Matrix} {Tape/Weave Type}

Prepreg identification is included as {Manufacturer} {Commercial Name} {Tape/weave type}. For fabric, information such as warp and fill fiber spacing is included when it is available. Fiber identification includes {Manufacturer} {Commercial Name} {Filament-Count} {Sizing} {Twist}. Resin identification is presented as {Manufacturer} {Commercial Name}.

❹ Glass transition temperature under dry and wet conditions is presented with the test method used to obtain these data (See Volume 1, Section 6.4.3).

❺ Basic processing information is presented. This includes the type of process, temperature, pressure, duration, and any other critical parameters for one or more processing steps.

❻ Any warning for limited data documentation is presented on each page of data presentation. On the first page of a data section, a warning is shown below the material identification block.

❼ The block below the material identification block presents various dates relevant to the fabrication and testing of the material. The date of data submittal determines the data documentation requirements that were used for the data set (Volume 1, Section 2.5.6) and the date of analysis determines the statistical analysis that was used (Volume 1, Section 8.3). Ranges of dates are presented where appropriate, such as a testing program which lasted several months.

❽ Lamina properties are summarized with the class of data provided for each property. The columns of the lamina property summary table define the environmental conditions. The first column contains room temperature ambient or dry data. Dry is used only if a drying procedure was used. Ambient refers to as-fabricated with subsequent storage in an ambient laboratory environment. The remaining columns are ordered from lowest to highest moisture content and within a given moisture content, from lowest to highest temperature. If there is enough space, a blank column separates the room temperature ambient/dry column from the other columns and each moisture condition from the others.

The rows of the lamina summary table identify the type test and direction. The basic mechanical properties are included in each summary table. If data are available, additional properties, such as the following, are included:

SB strength, 31-plane	G_{Ic}	CTE, 1-axis
SB strength, 23-plane	G_{IIc}	CTE, 2-axis
SB strength, 12-plane		CTE, 3-axis

For each test type and direction, the class of data for the strength, modulus, Poisson's ratio, and strain-to-failure are provided, in that order. For example, if the entry under RTA and Tension, 1-axis is FI-S. There is room temperature ambient data for longitudinal tension strength, modulus, and strain-to-failure, but not Poisson's ratio. The strength data are fully approved, the modulus data are interim, and the strain-to-failure data are screening. The classes of data approval are defined in Volume 1, Section 2.5.1. Fully approved data requires a minimum of thirty data points from at least five batches with more complete documentation; screening data represent a small number of data points from one batch, and interim data represent a minimum of fifteen specimens from at least three batches. Certain test methods, for example, short beam shear, result only in screening data. (Note that SB strength is the result of a short beam shear test, based on current terminology in ASTM D2344.)

Continuing on the second page of summary information (Table 1.4.2(a)):

① Any warning is placed at the top of this page.

② The box at the top of the second page of summary information presents basic physical parameters for the data set. The first data column contains nominal values, typically specification information.

③ The second data column presents the range of values for the data set submitted.

④ The last column presents the test method used to obtain these data. This information was not included in the early versions of data documentation requirements.

⑤ Laminate property data are summarized in the lower box in the same way as lamina property data are summarized on the previous page. Families of laminates are provided with properties listed below each laminate family. Specific lay-up information is provided in the detailed tables which follow. The type test and direction are included only if data are available and are based on Table 1.4.2(b).

TABLE 1.4.2(a) *Summary table format,* continued on next page.

X.X.X {Fiber} {Filament-Count}/{Matrix} {Tape/Weave Type}* ❶

MATERIAL:	{Fiber} {Filament-Count}/{Matrix} {Tape/Weave Type} ❸			❷
PREPREG:	{Manufacturer} {Commercial Name} {Tape/weave type}			
FIBER:	{Manufacturer} {Commercial Name} {Filament-Count} {Sizing} {Twist}	MATRIX:	{Manufacturer} {Commercial Name}	
T_g(dry):	XXX°F T_g(wet): XXX°F	T_g METHOD:	{Method} ❹	
PROCESSING:	{Type of Process}: {Temperature}, {Duration}, {Pressure} ❺			

*{Warning} ❻

Date of fiber manufacture	MM/YY	Date of testing	MM/YY
Date of resin manufacture	MM/YY	Date of data submittal	MM/YY
Date of prepreg manufacture	MM/YY	Date of analysis	MM/YY
Date of composite manufacture	MM/YY		❼

LAMINA PROPERTY SUMMARY ❽

	{RTA}		{Ambient/dry, coldest to hottest}					{Wet, coldest to hottest}		
Tension, 1-axis										
Tension, 2-axis										
Tension, 3-axis										
Compression, 1-axis										
Compression, 2-axis			Classes of approval are noted for each type test/direction/environmental condition combination							
Compression, 3-axis										
Shear, 12-plane										
Shear, 23-plane										
Shear, 31-plane										
{Additional type test/direction}										
.										
.										
.										

Classes of data: F - Fully approved, I - Interim, S - Screening in Strength/Modulus/Poisson's ratio/Strain-to-failure order.

TABLE 1.4.2(a) *Summary table format,* concluded.

Warning ①

		Nominal ②	As Submitted ③	Test Method ④
Fiber Density	(g/cm^3)	X.XX	{Minimum} - {Maximum}	{Method}
Resin Density	(g/cm^3)	X.XX	{Minimum} - {Maximum}	{Method}
Composite Density	(g/cm^3)	X.XX	{Minimum} - {Maximum}	{Method}
Fiber Areal Weight	(g/m^2)	XXX	{Minimum} - {Maximum}	{Method}
Ply Thickness	(in)	0.0XXX	{Minimum} - {Maximum}	{Method}

LAMINATE PROPERTY SUMMARY ⑤

	{RTA}		{Ambient/dry, coldest to hottest}					{Wet, coldest to hottest}		
{Laminate Family}										
{Type test/direction}										
.										
.										
.			Classes of approval are noted for each type test/direction/environmental condition combination							

Classes of data: F - Fully approved, I - Interim, S - Screening in Strength/Modulus/Poisson's ratio/Strain-to-failure order

TABLE 1.4.2(b) *Laminate type test and directions*

Type Test		Direction	
Tension	Compression After Impact	x-axis	xy-plane
Compression	Bearing	y-axis	yz-plane
Shear	CTE	z-axis	zx-plane
Open Hole Tension			
Open Hole Compression			

Unless otherwise noted, the x-axis corresponds to the 0 direction of the laminate lay-up. Data included for this material are indicated by the class of approval, identified in the footnote.

The format for a data table containing normalized material property information is shown in Table 1.4.2(c). Requirements and procedures for normalization are found in Volume 1, Section 2.5.7 and 2.4.3.

❶ Warnings are shown on each page for data sets which do not meet the data documentation requirements. Many of the data sets were submitted before the establishment of the data documentation requirements. Data sets which do not meet the first version of data documentation requirements or the data documentation requirements that were current when the data were submitted will not receive full approval.

❷ At the top right corner of each page is a box with a heavy border. This box contains information which identifies the data set, the type of test for which results are shown, specimen orientation, test conditions, and the classes of data.

{Table Number} *{Fiber Class}/{Matrix Class} {FAW}-{Form}* *{Fiber Name}/{Matrix Name}* *{Test Type}, {Direction}* *{Lay-up}* *{Test Temperature}/{Moisture Content}* *{Classes of Data Approval}*	- FAW, fiber areal weight - repeated for each data column

❸ Material identification is provided for the composite material as

{Fiber} {Filament-Count}/{Matrix} {Tape/Weave Type}

The range of physical parameters, resin content, fiber volume, ply thickness, composite density, and void content are presented for the data on this particular page.

❹ The test method is identified with the organization, number, and date. For compression after impact, the nominal impact energy level used for the test is appended to the test method, since alternate levels are often used.

❺ The method of calculating the modulus is presented for mechanical property data. This includes the calculation method, and the location or range of measurements used for the calculation.

❻ The normalization method is presented for data that have been normalized (See Volume 1, Section 2.4.3). The fiber volume to which the data are normalized is also included. This value is typically

60% for unidirectional tape and 57% for weaves. Types of normalization as entered are:

Normalized by fiber volume to XX% (0.0XXX in. CPT)
Normalized by specimen thickness and batch fiber volume to XX% (0.0XXX in. CPT)
Normalized by specimen thickness and batch fiber areal weight to XX% fiber volume (0.0XXX in. CPT)

Corresponding cured ply thickness (CPT) values, based on a nominal fiber areal weight, are included for reference for each method.

7 At the top of each data column are the test conditions. Nominally dry conditions, for materials that are fabricated and stored under controlled conditions are noted. Wet conditions which are not conditioned to equilibrium are also noted. The source code provides a means for identifying data sets from the same source. No other source identification is provided.

8 Strength data and strain-to-failure data are presented in the handbook with a full set of statistical parameters. The class of data approval is indicated for each property/condition combination. B-values are presented only for fully approved data. A-basis values are presented for fully approved data which meet the batch and specimen number requirements for A-basis values. The distribution of method of analysis is presented. The constants, C_1 and C_2, correspond to the distribution. These are as follows:

	C_1	C_2
Weibull	scale parameter	shape parameter
Normal	mean	standard deviation
Lognormal	mean of the natural log of the data	standard deviation of the natural log of the data
Nonparametric	rank	Hanson-Koopmans coefficient
ANOVA	sample between-batch variance	sample within-batch variance

All statistical parameters are presented for normalized and as-measured strength data. All statistical parameters are presented for as-measured strain-to-failure data.

9 Modulus data are presented with only mean, minimum, maximum, coefficient of variation, batch size, and sample size. Values are presented for both normalized and as-measured data. Where available, Poisson's ratio data are presented with batch size and sample size information.

10 Footnotes are presented wherever additional information is pertinent. Information frequently presented in footnotes include conditioning parameters, reasons for not presenting B-values, and deviations from standard test methods.

Symbols for properties are presented with property directions as subscripts and property type, for example, tension (t), as superscripts. The example table shows symbols for lamina tension in the fiber direction.

Table 1.4.2(d) shows an example table for material properties that are not normalized. The basic table format and information is identical to the table format for normalized data. Only as-measured data are presented in each column of information. The statistical parameters are the same provided for normalized data. Table 1.4.2(e) presents the format for bearing data, Table 1.4.2(f) the format for notched laminate data, and Table 1.4.2(g) the format for bearing/bypass data.

TABLE 1.4.2(c) *Table format for normalized data.*

{Warning} ❶

MATERIAL: {Fiber} {Filament count}/{Matrix} {Tape/weave type} ❸ RESIN CONTENT: XX - XX wt% COMP: DENSITY: X.XX-X.XX g/cm^3 FIBER VOLUME: XX - XX vol % VOID CONTENT: 0.X to X.X % PLY THICKNESS: 0.0XXX - 0.0XXX in. TEST METHOD: ❹ MODULUS CALCULATION: ❺ {Organization} {Number} {Date} {Method}, XXXX - XXXX με NORMALIZED BY: {Method} to XX % (0.0XXX in. CPT) ❻						❷
Temperature (°F) Moisture Content (%) Equilibrium at T, RH Source Code	❼					
	Normalized	Measured	Normalized	Measured	Normalized	Measured
F_1^{tu} (ksi) — Mean Minimum Maximum C.V.(%) B-value Distribution C_1 C_2 No. Specimens No. Batches Approval Class	❽					
E_1^t (Msi) — Mean Minimum Maximum C.V.(%) No. Specimens No. Batches Approval Class	❾					
ν_{12}^t — Mean No. Specimens No. Batches Approval Class	❾					
ε_1^{tu} (με) — Mean Minimum Maximum C.V.(%) B-value Distribution C_1 C_2 No. Specimens No. Batches Approval Class	❽					

❿

TABLE 1.4.2(d) *Table format for as-measured data.*

{Warning} ❶

MATERIAL: {Fiber} {Filament count}/{Matrix} {Tape/weave type} ❸					❷
RESIN CONTENT: XX - XX wt% FIBER VOLUME: XX - XX vol % PLY THICKNESS: 0.0XXX - 0.0XXX in.		COMP: DENSITY: X.XX-X.XX g/cm^3 VOID CONTENT: 0.X to X.X %			
TEST METHOD: ❹ {Organization} {Number} {Date}		MODULUS CALCULATION: ❺ {Method}, XXXX - XXXX $\mu\varepsilon$			
NORMALIZED BY: Not normalized ❻					
Temperature (°F) Moisture Content (%) Equilibrium at T, RH Source Code	❼				
F(1) (ksi) — Mean Minimum Maximum C.V.(%) B-value Distribution C_1 C_2 No. Specimens No. Batches Approval Class	❽				
(Msi) — Mean Minimum Maximum C.V.(%) No. Specimens No. Batches Approval Class	❾				
Mean No. Specimens No. Batches Approval Class	❾				
($\mu\varepsilon$) — Mean Minimum Maximum C.V.(%) B-value Distribution C_1 C_2 No. Specimens No. Batches Approval Class	❽				

❿

TABLE 1.4.2(e) *Format for bearing strength property table.*

MATERIAL:	{Fiber} {Fil. Count} / {Matrix} {tape/weave type}			**{Table Number} {Material Class}/{Form} {Material Designation} {Property and Direction} {Laminate Family} {Test Temp/Moisture Conditions} {Data Approval Status}**
RESIN CONTENT:	XX-XX wt%	COMP: DENSITY:	0.0XX lb/in³	
FIBER VOLUME:	XX-XX %	VOID CONTENT:	X.X - X.X %	
PLY THICKNESS:	0.00XX - 0.00XX in.			
TEST METHOD:		{Org. Method - Date}		
TYPE OF BEARING TEST:		{material/joint}		
JOINT CONFIGURATION				
Member 1 (t,w,layup):		{thickness, width, layup}		
Member 2 (t,w,layup):		{thickness, width, layup}		
FASTENER TYPE:		{ }		
TORQUE:		{ }		
NORMALIZED BY:		Not normalized		
Failure Description (3):				

Temperature (°F) Moisture Content (%) Equilibrium at T, RH (°F, %) Source Code		75 (1)		ET (2)	
Diameter		0.25	D2	0.25	D2
F^{bru}	Mean				
	Minimum				
	Maximum				
	C.V. (%)				
(ksi)	No. Specimens				
	No. Batches				
F^{br}_{offset}	Mean				
	Minimum				
	Maximum				
	C.V. (%)				
(ksi)	No. Specimens				
	No. Batches				
F^{br}_{pl}	Mean				
	Minimum				
	Maximum				
	C.V. (%)				
(ksi)	No. Specimens				
	No. Batches				

(1) As fabricated.
(2) Conditioned at XXX°F, XX% RH for XX days.
(3) See Table 8.3.2.1 for failure mode description.

TABLE 1.4.2(f) *Format for notched laminate strength property table.*

MATERIAL:	{Fiber} {Fil. Count} / {Matrix} {tape/weave type}			**{Table Number}** **{Material Class}/{Form}** **{Material Designation}** **{Property and Direction}** **{Laminate Family}** **{Test Temp/Moisture Conditions}** **{Data Approval Status}**
RESIN CONTENT: FIBER VOLUME:	XX-XX wt% XX-XX %	COMP: DENSITY: VOID CONTENT:	0.0XX lb/in^3 X.X - X.X %	
PLY THICKNESS:	0.00XX - 0.00XX in.			
TEST METHOD:		{Org. Method - Date}		
SPECIMEN GEOMETRY: FASTENER TYPE: TORQUE:		{thickness} { } { }		
NORMALIZED BY:		Not normalized		

Temperature (°F) Moisture Content (%) Equilibrium at T, RH (°F, %) Source Code		75 (1)			ET (2)		
Width to Diameter Ratio		W1/D	W2/D	W3/D	W1/D	W2/D	W3/D
Diameter = 0.25	Mean						
	Minimum						
	Maximum						
F_x^{cu}							
	C.V. (%)						
(ksi)	No. Specimens						
	No. Batches						
Diameter = D2	Mean						
	Minimum						
	Maximum						
F_x^{cu}							
	C.V. (%)						
(ksi)	No. Specimens						
	No. Batches						
Diameter = D3	Mean						
	Minimum						
	Maximum						
F_x^{cu}	C.V. (%)						
(ksi)	No. Specimens						
	No. Batches						

(1) As fabricated.
(2) Conditioned at XXX°F, XX% RH for XX days.

TABLE 1.4.2(g) *Format for bearing/bypass property table.*

MATERIAL:	{Fiber} {Fil. Count} / {Matrix} {tape/weave type}			**{Table Number} {Material Class}/{Form} {Material Designation} {Property and Direction} {Laminate Family} {Test Temp/Moisture Conditions} {Data Approval Status}**
RESIN CONTENT:	XX-XX wt%	COMP: DENSITY:	0.0XX lb/in^3	
FIBER VOLUME:	XX-XX %	VOID CONTENT:	X.X - X.X %	
PLY THICKNESS:	0.00XX - 0.00XX in.			
TEST METHOD:		{Org. Method - Date}		
SPECIMEN GEOMETRY:		{thickness}		
FASTENER TYPE:		{ }		
TORQUE:		{ }		
NORMALIZED BY:		Not normalized		

Temperature (°F) Moisture Content (%) Equilibrium at T, RH (°F, %) Source Code		75 (1)	
Bearing/Bypass Ratio		0.50	0.75
F^{cu}_{byp}	Mean		
(ksi)	Minimum Maximum C.V. (%)		
F^{br}	Mean		
(ksi)	Minimum Maximum C.V. (%)		
	No. Specimens No. Batches		
F^{tu}_{byp}	Mean		
(ksi)	Minimum Maximum C.V. (%)		
F^{br}	Mean		
(ksi)	Minimum Maximum C.V. (%)		
	No. Specimens No. Batches		

(1) As fabricated.

1.4.3 Sample graphs

Sample graphs for the stress-strain curves are given in Volume 1, Section 8.4.4. Figure 1.4.3(a) shows a graph of the tangent modulus and the secant modulus as a function of strain for the data of Volume 1, Figure 8.4.4(a).

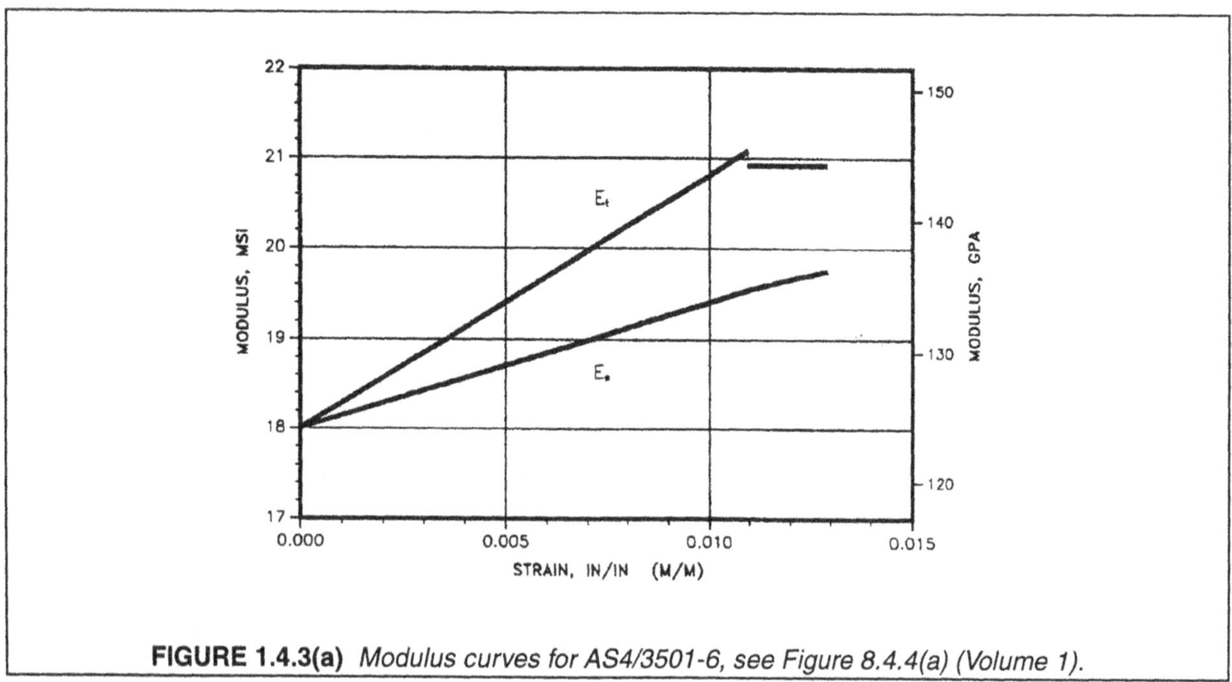

FIGURE 1.4.3(a) *Modulus curves for AS4/3501-6, see Figure 8.4.4(a) (Volume 1).*

For thermal properties, room temperature values will be listed in the data summary and additional values of the thermal expansion coefficient will be presented graphically as a function of temperature. The specific heat and the thermal conductivity will be presented in a similar fashion. When data are adequate to present curves showing specific heat, thermal conductivity, and mean coefficient of thermal expansion over a range of temperatures, graphical presentation is used in addition to tabular presentation. A smooth curve is drawn through the plotted points to depict the overall trend of the data. The smooth curve formats for specific heat, thermal conductivity, and thermal expansion are presented along with representative data in Figures 1.4.3(b) - (d), respectively. The reference temperature for thermal expansion should be shown on the figure. In Figure 1.4.3(d), the reference temperature of 70°F (21°C, 294 K) indicates that the mean coefficient of thermal expansion between 70°F (21°C, 294 K) and the indicated temperature is plotted.

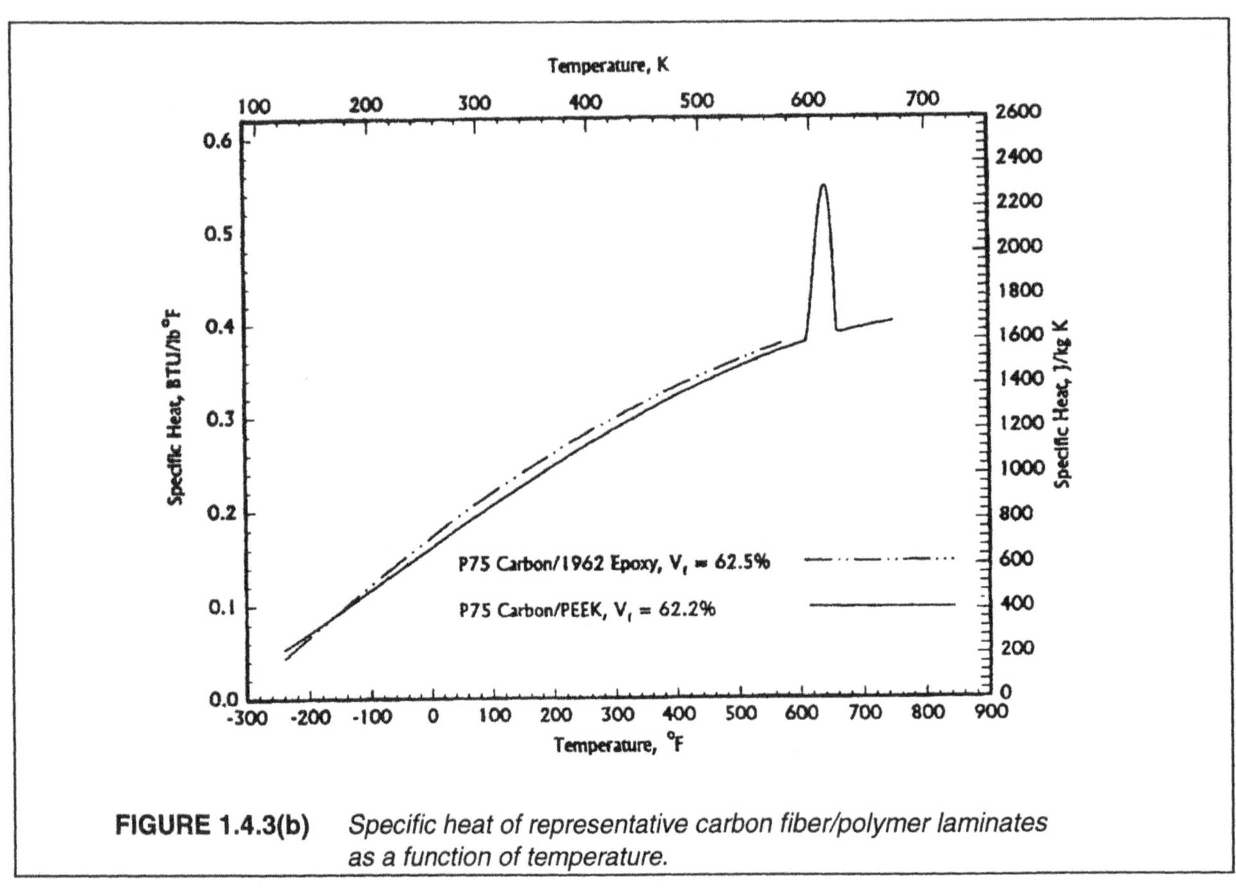

FIGURE 1.4.3(b) *Specific heat of representative carbon fiber/polymer laminates as a function of temperature.*

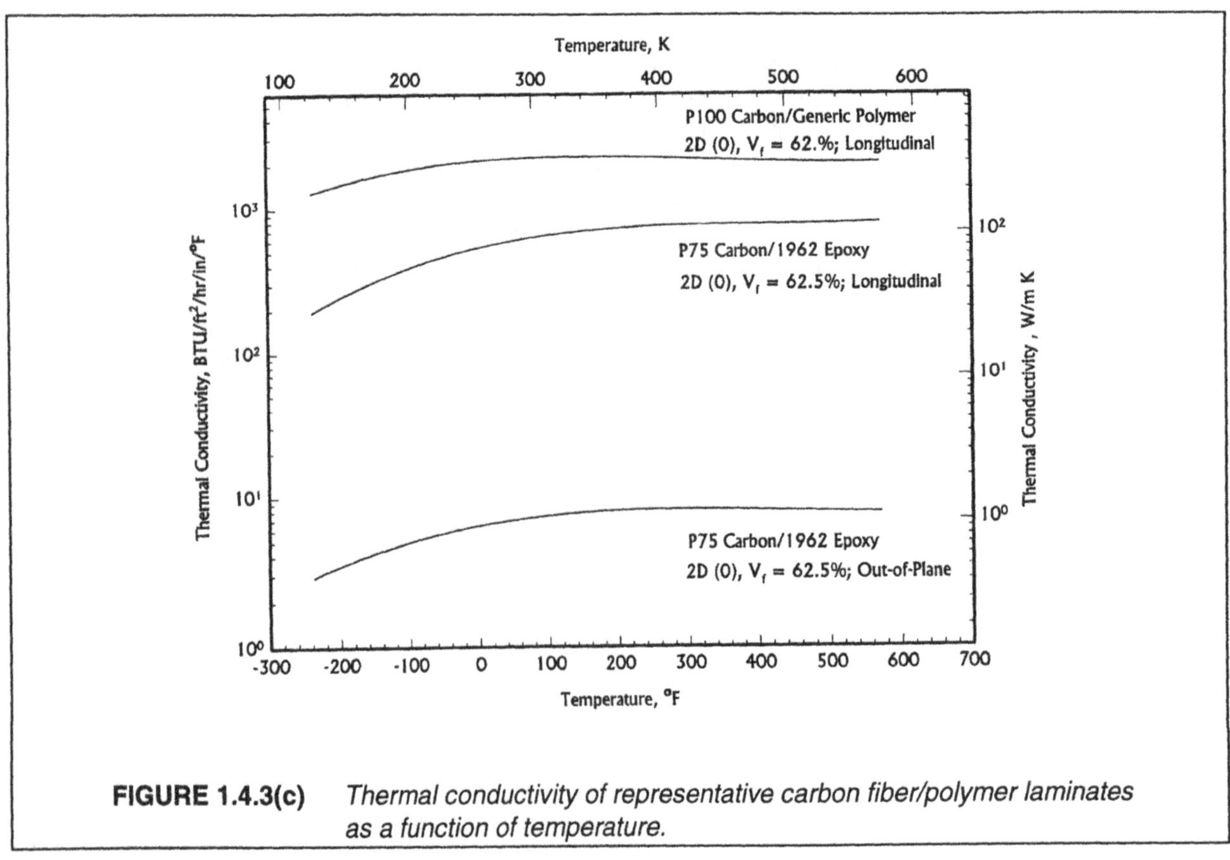

FIGURE 1.4.3(c) *Thermal conductivity of representative carbon fiber/polymer laminates as a function of temperature.*

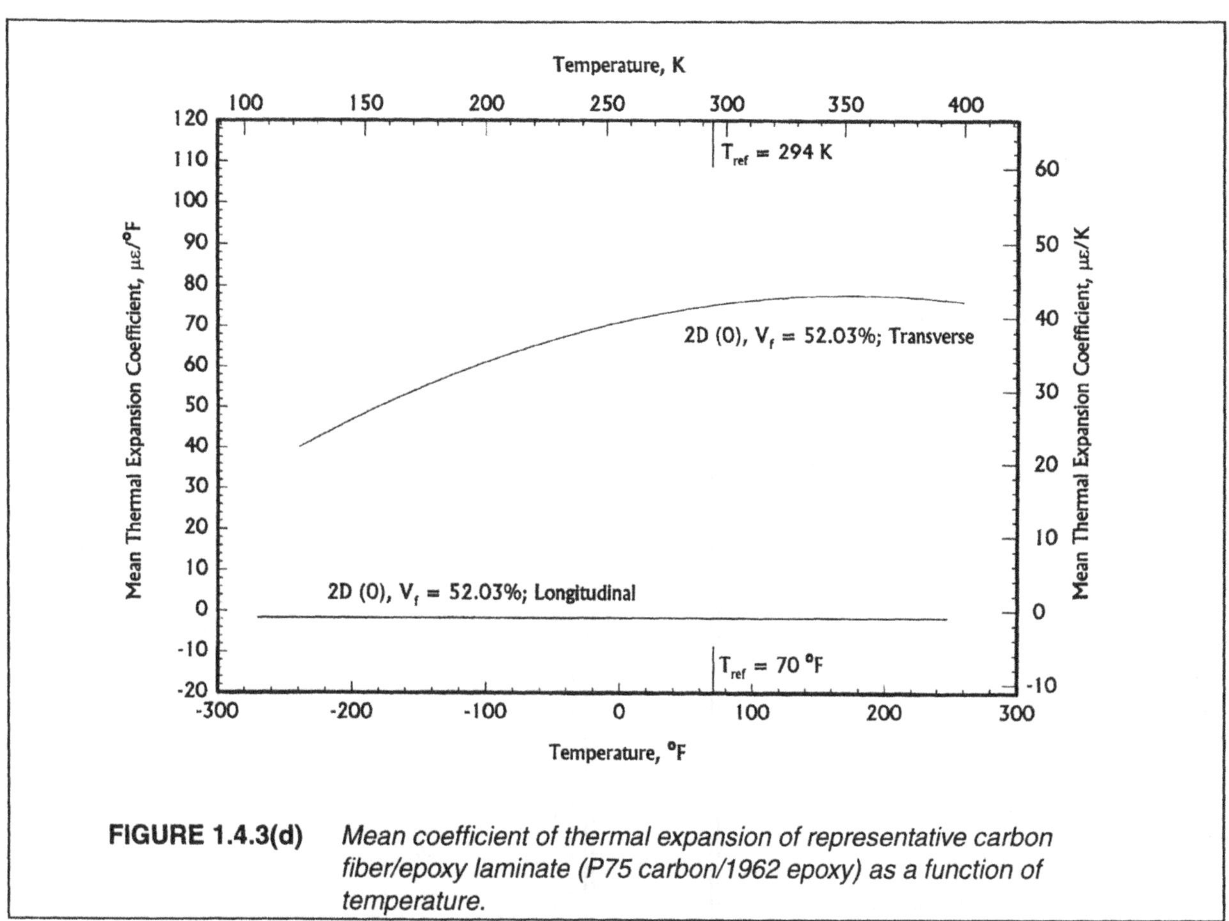

FIGURE 1.4.3(d) *Mean coefficient of thermal expansion of representative carbon fiber/epoxy laminate (P75 carbon/1962 epoxy) as a function of temperature.*

1.5 MATERIALS SYSTEMS

1.5.1 Materials system codes

The materials systems codes which are used in the handbook consist of a fiber system code and a matrix material code separated by a virgule (/). The codes for the fiber and matrix materials appear in Tables 1.5.1(a) and (b).

TABLE 1.5.1(a) *Fiber system codes.*

AlO	Alumina
Ar	Aramid
B	Boron
C	Carbon
DGl	D-Glass
EGl	E-Glass
Gl	Glass
Gr	Graphite
Li	Lithium
PAN	Polyacrylonitrile
PBT	Polybenzothiazole
Q	Quartz
Si	Silicon
SiC	Silicon carbide
SGl	S-Glass
Ti	Titanium
W	Tungsten

TABLE 1.5.1(b) *Matrix material codes.*

BMI	Bismaleimide
EP	Epoxy
PAI	Polyamide-imide
PEEK	Polyetheretherketone
PEI	Polyetherimide
PES	Polyethersulfone
PI	Polyimide
PPS	Polyphenylene sulfide
PSU	Polysulfone
SI	Silicone
TP	Thermoplastic
TPES	Thermoplastic polyester
	Fluorocarbon
	Phenolic
	Polybenzimidazole

1.5.2 Index of materials

This section is reserved for future use.

1.6 MATERIAL ORIENTATION CODES

1.6.1 Laminate orientation codes

The purpose of a laminate orientation code is to provide a simple, easily understood method of describing the lay-up of a laminate. The laminate orientation code is based largely on the code used in the Advanced Composites Design Guide (Reference 1.6.1(a)). The following information and the examples in Figure 1.6.1 describe the laminate orientation code used in MIL-HDBK-17.

1. The orientation of each lamina with respect to the x-axis is indicated by the angle between the fiber direction and the x-axis. Positive angles are measured counter-clockwise from the x-axis when looking toward the lay-up surface (right-hand rule).
2. When indicating the lay-up of a weave, the angle is measured between the warp direction and the x-axis.
3. Orientations of successive laminae with different absolute values are separated by a virgule (/).
4. Two or more adjacent laminae with the same orientation are indicated by adding a subscript, to the angle of the first such lamina, equal to the number of repetitions of laminae with that orientation.
5. Laminae are listed in order from the first laid up to the last. Brackets are used to indicate the beginning and the end of the code.
6. A subscript of 's' is used if the first half of the lay-up is indicated and the second half is symmetric with the first. When a symmetric lay-up with an odd number of laminae is shown, the layer which is not repeated is indicated by overlining the angle of that lamina.
7. A repeated set of laminae are enclosed in parentheses and the number of repetitions of the set indicated by a subscript.
8. The convention used for indicating materials is no subscript for a tape ply and a subscript "f" for a weave.
9. The laminate code for a hybrid has the different materials contained in the laminate indicated by subscripts on the laminae.
10. Since the majority of computer programs do not permit the use of subscripts and superscripts, the following modifications are recommended based on ASTM Committee E-49 guidelines (Reference 1.6.1(b)).

    a. Subscript information will be preceded by a colon (:), e.g., [90/0:2/45]:s.
    b. A bar over a ply (designating a non-repeated ply in a symmetric laminate) should be indicated by a backslash (\) after the ply, e.g., [0/45/90\]:s.

1.6.2 Braiding, orientation codes

This section is reserved for future use.

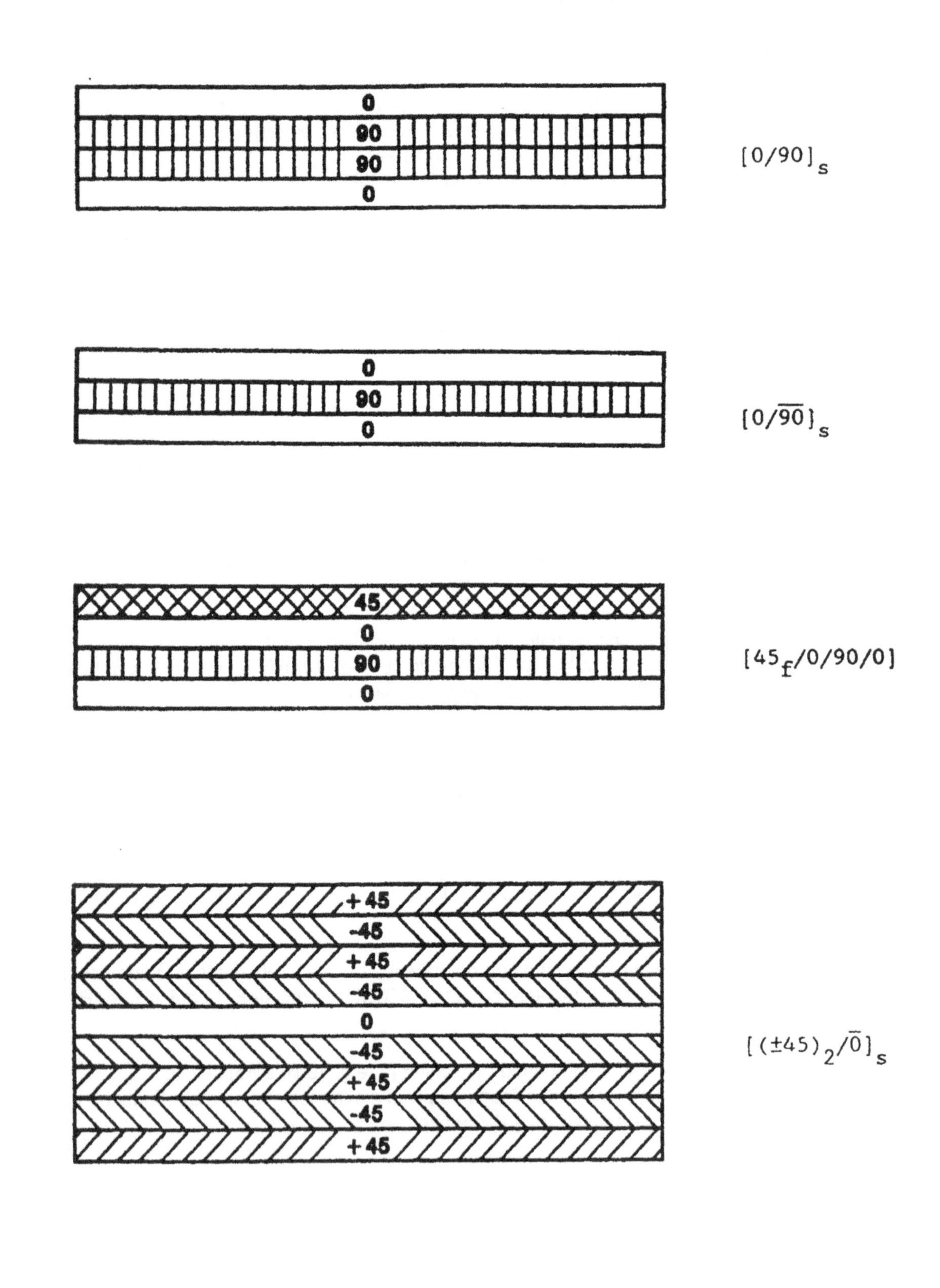

FIGURE 1.6.1 *Example laminate orientation codes.*

1.7 SYMBOLS, ABBREVIATIONS, AND SYSTEMS OF UNITS

This section defines the symbols and abbreviations which are used within MIL-HDBK-17 and describes the system of units which is maintained. Common usage is maintained where possible. References 1.7(a) - (c) served as primary sources for this information.

1.7.1 Symbols and abbreviations

The symbols and abbreviations used in this document are defined in this section with the exception of statistical symbols. These latter symbols are defined in Chapter 8. The lamina/laminate coordinate axes used for all properties and a summary of the mechanical property notation are shown in Figure 1.7.1.

- The symbols f and m, when used as either subscripts or superscripts, always denote fiber and matrix, respectively.
- The type of stress (for example, cy - compression yield) is always used in the superscript position.
- Direction indicators (for example, x, y, z, 1, 2, 3, etc.) are always used in the subscript position.
- Ordinal indicators of laminae sequence (e.g., 1, 2, 3, etc.) are used in the superscript position and must be parenthesized to distinguish them from mathematical exponents.
- Other indicators may be used in either subscript or superscript position, as appropriate for clarity.
- Compound symbols (such as, basic symbols plus indicators) which deviate from these rules are shown in their specific form in the following list.

The following general symbols and abbreviations are considered standard for use in MIL-HDBK-17. Where exceptions are made, they are noted in the text and tables.

A	- (1) area (m^2,in^2) - (2) ratio of alternating stress to mean stress - (3) A-basis for mechanical property values
a	- (1) length dimension (mm,in) - (2) acceleration (m/sec^2,ft/sec^2) - (3) amplitude - (4) crack or flaw dimension (mm,in)
B	- (1) B-basis for mechanical property values - (2) biaxial ratio
Btu	- British thermal unit(s)
b	- width dimension (mm,in), e.g., the width of a bearing or compression panel normal to load, or breadth of beam cross-section
C	- (1) specific heat (kJ/kg °C,Btu/lb °F) - (2) Celsius
CF	- centrifugal force (N,lbf)
CPF	- crossply factor
CPT	- cured ply thickness (mm, in.)
CG	- (1) center of mass, "center of gravity" - (2) area or volume centroid
℄	- centerline
c	- column buckling end-fixity coefficient

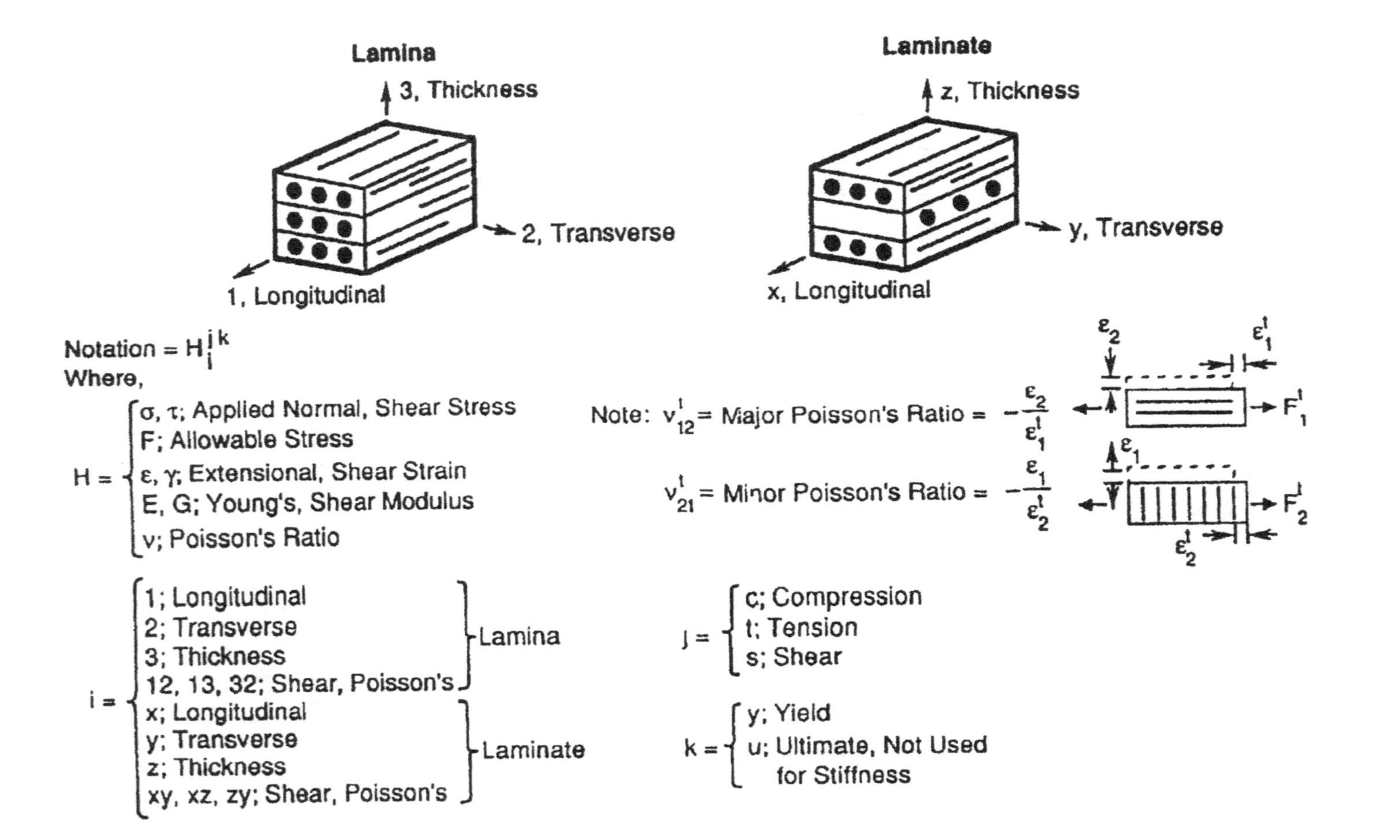

FIGURE 1.7.1 *Mechanical property notation.*

$\bar{c}$ - honeycomb sandwich core depth (mm,in)
cpm - cycles per minute
D - (1) diameter (mm,in)
- (2) hole or fastener diameter (mm,in)
- (3) plate stiffness (N-m,lbf-in)
d - mathematical operator denoting differential
E - modulus of elasticity in tension, average ratio of stress to strain for stress below proportional limit (GPa,Msi)
E' - storage modulus (GPa,Msi)
E" - loss modulus (GPa,Msi)
E_c - modulus of elasticity in compression, average ratio of stress to strain for stress below proportional limit (GPa,Msi)
E_c' - modulus of elasticity of honeycomb core normal to sandwich plane (GPa,Msi)
E^{sec} - secant modulus (GPa,Msi)
E^{tan} - tangent modulus (GPa,Msi)
e - minimum distance from a hole center to the edge of the sheet (mm,in)
e/D - ratio of edge distance to hole diameter (bearing strength)
F - (1) stress (MPa,ksi)
- (2) Fahrenheit
F^b - bending stress (MPa,ksi)
F^{ccr} - crushing or crippling stress (upper limit of column stress for failure) (MPa,ksi)
F^{su} - ultimate stress in pure shear (this value represents the average shear stress over the cross-section) (MPa,ksi)
FAW - fiber areal weight (g/m^2, lb/in^2)
FV - fiber volume (%)
f - (1) internal (or calculated) stress (MPa,ksi)
- (2) stress applied to the gross flawed section (MPa,ksi)
- (3) creep stress (MPa,ksi)
f^c - internal (or calculated) compressive stress (MPa,ksi)
f_c - (1) maximum stress at fracture (MPa,ksi)
- (2) gross stress limit (for screening elastic fracture data (MPa,ksi)
ft - foot, feet
G - modulus of rigidity (shear modulus) (GPa,Msi)
GPa - gigapascal(s)
g - (1) gram(s)
- (2) acceleration due to gravity (m/s^2,ft/s^2)
H/C - honeycomb (sandwich)
h - height dimension (mm,in) e.g. the height of a beam cross-section
hr - hour(s)
I - area moment of inertia (mm^4,in^4)
i - slope (due to bending) of neutral plane in a beam, in radians
in. - inch(es)
J - (1) torsion constant (= I_p for round tubes) (m^4,in^4)
- (2) Joule
K - (1) Kelvin
- (2) stress intensity factor ($MPa\sqrt{m}$,$ksi\sqrt{in}$)
- (3) coefficient of thermal conductivity (W/m °C, Btu/ft^2/hr/in/°F)
- (4) correction factor
- (5) dielectric constant
K_{app} - apparent plane strain fracture toughness or residual strength ($MPa\sqrt{m}$,$ksi\sqrt{in}$)
K_c - critical plane strain fracture toughness, a measure of fracture toughness at point of crack growth instability ($MPa\sqrt{m}$,$ksi\sqrt{in}$)
K_{Ic} - plane strain fracture toughness ($MPa\sqrt{m}$,$ksi\sqrt{in}$)
K_N - empirically calculated fatigue notch factor

K_s - plate or cylinder shear buckling coefficient
K_t - (1) theoretical elastic stress concentration factor
- (2) t_w/c ratio in H/C sandwich
Kv - dielectric strength (KV/mm, V/mil)
K_x, K_y - plate or cylinder compression buckling coefficient
k - strain at unit stress (m/m,in/in)
L - cylinder, beam, or column length (mm,in)
L' - effective column length (mm,in)
lb - pound
M - applied moment or couple (N-m,in-lbf)
Mg - megagram(s)
MPa - megapascal(s)
MS - military standard
M.S. - margin of safety
MW - molecular weight
MWD - molecular weight distribution
m - (1) mass (kg,lb)
- (2) number of half wave lengths
- (3) metre
- (4) slope
N - (1) number of fatigue cycles to failure
- (2) number of laminae in a laminate
- (3) distributed in-plane forces on a panel (lbf/in)
- (4) Newton
- (5) normalized
NA - neutral axis
n - (1) number of times in a set
- (2) number of half or total wavelengths
- (3) number of fatigue cycles endured
P - (1) applied load (N,lbf)
- (2) exposure parameter
- (3) probability
- (4) specific resistance (Ω)
P^u - test ultimate load, (N,lb per fastener)
P^y - test yield load, (N,lb per fastener)
p - normal pressure (Pa,psi)
psi - pounds per square inch
Q - area static moment of a cross-section (mm^3,in^3)
q - shear flow (N/m,lbf/in)
R - (1) algebraic ratio of minimum load to maximum load in cyclic loading
- (2) reduced ratio
RA - reduction of area
R.H. - relative humidity
RMS - root-mean-square
RT - room temperature
r - (1) radius (mm,in)
- (2) root radius (mm,in)
- (3) reduced ratio (regression analysis)
S - (1) shear force (N,lbf)
- (2) nominal stress in fatigue (MPa,ksi)
- (3) S-basis for mechanical property values
S_a - stress amplitude in fatigue (MPa,ksi)
S_e - fatigue limit (MPa,ksi)
S_m - mean stress in fatigue (MPa,ksi)

S_{max}	- highest algebraic value of stress in the stress cycle (MPa,ksi)
S_{min}	- lowest algebraic value of stress in the stress cycle (MPa,ksi)
S_R	- algebraic difference between the minimum and maximum stresses in one cycle (MPa,ksi)
S.F.	- safety factor
s	- (1) arc length (mm,in)
	- (2) H/C sandwich cell size (mm,in)
T	- (1) temperature (°C,°F)
	- (2) applied torsional moment (N-m,in-lbf)
T_d	- thermal decomposition temperature (°C,°F)
T_F	- exposure temperature (°C,°F)
T_g	- glass transition temperature(°C,°F)
T_m	- melting temperature (°C,°F)
t	- (1) thickness (mm,in)
	- (2) exposure time (s)
	- (3) elapsed time (s)
V	- (1) volume (mm^3,in^3)
	- (2) shear force (N,lbf)
W	- (1) weight (N,lbf)
	- (2) width (mm,in)
	- (3) Watt
x	- distance along a coordinate axis
Y	- nondimensional factor relating component geometry and flaw size
y	- (1) deflection (due to bending) of elastic curve of a beam (mm,in)
	- (2) distance from neutral axis to given point
	- (3) distance along a coordinate axis
Z	- section modulus, I/y (mm^3,in^3)
α	- coefficient of thermal expansion (m/m/°C,in/in/°F)
γ	- shear strain (m/m,in/in)
Δ	- difference (used as prefix to quantitative symbols)
δ	- elongation or deflection (mm,in)
ε^e	- strain (m/m,in/in)
ε^p	- elastic strain (m/m,in/in)
ε	- plastic strain (m/m,in/in)
μ	- permeability
η	- plasticity reduction factor
$[\eta]$	- intrinsic viscosity
η^*	- dynamic complex viscosity
ν	- Poisson's ratio
ρ	- (1) density (kg/m^3,lb/in^3)
	- (2) radius of gyration (mm,in)
ρ_c'	- H/C sandwich core density (kg/m^3,lb/in^3)
Σ	- total, summation
σ	- standard deviation
σ_{ij}, τ_{ij}	- stress in j direction on surface whose outer normal is in i direction (i, j = 1, 2, 3 or x, y, z) (MPa,ksi)
T	- applied shear stress (MPa,ksi)
ω	- angular velocity (radians/s)
∞	- infinity

1.7.1.1 Constituent properties

The following symbols apply specifically to the constituent properties of a typical composite material.

E^f - Young's modulus of filament material (MPa,ksi)
E^m - Young's modulus of matrix material (MPa,ksi)
E^g_x - Young's modulus of impregnated glass scrim cloth in the filament direction or in the warp direction of a fabric (MPa,ksi)
E^g_y - Young's modulus of impregnated glass scrim cloth transverse to the filament direction or to the warp direction in a fabric (MPa,ksi)
G^f - shear modulus of filament material (MPa,ksi)
G^m - shear modulus of matrix (MPa,ksi)
G^g_{xy} - shear modulus of impregnated glass scrim cloth (MPa,ksi)
G'_{cx} - shear modulus of sandwich core along X-axis (MPa,ksi)
G'_{cy} - shear modulus of sandwich core along Y-axis (MPa,ksi)
ℓ - filament length (mm,in)
α^f - coefficient of thermal expansion for filament material (m/m/°C,in/in/°F)
α^m - coefficient of thermal expansion for matrix material (m/m/°C,in/in/°F)
α^g_x - coefficient of thermal expansion of impregnated glass scrim cloth in the filament direction or in the warp direction of a fabric (m/m/°C,in/in/°F)
α^g_y - coefficient of thermal expansion of impregnated glass scrim cloth transverse to the filament direction or to the warp direction in a fabric (m/m/°C,in/in/°F)
ν^f - Poisson's ratio of filament material
ν^m - Poisson's ratio of matrix material
ν^g_{xy} - glass scrim cloth Poisson's ratio relating to contraction in the transverse (or fill) direction as a result of extension in the longitudinal (or warp) direction
ν^g_{yx} - glass scrim cloth Poisson's ratio relating to contraction in the longitudinal (or warp) direction as a result of extension in the transverse (or fill) direction
σ - applied axial stress at a point, as used in micromechanics analysis (MPa,ksi)
τ - applied shear stress at a point, as used in micromechanics analysis (MPa,ksi)

1.7.1.2 Laminae and laminates

The following symbols, abbreviations, and notations apply to composite laminae and laminates. At the present time the focus in MIL-HDBK-17 is on laminae properties. However, commonly used nomenclature for both laminae and laminates are included here to avoid potential confusion.

A_{ij} (i,j = 1,2,6) - extensional rigidities (N/m,lbf/in)
B_{ij} (i,j = 1,2,6) - coupling matrix (N,lbf)
C_{ij} (i,j = 1,2,6) - elements of stiffness matrix (Pa,psi)
D_x, D_y - flexural rigidities (N-m,lbf-in)
D_{xy} - twisting rigidity (N-m,lbf-in)
D_{ij} (i,j = 1,2,6) - flexural rigidities (N-m,lbf-in)
E_1 - Young's modulus of lamina parallel to filament or warp direction (GPa,Msi)
E_2 - Young's modulus of lamina transverse to filament or warp direction (GPa,Msi)
E_x - Young's modulus of laminate along x reference axis (GPa,Msi)
E_y - Young's modulus of laminate along y reference axis (GPa,Msi)
G_{12} - shear modulus of lamina in 12 plane (GPa,Msi)
G_{xy} - shear modulus of laminate in xy reference plane (GPa,Msi)
h_i - thickness of ith ply or lamina (mm,in)

M_x, M_y, M_{xy} - bending and twisting moment components (N-m/m, in-lbf/in in plate and shell analysis)

n_f - number of filaments per unit length per lamina

Q_x, Q_y - shear force parallel to z axis of sections of a plate perpendicular to x and y axes, respectively (N/m,lbf/in)

Q_{ij} (i,j = 1,2,6) - reduced stiffness matrix (Pa,psi)

u_x, u_y, u_z - components of the displacement vector (mm,in)

u_x^o, u_y^o, u_z^o - components of the displacement vector at the laminate's midsurface (mm,in)

V_v - void content (% by volume)

V_f - filament content or fiber volume (% by volume)

V_g - glass scrim cloth content (% by volume)

V_m - matrix content (% by volume)

V_x, V_y - edge or support shear force (N/m,lbf/in)

W_f - filament content (% by weight)

W_g - glass scrim cloth content (% by weight)

W_m - matrix content (% by weight)

W_s - weight of laminate per unit surface area (N/m^2,lbf/in^2)

α_1 - lamina coefficient of thermal expansion along 1 axis (m/m/°C,in/in/°F)

α_2 - lamina coefficient of thermal expansion along 2 axis (m/m/°C,in/in/°F)

α_x - laminate coefficient of thermal expansion along general reference x axis (m/m/°C, in/in/°F)

α_y - laminate coefficient of thermal expansion along general reference y axis (m/m/°C, in/in/°F)

α_{xy} - laminate shear distortion coefficient of thermal expansion (m/m/°C,in/in/°F)

θ - angular orientation of a lamina in a laminate, i.e., angle between 1 and x axes (°)

λ_{xy} - product of ν_{xy} and ν_{yx}

ν_{12} - Poisson's ratio relating contraction in the 2 direction as a result of extension in the 1 direction[1]

ν_{21} - Poisson's ratio relating contraction in the 1 direction as a result of extension in the 2 direction[1]

ν_{xy} - Poisson's ratio relating contraction in the y direction as a result of extension in the x direction[1]

ν_{yx} - Poisson's ratio relating contraction in the x direction as a result of extension in the y direction[1]

ρ_c - density of a single lamina (kg/m^3,lb/in^3)

$\bar{\rho}_c$ - density of a laminate (kg/m^3,lb/in^3)

ϕ - (1) general angular coordinate, (°)
- (2) angle between x and load axes in off-axis loading (°)

1.7.1.3 Subscripts

The following subscript notations are considered standard in MIL-HDBK-17.

1, 2, 3 - laminae natural orthogonal coordinates (1 is filament or warp direction)

A - axial

a - (1) adhesive
- (2) alternating

app - apparent

byp - bypass

[1]The convention for Poisson's ratio should be checked before comparing different sources as different conventions are used.

c - composite system, specific filament/matrix composition. Composite as a whole, contrasted to individual constituents. Also, sandwich core when used in conjunction with prime (')
- (4) critical
cf - centrifugal force
e - fatigue or endurance
eff - effective
eq - equivalent
f - filament
g - glass scrim cloth
H - hoop
i - i^{th} position in a sequence
L - lateral
m - (1) matrix
- (2) mean
max - maximum
min - minimum
n - (1) n^{th} (last) position in a sequence
- (2) normal
p - polar
s - symmetric
st - stiffener
T - transverse
t - value of parameter at time t
x, y, z - general coordinate system
Σ - total, or summation
o - initial or reference datum
( ) - format for indicating specific, temperature associated with term in parentheses. RT - room temperature (21°C,70°F); all other temperatures in °F unless specified.

1.7.1.4 Superscripts

The following superscript notations are considered standard in MIL-HDBK-17.

b - bending
br - bearing
c - (1) compression
- (2) creep
cc - compression crippling
cr - compression buckling
e - elastic
f - filament
flex - flexure
g - glass scrim cloth
is - interlaminar shear
(i) - i^{th} ply or lamina
lim - limit, used to indicate limit loading
m - matrix
ohc - open hole compression
oht - open hole tension
p - plastic
pl - proportional limit
rup - rupture
s - shear
scr - shear buckling

sec - secant (modulus)
so - offset shear
T - temperature or thermal
t - tension
tan - tangent (modulus)
u - ultimate
y - yield
' - secondary (modulus), or denotes properties of H/C core when used with subscript c
CAI - compression after impact

1.7.1.5 *Acronyms*

The following acronyms are used in MIL-HDBK-17.

AA - atomic absorption
AES - Auger electron spectroscopy
AIA - Aerospace Industries Association
ANOVA - analysis of variance
ARL - US Army Research Laboratory - Materials Directorate
ASTM - American Society for Testing and Materials
BMI - bismaleimide
BVID - barely visible impact damage
CAI - compression after impact
CCA - composite cylinder assemblage
CLS - crack lap shear
CMCS - Composite Motorcase Subcommittee (JANNAF)
CPT - cured ply thickness
CTA - cold temperature ambient
CTD - cold temperature dry
CTE - coefficient of thermal expansion
CV - coefficient of variation
CVD - chemical vapor deposition!
DCB - double cantilever beam
DDA - dynamic dielectric analysis
DLL - design limit load
DMA - dynamic mechanical analysis
DOD - Department of Defense
DSC - differential scanning calorimetry
DTA - differential thermal analysis
DTRC - David Taylor Research Center
ENF - end notched flexure
EOL - end-of-life
ESCA - electron spectroscopy for chemical analysis
ESR - electron spin resonance
ETW - elevated temperature wet
FAA - Federal Aviation Administration
FFF - field flow fractionation
FMECA - Failure Modes Effects Criticality Analysis
FOD - foreign object damage
FTIR - Fourier transform infrared spectroscopy
FWC - finite width correction factor
GC - gas chromatography
GSCS - Generalized Self Consistent Scheme
HDT - heat distortion temperature

HPLC - high performance liquid chromatography
ICAP - inductively coupled plasma emission
IITRI - Illinois Institute of Technology Research Institute
IR - infrared spectroscopy
ISS - ion scattering spectroscopy
JANNAF - Joint Army, Navy, NASA, and Air Force
LC - liquid chromatography
LPT - laminate plate theory
LSS - laminate stacking sequence
MMB - mixed mode bending
MOL - material operational limit
MS - mass spectroscopy
MSDS - material safety data sheet
MTBF - Mean Time Between Failure
NAS - National Aerospace Standard
NASA - National Aeronautics and Space Administration
NDI - nondestructive inspection
NMR - nuclear magnetic resonance
PEEK - polyether ether ketone
RDS - rheological dynamic spectroscopy
RH - relative humidity
RT - room temperature
RTA - room temperature ambient
RTD - room temperature dry
RTM - resin transfer molding
SACMA - Suppliers of Advanced Composite Materials Association
SAE - Society of Automotive Engineers
SANS - small-angle neutron scattering spectroscopy
SEC - size-exclusion chromatography
SEM - scanning electron microscopy
SFC - supercritical fluid chromatography
SI - International System of Units (Le Système Interational d'Unités)
SIMS - secondary ion mass spectroscopy
TBA - torsional braid analysis
TEM - transmission electron microscopy
TGA - thermogravimetric analysis
TLC - thin-layer chromatography
TMA - thermal mechanical analysis
TOS - thermal oxidative stability
TVM - transverse microcrack
UDC - unidirectional fiber composite
VNB - V-notched beam
XPS - X-ray photoelectron spectroscopy

1.7.2 System of units

To comply with Department of Defense Instructive 5000.2, Part 6, Section M, "Use of the Metric System," dated February 23, 1991, the data in MIL-HDBK-17 are generally presented in both the International System of Units (SI units) and the U. S. Customary (English) system of units. ASTM E380, Standard for Metric Practice, provides guidance for the application for SI units which are intended as a basis for world-wide standardization of measurement units (Reference 1.7.2(a)). Further guidelines on the use of the SI system of units and conversion factors are contained in the following publications (References 1.7.2(b) - (e)):

(1) DARCOM P 706-470, *Engineering Design Handbook: Metric Conversion Guide*, July 1976.

(2) NBS Special Publication 330, "The International System of Units (SI)," National Bureau of Standards, 1986 edition.

(3) NBS Letter Circular LC 1035, "Units and Systems of Weights and Measures, Their Origin, Development, and Present Status," National Bureau of Standards, November 1985.

(4) NASA Special Publication 7012, "The International System of Units Physical Constants and Conversion Factors", 1964.

English to SI conversion factors pertinent to MIL-HDBK-17 data are contained in Table 1.7.2.

TABLE 1.7.2 *English to SI conversion factors.*

To convert from	to	Multiply by
Btu (thermochemical)/in^2-s	watt/meter^2 (W/m^2)	1.634 246 E+06
Btu-in/(s-ft^2-°F)	W/(m K)	5.192 204 E+02
degree Fahrenheit	degree Celsius (°C)	T = (T - 32)/1.8
degree Fahrenheit	kelvin (K)	T = (T + 459.67)/1.8
foot	meter (m)	3.048 000 E-01
ft^2	m^2	9.290 304 E-02
foot/second	meter/second (m/s)	3.048 000 E-01
ft/s^2	m/s^2	3.048 000 E-01
inch	meter (m)	2.540 000 E-02
in.^{2}	meter^2 (m^2)	6.451 600 E-04
in.^{3}	m^3	1.638 706 E-05
kilogram-force (kgf)	newton (N)	9.806 650 E+00
kgf/m^2	pascal (Pa)	9.806 650 E+00
kip (1000 lbf)	newton (N)	4.448 222 E+03
ksi (kip/in^2)	MPa	6.894 757 E+00
lbf-in	N-m	1.129 848 E-01
lbf-ft	N-m	1.355 818 E+00
lbf/in^2 (psi)	pascal (Pa)	6.894 757 E+03
lb/in^2	gm/m^2	7.030 696 E+05
lb/in^3	kg/m^3	2.767 990 E+04
Msi (10^6 psi)	GPa	6.894 757 E+00
pound-force (lbf)	newton (N)	4.488 222 E+00
pound-mass (lb avoirdupois)	kilogram (kg)	4.535 924 E-01
torr	pascal (Pa)	1.333 22 E+02

* The letter "E" following the conversion factor stands for exponent and the two digits after the letter "E" indicate the power of 10 by which the number is to be multiplied.

1.8 DEFINITIONS

The following definitions are used within MIL-HDBK-17. This glossary of terms is not totally comprehensive but it does represent nearly all commonly used terms. Where exceptions are made, they are noted in the text and tables. For ease of identification the definitions have been organized alphabetically.

A-Basis (or A-Value) -- A statistically-based material property; a 95% lower confidence bound on the first percentile of a specified population of measurements. Also a 95% lower tolerance bound for the upper 99% of a specified population.

A-Stage -- An early stage in the reaction of thermosetting resins in which the material is still soluble in certain liquids and may be liquid or capable of becoming liquid upon heating. (Sometimes referred to as **resol**.)

Absorption -- A process in which one material (the absorbent) takes in or absorbs another (the absorbate).

Accelerator -- A material which, when mixed with a catalyzed resin, will speed up the chemical reaction between the catalyst and the resin.

Accuracy -- The degree of conformity of a measured or calculated value to some recognized standard or specified value. Accuracy involves the systematic error of an operation.

Addition Polymerization -- Polymerization by a repeated addition process in which monomers are linked together to form a polymer without splitting off of water or other simple molecules.

Adhesion -- The state in which two surfaces are held together at an interface by forces or interlocking action or both.

Adhesive -- A substance capable of holding two materials together by surface attachment. In the handbook, the term is used specifically to designate structural adhesives, those which produce attachments capable of transmitting significant structural loads.

ADK -- Notation used for the k-sample Anderson-Darling statistic, which is used to test the hypothesis that k batches have the same distribution.

Aliquot -- A small, representative portion of a larger sample.

Aging -- The effect, on materials, of exposure to an environment for a period of time; the process of exposing materials to an environment for an interval of time.

Ambient -- The surrounding environmental conditions such as pressure or temperature.

Anelasticity -- A characteristic exhibited by certain materials in which strain is a function of both stress and time, such that, while no permanent deformations are involved, a finite time is required to establish equilibrium between stress and strain in both the loading and unloading directions.

Angleply -- Same as **Crossply**.

Anisotropic -- Not isotropic; having mechanical and/or physical properties which vary with direction relative to natural reference axes inherent in the material.

Aramid -- A manufactured fiber in which the fiber-forming substance consisting of a long-chain synthetic aromatic polyamide in which at least 85% of the amide (-CONH-) linkages are attached directly to two aromatic rings.

Areal Weight of Fiber -- The weight of fiber per unit area of prepreg. This is often expressed as grams per square meter. See Table 1.7.2 for conversion factors.

Artificial Weathering -- Exposure to laboratory conditions which may be cyclic, involving changes in temperature, relative humidity, radiant energy and any other elements found in the atmosphere in various geographical areas.

Aspect Ratio -- In an essentially two-dimensional rectangular structure (e.g., a panel), the ratio of the long dimension to the short dimension. However, in compression loading, it is sometimes considered to be the ratio of the load direction dimension to the transverse dimension. Also, in fiber micro-mechanics, it is referred to as the ratio of length to diameter.

Autoclave -- A closed vessel for producing an environment of fluid pressure, with or without heat, to an enclosed object which is undergoing a chemical reaction or other operation.

Autoclave Molding -- A process similar to the pressure bag technique. The lay-up is covered by a pressure bag, and the entire assembly is placed in an autoclave capable of providing heat and pressure for curing the part. The pressure bag is normally vented to the outside.

Axis of Braiding -- The direction in which the braided form progresses.

B-Basis (or B-Value) -- A statistically-based material property; a 95% lower confidence bound on the tenth percentile of a specified population of measurements. Also a 95% lower tolerance bound for the upper 90% of a specified population. (See Volume 1, Section 8.1.4)

B-Stage -- An intermediate stage in the reaction of a thermosetting resin in which the material softens when heated and swells when in contact with certain liquids but does not entirely fuse or dissolve. Materials are usually precured to this stage to facilitate handling and processing prior to final cure. (Sometimes referred to as **resitol.**)

Bag Molding -- A method of molding or laminating which involves the application of fluid pressure to a flexible material which transmits the pressure to the material being molded or bonded. Fluid pressure usually is applied by means of air, steam, water or vacuum.

Balanced Laminate -- A composite laminate in which all identical laminae at angles other than 0 degrees and 90 degrees occur only in $\pm$ pairs (not necessarily adjacent).

Batch (or Lot) -- For fibers and resins, a quantity of material formed during the same process and having identical characteristics throughout. For prepregs, laminae, and laminates, material made from one batch of fiber and one batch of resin.

Bearing Area -- The product of the pin diameter and the specimen thickness.

Bearing Load -- A compressive load on an interface.

Bearing Yield Strength -- The bearing stress at which a material exhibits a specified limiting deviation from the proportionality of bearing stress to bearing strain.

Bend Test -- A test of ductility by bending or folding, usually with steadily applied forces. In some instances the test may involve blows to a specimen having a cross section that is essentially uniform over a length several times as great as the largest dimension of the cross section.

Binder -- A bonding resin used to hold strands together in a mat or preform during manufacture of a molded object.

Binomial Random Variable -- The number of successes in independent trials where the probability of success is the same for each trial.

Birefringence -- The difference between the two principal refractive indices (of a fiber) or the ratio between the retardation and thickness of a material at a given point.

Bleeder Cloth -- A nonstructural layer of material used in the manufacture of composite parts to allow the escape of excess gas and resin during cure. The bleeder cloth is removed after the curing process and is not part of the final composite.

Bobbin -- A cylinder or slightly tapered barrel, with or without flanges, for holding tows, rovings, or yarns.

Bond -- The adhesion of one surface to another, with or without the use of an adhesive as a bonding agent.

Braid -- A system of three or more yarns which are interwoven in such a way that no two yarns are twisted around each other.

Braid Angle -- The acute angle measured from the axis of braiding.

Braid, Biaxial -- Braided fabric with two-yarn systems, one running in the $+\theta$ direction, the other in the $-\theta$ direction as measured from the axis of braiding.

Braid Count -- The number of braiding yarn crossings per inch measured along the axis of a braided fabric.

Braid, Diamond -- Braided fabric with an over one, under one weave pattern, (1 x 1).

Braid, Flat -- A narrow bias woven tape wherein each yarn is continuous and is intertwined with every other yarn in the system without being intertwined with itself.

Braid, Hercules -- A braided fabric with an over three, under three weave pattern, (3 x 3).

Braid, Jacquard -- A braided design made with the aid of a jacquard machine, which is a shedding mechanism by means of which a large number of ends may be controlled independently and complicated patterns produced.

Braid, Regular -- A braided fabric with an over two, under two weave pattern (2 x 2).

Braid, Square -- A braided pattern in which the yarns are formed into a square pattern.

Braid, Two-Dimensional -- Braided fabric with no braiding yarns in the through thickness direction.

Braid, Three-Dimensional -- Braided fabric with one or more braiding yarns in the through thickness direction.

Braid, Triaxial -- A biaxial braided fabric with laid in yarns running in the axis of braiding.

Braiding -- A textile process where two or more strands, yarns or tapes are intertwined in the bias direction to form an integrated structure.

Broadgoods -- A term loosely applied to prepreg material greater than about 12 inches in width, usually furnished by suppliers in continuous rolls. The term is currently used to designate both collimated uniaxial tape and woven fabric prepregs.

Buckling (Composite) -- A mode of structural response characterized by an out-of-plane material deflection due to compressive action on the structural element involved. In advanced composites, buckling may take the form not only of conventional general instability and local instability but also a micro-instability of individual fibers.

Bundle -- A general term for a collection of essentially parallel filaments or fibers.

C-Stage -- The final stage of the curing reaction of a thermosetting resin in which the material has become practically infusable and insoluble. (Normally considered fully cured and sometimes referred to as **resite**.)

Capstan -- A friction type take-up device which moves braided fabric away from the fell. The speed of which determines the braid angle.

Carbon Fibers -- Fibers produced by the pyrolysis of organic precursor fibers such as rayon, polyacrylonitrile (PAN), and pitch in an inert atmosphere. The term is often used interchangeably with "graphite"; however, carbon fibers and graphite fibers differ in the temperature at which the fibers are made and heat-treated, and the amount of carbon produced. Carbon fibers typically are carbonized at about 2400°F (1300°C) and assay at 93 to 95% carbon, while graphite fibers are graphitized at 3450 to 5450°F (1900 to 3000°C) and assay at more than 99% elemental carbon.

Carrier -- A mechanism for carrying a package of yarn through the braid weaving motion. A typical carrier consists of a bobbin spindle, a track follower, and a tensioning device.

Caul Plates -- Smooth metal plates, free of surface defects, the same size and shape as a composite lay-up, used immediately in contact with the lay-up during the curing process to transmit normal pressure and to provide a smooth surface on the finished laminate.

Censoring -- Data is right (left) censored at M, if, whenever an observation is less than or equal to M (greater than or equal to M), the actual value of the observation is recorded. If the observation exceeds (is less than) M, the observation is recorded as M.

Chain-Growth Polymerization -- One of the two principal polymerization mechanisms. In chain-growth polymerization, the reactive groups are continuously regenerated during the growth process. Once started, the polymer molecule grows rapidly by a chain of reactions emanating from a particular reactive initiator which may be a free radical, cation or anion.

Chromatogram -- A plot of detector response against peak volume of solution (eluate) emerging from the system for each of the constituents which have been separated.

Circuit -- One complete traverse of the fiber feed mechanism of a winding machine; one complete traverse of a winding band from one arbitrary point along the winding path to another point on a plane through the starting point and perpendicular to the axis.

Cocuring -- The act of curing a composite laminate and simultaneously bonding it to some other prepared surface during the same cure cycle (see **Secondary Bonding**).

Coefficient of Linear Thermal Expansion -- The change in length per unit length resulting from a one-degree rise in temperature.

Coefficient of Variation -- The ratio of the population (or sample) standard deviation to the population (or sample) mean.

Collimated -- Rendered parallel.

Compatible -- The ability of different resin systems to be processed in contact with each other without degradation of end product properties. (See **Compatible**, Volume 1, Section 8.1.4)

Composite Class -- As used in the handbook, a major subdivision of composite construction in which the class is defined by the fiber system and the matrix class, e.g., organic-matrix filamentary laminate.

Composite Material -- Composites are considered to be combinations of materials differing in composition or form on a macroscale. The constituents retain their identities in the composite; that is, they do not dissolve or otherwise merge completely into each other although they act in concert. Normally, the components can be physically identified and exhibit an interface between one another.

Compound -- An intimate mixture of polymer or polymers with all the materials necessary for the finished product.

Condensation Polymerization -- This is a special type of step-growth polymerization characterized by the formation of water or other simple molecules during the stepwise addition of reactive groups.

Confidence Coefficient -- See **Confidence Interval.**

Confidence Interval -- A confidence interval is defined by a statement of one of the following forms:

(1) $P\{a<\theta\} \leq 1-\alpha$
(2) $P\{\theta<b\} \leq 1-\alpha$
(3) $P\{a<\theta<b\} \leq 1-\alpha$

where $1-\alpha$ is called the confidence coefficient. A statement of type (1) or (2) is called a one-sided confidence interval and a statement of type (3) is called a two-sided confidence interval. In (1) a is a lower confidence limit and in (2) b is an upper confidence limit. With probability at least $1-\alpha$, the confidence interval will contain the parameter θ.

Constituent -- In general, an element of a larger grouping. In advanced composites, the principal constituents are the fibers and the matrix.

Continuous Filament -- A yarn or strand in which the individual filaments are substantially the same length as the strand.

Coupling Agent -- Any chemical substance designed to react with both the reinforcement and matrix phases of a composite material to form or promote a stronger bond at the interface. Coupling agents are applied to the reinforcement phase from an aqueous or organic solution or from a gas phase, or added to the matrix as an integral blend.

Coverage -- The measure of the fraction of surface area covered by the braid.

Crazing -- Apparent fine cracks at or under the surface of an organic matrix.

Creel -- A framework arranged to hold tows, rovings, or yarns so that many ends can be withdrawn smoothly and evenly without tangling.

Creep -- The time dependent part of strain resulting from an applied stress.

Creep, Rate Of -- The slope of the creep-time curve at a given time.

Crimp -- The undulations induced into a braided fabric via the braiding process.

Crimp Angle -- The maximum acute angle of a single braided yarn's direction measured from the average axis of tow.

Crimp Exchange -- The process by which a system of braided yarns reaches equilibrium when put under tension or compression.

Critical Value(s) -- When testing a one-sided statistical hypothesis, a critical value is the value such that, if the test statistic is greater than (less than) the critical value, the hypothesis is rejected. When testing a two-sided statistical hypothesis, two critical values are determined. If the test statistic is either less than the smaller critical value or greater than the larger critical value, then the hypothesis is rejected. In both cases, the critical value chosen depends on the desired risk (often 0.05) of rejecting the hypothesis when it is true.

Crossply -- Any filamentary laminate which is not uniaxial. Same as Angleply. In some references, the term crossply is used to designate only those laminates in which the laminae are at right angles to one another, while the term angleply is used for all others. In the handbook, the two terms are used synonymously. The reservation of a separate terminology for only one of several basic orientations is unwarranted because a laminate orientation code is used.

Cumulative Distribution Function -- See Volume 1, Section 8.1.4.

Cure -- To change the properties of a thermosetting resin irreversibly by chemical reaction, i.e., condensation, ring closure, or addition. Cure may be accomplished by addition of curing (cross-linking) agents, with or without catalyst, and with or without heat. Cure may occur also by addition, such as occurs with anhydride cures for epoxy resin systems.

Cure Cycle -- The schedule of time periods at specified conditions to which a reacting thermosetting material is subjected in order to reach a specified property level.

Cure Stress -- A residual internal stress produced during the curing cycle of composite structures. Normally, these stresses originate when different components of a lay-up have different thermal coefficients of expansion.

Debond -- A deliberate separation of a bonded joint or interface, usually for repair or rework purposes. (See **Disbond, Unbond**).

Deformation -- The change in shape of a specimen caused by the application of a load or force.

Degradation -- A deleterious change in chemical structure, physical properties or appearance.

Delamination -- The separation of the layers of material in a laminate. This may be local or may cover a large area of the laminate. It may occur at any time in the cure or subsequent life of the laminate and may arise from a wide variety of causes.

Denier -- A direct numbering system for expressing linear density, equal to the mass in grams per 9000 meters of yarn, filament, fiber, or other textile strand.

Density -- The mass per unit volume.

Desorption -- A process in which an absorbed or adsorbed material is released from another material. Desorption is the reverse of absorption, adsorption, or both.

Deviation -- Variation from a specified dimension or requirement, usually defining the upper and lower limits.

Dielectric Constant -- The ratio of the capacity of a condenser having a dielectric constant between the plates to that of the same condenser when the dielectric is replaced by a vacuum; a measure of the electrical charge stored per unit volume at unit potential.

Dielectric Strength -- The average potential per unit thickness at which failure of the dielectric material occurs.

Disbond -- An area within a bonded interface between two adherends in which an adhesion failure or separation has occurred. It may occur at any time during the life of the structure and may arise from a wide variety of causes. Also, colloquially, an area of separation between two laminae in the finished laminate (in this case the term "delamination" is normally preferred.) (See **Debond, Unbond, Delamination**.)

Distribution -- A formula which gives the probability that a value will fall within prescribed limits. (See **Normal**, **Weibull**, and **Lognormal Distributions**, also Volume 1, Section 8.1.4).

Dry -- a material condition of moisture equilibrium with a surrounding environment at 5% or lower relative humidity.

Dry Fiber Area -- Area of fiber not totally encapsulated by resin.

Ductility -- The ability of a material to deform plastically before fracturing.

Elasticity -- The property of a material which allows it to recover its original size and shape immediately after removal of the force causing deformation.

Elongation -- The increase in gage length or extension of a specimen during a tension test, usually expressed as a percentage of the original gage length.

Eluate -- The liquid emerging from a column (in liquid chromatography).

Eluent -- The mobile phase used to sweep or elute the sample (solute) components into, through, and out of the column.

End -- A single fiber, strand, roving or yarn being or already incorporated into a product. An end may be an individual warp yarn or cord in a woven fabric. In referring to aramid and glass fibers, an end is usually an untwisted bundle of continuous filaments.

Epoxy Equivalent Weight -- The number of grams of resin which contain one chemical equivalent of the epoxy group.

Epoxy Resin -- Resins which may be of widely different structures but are characterized by the presence of the epoxy group. (The epoxy or epoxide group is usually present as a glycidyl ether, glycidyl amine, or as part of an aliphatic ring system. The aromatic type epoxy resins are normally used in composites.)

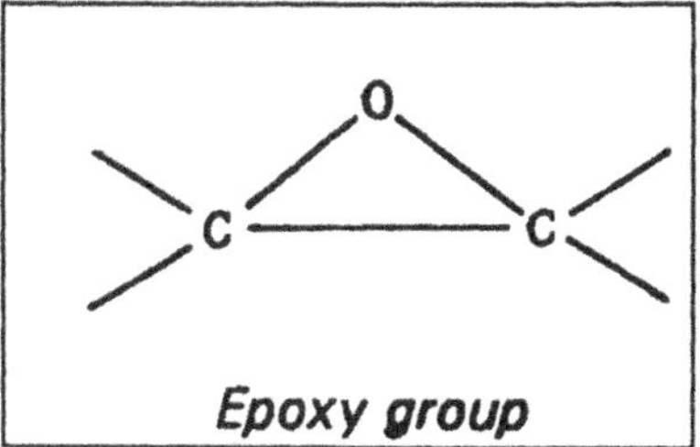

Epoxy group

Extensometer -- A device for measuring linear strain.

F-Distribution -- See Volume 1, Section 8.1.4.

Fabric, Nonwoven -- A textile structure produced by bonding or interlocking of fibers, or both, accomplished by mechanical, chemical, thermal, or solvent means, and combinations thereof.

Fabric, Woven -- A generic material construction consisting of interlaced yarns or fibers, usually a planar structure. Specifically, as used in this handbook, a cloth woven in an established weave pattern from advanced fiber yarns and used as the fibrous constituent in an advanced composite lamina. In a fabric lamina, the warp direction is considered the longitudinal direction, analogous to the filament direction in a filamentary lamina.

Fell -- The point of braid formation, which is defined as the point at which the yarns in a braid system cease movement relative to each other.

Fiber -- A general term used to refer to filamentary materials. Often, fiber is used synonymously with filament. It is a general term for a filament of finite length. A unit of matter, either natural or manmade, which forms the basic element of fabrics and other textile structures.

Fiber Content -- The amount of fiber present in a composite. This is usually expressed as a percentage volume fraction or weight fraction of the composite.

Fiber Count -- The number of fibers per unit width of ply present in a specified section of a composite.

Fiber Direction -- The orientation or alignment of the longitudinal axis of the fiber with respect to a stated reference axis.

Fiber System -- The type and arrangement of fibrous material which comprises the fiber constituent of an advanced composite. Examples of fiber systems are collimated filaments or filament yarns, woven fabric, randomly oriented short-fiber ribbons, random fiber mats, whiskers, etc.

Fiber Volume (Fraction) -- See fiber content.

Filament -- The smallest unit of a fibrous material. The basic units formed during spinning and which are gathered into strands of fiber, (for use in composites). Filaments usually are of extreme length and of very small diameter. Filaments normally are not used individually. Some textile filaments can function as a yarn when they are of sufficient strength and flexibility.

Filamentary Composite -- A composite material reinforced with continuous fibers.

Filament winding -- See **Winding.**

Filament Wound -- Pertaining to an object created by the filament winding method of fabrication.

Fill (Filling) -- In a woven fabric, the yarn running from selvage to selvage at right angles to the warp.

Filler -- A relatively inert substance added to a material to alter its physical, mechanical, thermal, electrical, and other properties or to lower cost. Sometimes the term is used specifically to mean particulate additives.

Finish (or Size System) -- A material, with which filaments are treated, which contains a coupling agent to improve the bond between the filament surface and the resin matrix in a composite material. In addition, finishes often contain ingredients which provide lubricity to the filament surface, preventing abrasive damage during handling, and a binder which promotes strand integrity and facilitates packing of the filaments.

Fixed Effect -- A systematic shift in a measured quantity due to a particular level change of a treatment or condition. (See Volume 1, Section 8.1.4.)

Flash -- Excess material which forms at the parting line of a mold or die, or which is extruded from a closed mold.

Former Plate -- A die attached to a braiding machine which helps to locate the fell.

Fracture Ductility -- The true plastic strain at fracture.

Gage Length -- the original length of that portion of the specimen over which strain or change of length is determined.

Gel -- The initial jelly-like solid phase that develops during formation of a resin from a liquid. Also, a semi-solid system consisting of a network of solid aggregates in which liquid is held.

Gel Coat -- A quick-setting resin used in molding processes to provide an improved surface for the composite; it is the first resin applied to the mold after the mold-release agent.

Gel Point -- The stage at which a liquid begins to exhibit pseudo-elastic properties. (This can be seen from the inflection point on a viscosity-time plot.)

Gel Time -- The period of time from a pre-determined starting point to the onset of gelation (gel point) as defined by a specific test method.

Glass -- An inorganic product of fusion which has cooled to a rigid condition without crystallizing. In the handbook, all reference to glass will be to the fibrous form as used in filaments, woven fabric, yarns, mats, chopped fibers, etc.

Glass Cloth -- Conventionally-woven glass fiber material (see **Scrim**).

Glass Fibers -- A fiber spun from an inorganic product of fusion which has cooled to a rigid condition without crystallizing.

Glass Transition -- The reversible change in an amorphous polymer or in amorphous regions of a partially crystalline polymer from (or to) a viscous or rubbery condition to (or from) a hard and relatively brittle one.

Glass Transition Temperature -- The approximate midpoint of the temperature range over which the glass transition takes place.

Graphite Fibers -- See **Carbon Fibers**.

Greige -- Fabric that has received no finish.

Hand Lay-up -- A process in which components are applied either to a mold or a working surface, and the successive plies are built up and worked by hand.

Hardness -- Resistance to deformation; usually measured by indention. Types of standard tests include Brinell, Rockwell, Knoop, and Vickers.

Heat Cleaned -- Glass or other fibers which have been exposed to elevated temperatures to remove preliminary sizings or binders which are not compatible with the resin system to be applied.

Heterogeneous -- Descriptive term for a material consisting of dissimilar constituents separately identifiable; a medium consisting of regions of unlike properties separated by internal boundaries. (Note that all nonhomogeneous materials are not necessarily heterogeneous).

Homogeneous -- Descriptive term for a material of uniform composition throughout; a medium which has no internal physical boundaries; a material whose properties are constant at every point, in other words, constant with respect to spatial coordinates (but not necessarily with respect to directional coordinates).

Horizontal Shear -- Sometimes used to indicate interlaminar shear. This is not an approved term for use in this handbook.

Humidity, Relative -- The ratio of the pressure of water vapor present to the pressure of saturated water vapor at the same temperature.

Hybrid -- A composite laminate comprised of laminae of two or more composite material systems. Or, a combination of two or more different fibers such as carbon and glass or carbon and aramid into a structure (tapes, fabrics and other forms may be combined).

Hygroscopic -- Capable of absorbing and retaining atmospheric moisture.

Hysteresis -- The energy absorbed in a complete cycle of loading and unloading.

Inclusion -- A physical and mechanical discontinuity occurring within a material or part, usually consisting of solid, encapsulated foreign material. Inclusions are often capable of transmitting some structural stresses and energy fields, but in a noticeably different manner from the parent material.

Integral Composite Structure -- Composite structure in which several structural elements, which would conventionally be assembled by bonding or with mechanical fasteners after separate fabrication, are instead laid up and cured as a single, complex, continuous structure; e.g., spars, ribs, and one stiffened cover of a wing box fabricated as a single integral part. The term is sometimes applied more loosely to any composite structure not assembled by mechanical fasteners.

Interface -- The boundary between the individual, physically distinguishable constituents of a composite.

Interlaminar -- Between the laminae of a laminate.

Discussion: describing objects (e.g., voids), events (e.g., fracture), or fields (e.g., stress).

Interlaminar Shear -- Shearing force tending to produce a relative displacement between two laminae in a laminate along the plane of their interface.

Intermediate Bearing Stress -- The bearing stress at the point on the bearing load-deformation curve where the tangent is equal to the bearing stress divided by a designated percentage (usually 4%) of the original hole diameter.

Intralaminar -- Within the laminae of a laminate.

Discussion: describing objects (for example, voids), event (for example, fracture), or fields (for example, stress).

Isotropic -- Having uniform properties in all directions. The measured properties of an isotropic material are independent of the axis of testing.

Jammed State -- The state of a braided fabric under tension or compression where the deformation of the fabric is dominated by the deformation properties of the yarn.

Knitting -- A method of constructing fabric by interlocking series of loops of one or more yarns.

Knuckle Area -- The area of transition between sections of different geometry in a filament wound part.

k-Sample Data -- A collection of data consisting of values observed when sampling from k batches.

Laid-In Yarns -- A system of longitudinal yarns in a triaxial braid which are inserted between the bias yarns.

Lamina -- A single ply or layer in a laminate.

Discussion: For filament winding, a lamina is a layer.

Laminae -- Plural of lamina.

Laminate -- for fiber-reinforced composites, a consolidated collection of laminae (plies) with one or more orientations with respect to some reference direction.

Laminate Orientation -- The configuration of a crossplied composite laminate with regard to the angles of crossplying, the number of laminae at each angle, and the exact sequence of the lamina lay-up.

Lattice Pattern -- A pattern of filament winding with a fixed arrangement of open voids.

Lay-up -- A process of fabrication involving the assembly of successive layers of resin-impregnated material.

Lognormal Distribution -- A probability distribution for which the probability that an observation selected at random from this population falls between a and b ($0 < a < b < B$) is given by the area under the normal distribution between $\log a$ and $\log b$. The common (base 10) or the natural (base e) logarithm may be used. (See Volume 1, Section 8.1.4.)

Lower Confidence Bound -- See **Confidence Interval**.

Macro -- In relation to composites, denotes the gross properties of a composite as a structural element but does not consider the individual properties or identity of the constituents.

Macrostrain -- The mean strain over any finite gage length of measurement which is large in comparison to the material's interatomic distance.

Mandrel -- A form fixture or male mold used for the base in the production of a part by lay-up, filament winding or braiding.

Mat -- A fibrous material consisting of randomly oriented chopped or swirled filaments loosely held together with a binder.

Material Acceptance -- The testing of incoming material to ensure that it meets requirements.

Material Qualification -- The procedures used to accept a material by a company or organization for production use.

Material System -- A specific composite material made from specifically identified constituents in specific geometric proportions and arrangements and possessed of numerically defined properties.

Material System Class -- As used in this handbook, a group consisting of material systems categorized by the same generic constituent materials, but without defining the constituents uniquely; e.g., the carbon/epoxy class.

Material Variability -- A source of variability due to the spatial and consistency variations of the material itself and due to variation in its processing. (See Volume 1, Section 8.1.4.)

Matrix -- The essentially homogeneous material in which the fiber system of a composite is embedded.

Matrix Content -- The amount of matrix present in a composite expressed either as percent by weight or percent by volume. Discussion: For polymer matrix composites this is called resin content, which is usually expressed as percent by weight

Mean -- See **Sample Mean** and **Population Mean**.

Mechanical Properties -- The properties of a material that are associated with elastic and inelastic reaction when force is applied, or the properties involving the relationship between stress and strain.

Median -- See **Sample Median** and **Population Median**.

Micro -- In relation to composites, denotes the properties of the constituents, i.e., matrix and reinforcement and interface only, as well as their effects on the composite properties.

Microstrain -- The strain over a gage length comparable to the material's interatomic distance.

Modulus, Chord -- The slope of the chord drawn between any two specified points on the stress-strain curve.

Modulus, initial -- The slope of the initial straight portion of a stress-strain curve.

Modulus, Secant -- The slope of the secant drawn from the origin to any specified point on the stress-strain curve.

Modulus, Tangent -- The ratio of change in stress to change in strain derived from the tangent to any point on a stress-strain curve.

Modulus, Young's -- The ratio of change in stress to change in strain below the elastic limit of a material. (Applicable to tension and compression).

Modulus of Rigidity (also Shear Modulus or Torsional Modulus) -- The ratio of stress to strain below the proportional limit for shear or torsional stress.

Modulus of Rupture, in Bending -- The maximum tensile or compressive stress (whichever causes failure) value in the extreme fiber of a beam loaded to failure in bending. The value is computed from the flexure equation:

$$F^b = \frac{Mc}{I} \qquad 1.8(a)$$

where M = maximum bending moment computed from the maximum load and the original moment arm,
c = initial distance from the neutral axis to the extreme fiber where failure occurs,
I = the initial moment of inertia of the cross section about its neutral axis.

Modulus of Rupture, in Torsion -- The maximum shear stress in the extreme fiber of a member of circular cross section loaded to failure in torsion calculated from the equation:

$$F^s = \frac{Tr}{J} \qquad 1.8(b)$$

where T = maximum twisting moment,
r = original outer radius,
J = polar moment of inertia of the original cross section.

Moisture Content -- The amount of moisture in a material determined under prescribed condition and expressed as a percentage of the mass of the moist specimen, i.e., the mass of the dry substance plus the moisture present.

Moisture Equilibrium -- The condition reached by a sample when it no longer takes up moisture from, or gives up moisture to, the surrounding environment.

Mold Release Agent -- A lubricant applied to mold surfaces to facilitate release of the molded article.

Molded Edge -- An edge which is not physically altered after molding for use in final form and particularly one which does not have fiber ends along its length.

Molding -- The forming of a polymer or composite into a solid mass of prescribed shape and size by the application of pressure and heat.

Monolayer -- The basic laminate unit from which crossplied or other laminates are constructed.

Monomer -- A compound consisting of molecules each of which can provide one or more constitutional units.

NDE -- Nondestructive evaluation. Broadly considered synonymous with NDI.

NDI -- Nondestructive inspection. A process or procedure for determining the quality or characteristics of a material, part, or assembly without permanently altering the subject or its properties.

NDT -- Nondestructive testing. Broadly considered synonymous with NDI.

Necking -- A localized reduction in cross-sectional area which may occur in a material under tensile stress.

Negatively Skewed -- A distribution is said to be negatively skewed if the distribution is not symmetric and the longest tail is on the left.

Nominal Specimen Thickness -- The nominal ply thickness multiplied by the number of plies.

Nominal Value -- A value assigned for the purpose of a convenient designation. A nominal value exists in name only.

Normal Distribution -- A two parameter (μ,σ) family of probability distributions for which the probability that an observation will fall between a and b is given by the area under the curve

$$f(x) = \frac{1}{\sigma\sqrt{2\pi}}\exp\left[-\frac{(x-\mu)^2}{2\sigma^2}\right] \qquad 1.8(c)$$

between a and b. (See Volume 1, Section 8.1.4.)

Normalization -- A mathematical procedure for adjusting raw test values for fiber-dominated properties to a single (specified) fiber volume content.

Normalized Stress -- Stress value adjusted to a specified fiber volume content by multiplying the measured stress value by the ratio of specimen fiber volume to the specified fiber volume. This ratio may be obtained directly by experimentally measuring fiber volume, or indirectly by calculation using specimen thickness and fiber areal weight.

Observed Significance Level (OSL) -- The probability of observing a more extreme value of the test statistic when the null hypotheses is true.

Offset Shear Strength --- (from valid execution of a material property shear response test) the value of shear stress at the intersection between a line parallel to the shear chord modulus of elasticity and the shear stress/strain curve, where the line has been offset along the shear strain axis from the origin by a specified strain offset value.

Oligomer -- A polymer consisting of only a few monomer units such as a dimer, trimer, etc., or their mixtures.

One-Sided Tolerance Limit Factor -- See **Tolerance Limit Factor**.

Orthotropic -- Having three mutually perpendicular planes of elastic symmetry.

Oven Dry -- The condition of a material that has been heated under prescribed conditions of temperature and humidity until there is no further significant change in its mass.

PAN Fibers -- Reinforcement fiber derived from the controlled pyrolysis of poly(acrylonitrile) fiber.

Parallel Laminate -- A laminate of woven fabric in which the plies are aligned in the same position as originally aligned in the fabric roll.

Parallel Wound -- A term used to describe yarn or other material wound into a flanged spool.

Peel Ply -- A layer of resin free material used to protect a laminate for later secondary bonding.

pH -- A measure of acidity or alkalinity of a solution, with neutrality represented by a value of 7, with increasing acidity corresponding to progressively smaller values, and increasing alkalinity corresponding to progressively higher values.

Pick Count -- The number of filling yarns per inch or per centimeter of woven fabric.

Pitch Fibers -- Reinforcement fiber derived from petroleum or coal tar pitch.

Plastic -- A material that contains one or more organic polymers of large molecular weight, is solid in its finished state, and, at some state in its manufacture or processing into finished articles, can be shaped by flow.

Plasticizer -- A material of lower molecular weight added to a polymer to separate the molecular chains. This results in a depression of the glass transition temperature, reduced stiffness and brittleness, and improved processability. (Note, many polymeric materials do not need a plasticizer.)

Plied Yarn -- A yarn formed by twisting together two or more single yarns in one operation.

Poisson's Ratio -- The absolute value of the ratio of transverse strain to the corresponding axial strain resulting from uniformly distributed axial stress below the proportional limit of the material.

Polymer -- An organic material composed of molecules characterized by the repetition of one or more types of monomeric units.

Polymerization -- A chemical reaction in which the molecules of monomers are linked together to form polymers via two principal reaction mechanisms. Addition polymerizations proceed by chain growth and most condensation polymerizations through step growth.

Population -- The set of measurements about which inferences are to be made or the totality of possible measurements which might be obtained in a given testing situation. For example, "all possible ultimate tensile strength measurements for carbon/epoxy system A, conditioned at 95% relative humidity and room temperature". In order to make inferences about a population, it is often necessary to make assumptions about its distributional form. The assumed distributional form may also be referred to as the population. (See Volume 1, Section 8.1.4.)

Population Mean -- The average of all potential measurements in a given population weighted by their relative frequencies in the population. (See Volume 1, Section 8.1.4.)

Population Median -- That value in the population such that the probability of exceeding it is 0.5 and the probability of being less than it is 0.5. (See Volume 1, Section 8.1.4.)

Population Variance -- A measure of dispersion in the population.

Porosity -- A condition of trapped pockets of air, gas, or vacuum within a solid material, usually expressed as a percentage of the total nonsolid volume to the total volume (solid plus nonsolid) of a unit quantity of material.

Positively Skewed -- A distribution is said to be positively skewed if the distribution is not symmetric and the longest tail is on the right.

Postcure -- Additional elevated temperature cure, usually without pressure, to increase the glass transition temperature, to improve final properties, or to complete the cure.

Pot Life -- The period of time during which a reacting thermosetting composition remains suitable for its intended processing after mixing with a reaction initiating agent.

Precision -- The degree of agreement within a set of observations or test results obtained. Precision involves repeatability and reproducibility.

Precursor (for Carbon or Graphite Fiber) -- Either the PAN or pitch fibers from which carbon and graphite fibers are derived.

Preform -- An assembly of dry fabric and fibers which has been prepared for one of several different wet resin injection processes. A preform may be stitched or stabilized in some other way to hold its A shape. A commingled preform may contain thermoplastic fibers and may be consolidated by elevated temperature and pressure without resin injection.

Preply -- Layers of prepreg material, which have been assembled according to a user specified stacking sequence.

Prepreg -- Ready to mold or cure material in sheet form which may be tow, tape, cloth, or mat impregnated with resin. It may be stored before use.

Pressure -- The force or load per unit area.

Probability Density Function -- See Volume 1, Section 8.1.4.

Proportional Limit -- The maximum stress that a material is capable of sustaining without any deviation from the proportionality of stress to strain (also known as Hooke's law).

Quasi-Isotropic Laminate -- A balanced and symmetric laminate for which a constitutive property of interest, at a given point, displays isotropic behavior in the plane of the laminate.

Discussion: Common quasi-isotropic laminates are $(0/\pm 60)_s$ and $(0/\pm 45/90)_s$.

Random Effect -- A shift in a measured quantity due to a particular level change of an external, usually uncontrollable, factor. (See Volume 1, Section 8.1.4.)

Random Error -- That part of the data variation that is due to unknown or uncontrolled factors and that affects each observation independently and unpredictably. (See Volume 1, Section 8.1.4.)

Reduction of Area -- The difference between the original cross sectional area of a tension test specimen and the area of its smallest cross section, usually expressed as a percentage of the original area.

Refractive Index - The ratio of the velocity of light (of specified wavelength) in air to its velocity in the substance under examination. Also defined as the sine of the angle of incidence divided by the sine of the angle of refraction as light passes from air into the substance.

Reinforced Plastic -- A plastic with relatively high stiffness or very high strength fibers embedded in the composition. This improves some mechanical properties over that of the base resin.

Release Agent -- See **Mold Release Agent**.

Resilience -- A property of a material which is able to do work against restraining forces during return from a deformed condition.

Resin -- An organic polymer or prepolymer used as a matrix to contain the fibrous reinforcement in a composite material or as an adhesive. This organic matrix may be a thermoset or a thermoplastic, and may contain a wide variety of components or additives to influence; handleability, processing behavior and ultimate properties.

Resin Content -- See **Matrix content.**

Resin Starved Area -- Area of composite part where the resin has a non-continuous smooth coverage of the fiber.

Resin System -- A mixture of resin, with ingredients such as catalyst, initiator, diluents, etc. required for the intended processing and final product.

Room Temperature Ambient (RTA) -- 1) an environmental condition of 73±5°F (23±3°C) at ambient laboratory relative humidity; 2) a material condition where, immediately following consolidation/cure, the material is stored at 73±5°F (23±3°C) and at a maximum relative humidity of 60%.

Roving -- A number of strands, tows, or ends collected into a parallel bundle with little or no twist. In spun yarn production, an intermediate state between sliver and yarn.

S-Basis (or S-Value) -- The mechanical property value which is usually the specified minimum value of the appropriate government specification or SAE Aerospace Material Specification for this material.

Sample -- A small portion of a material or product intended to be representative of the whole. Statistically, a sample is the collection of measurements taken from a specified population. (See Volume 1, Section 8.1.4.)

Sample Mean -- The arithmetic average of the measurements in a sample. The sample mean is an estimator of the population mean. (See Volume 1, Section 8.1.4.)

Sample Median -- Order the observation from smallest to largest. Then the sample median is the value of the middle observation if the sample size is odd; the average of the two central observations if n is even. If the population is symmetric about its mean, the sample median is also an estimator of the population mean. (See Volume 1, Section 8.1.4.)

Sample Standard Deviation -- The square root of the sample variance. (See Volume 1, Section 8.1.4.)

Sample Variance -- The sum of the squared deviations from the sample mean, divided by $n-1$. (See Volume 1, Section 8.1.4.)

Sandwich Construction -- A structural panel concept consisting in its simplest form of two relatively thin, parallel sheets of structural material bonded to, and separated by, a relatively thick, light-weight core.

Saturation -- An equilibrium condition in which the net rate of absorption under prescribed conditions falls essentially to zero.

Scrim (also called **Glass Cloth, Carrier**) -- A low cost fabric woven into an open mesh construction, used in the processing of tape or other B-stage material to facilitate handling.

Secondary Bonding -- The joining together, by the process of adhesive bonding, of two or more already-cured composite parts, during which the only chemical or thermal reaction occurring is the curing of the adhesive itself.

Selvage or Selvedge -- The woven edge portion of a fabric parallel to the warp.

Set -- The strain remaining after complete release of the force producing the deformation.

Shear Fracture (for crystalline type materials) -- A mode of fracture resulting from translation along slip planes which are preferentially oriented in the direction of the shearing stress.

Shelf Life -- The length of time a material, substance, product, or reagent can be stored under specified environmental conditions and continue to meet all applicable specification requirements and/or remain suitable for its intended function.

Short Beam Strength (SBS) -- a test result from valid execution of ASTM test method D2344.

Significant -- Statistically, the value of a test statistic is significant if the probability of a value at least as extreme is less than or equal to a predetermined number called the significance level of the test.

Significant Digit -- Any digit that is necessary to define a value or quantity.

Size System -- See **Finish**.

Sizing -- A generic term for compounds which are applied to yarns to bind the fiber together and stiffen the yarn to provide abrasion-resistance during weaving. Starch, gelatin, oil, wax, and man-made polymers such as polyvinyl alcohol, polystyrene, polyacrylic acid, and polyacetatates are employed.

Skewness -- See **Positively Skewed, Negatively Skewed**.

Sleeving -- A common name for tubular braided fabric.

Slenderness Ratio -- The unsupported effective length of a uniform column divided by the least radius of gyration of the cross-sectional area.

Sliver -- A continuous strand of loosely assembled fiber that is approximately uniform in cross-sectional area and has no twist.

Solute -- The dissolved material.

Specific Gravity -- The ratio of the weight of any volume of a substance to the weight of an equal volume of another substance taken as standard at a constant or stated temperature. Solids and liquids are usually compared with water at 39°F (4°C).

Specific Heat -- The quantity of heat required to raise the temperature of a unit mass of a substance one degree under specified conditions.

Specimen -- A piece or portion of a sample or other material taken to be tested. Specimens normally are prepared to conform with the applicable test method.

Spindle -- A slender upright rotation rod on a spinning frame, roving frame, twister or similar machine.

Standard Deviation -- See **Sample Standard Deviation**.

Staple -- Either naturally occurring fibers or lengths cut from filaments.

Step-Growth Polymerization -- One of the two principal polymerization mechanisms. In sep-growth polymerization, the reaction grows by combination of monomer, oligomer, or polymer molecules through the consumption of reactive groups. Since average molecular weight increases with monomer consumption, high molecular weight polymers are formed only at high degrees of conversion.

Strain -- the per unit change, due to force, in the size or shape of a body referred to its original size or shape. Strain is a nondimensional quantity, but it is frequently expressed in inches per inch, meters per meter, or percent.

Strand -- Normally an untwisted bundle or assembly of continuous filaments used as a unit, including slivers, tow, ends, yarn, etc. Sometimes a single fiber or filament is called a strand.

Strength -- the maximum stress which a material is capable of sustaining.

Stress -- The intensity at a point in a body of the forces or components of forces that act on a given plane through the point. Stress is expressed in force per unit area (pounds-force per square inch, megapascals, etc.).

Stress Relaxation -- The time dependent decrease in stress in a solid under given constraint conditions.

Stress-Strain Curve (Diagram) -- A graphical representation showing the relationship between the change in dimension of the specimen in the direction of the externally applied stress and the magnitude of the applied stress. Values of stress usually are plotted as ordinates (vertically) and strain values as abscissa (horizontally).

Structural Element -- a generic element of a more complex structural member (for example, skin, stringer, shear panels, sandwich panels, joints, or splices).

Structured Data -- See Volume 1, Section 8.1.4.

Surfacing Mat -- A thin mat of fine fibers used primarily to produce a smooth surface on an organic matrix composite.

Symmetrical Laminate -- A composite laminate in which the sequence of plies below the laminate midplane is a mirror image of the stacking sequence above the midplane.

Tack -- Stickiness of the prepreg.

Tape -- Prepreg fabricated in widths up to 12 inches wide for carbon and 3 inches for boron. Cross stitched carbon tapes up to 60 inches wide are available commercially in some cases.

Tenacity -- The tensile stress expressed as force per unit linear density of the unstrained specimen i.e., grams-force per denier or grams-force per tex.

Tex -- A unit for expressing linear density equal to the mass or weight in grams of 1000 meters of filament, fiber, yarn or other textile strand.

Thermal Conductivity -- Ability of a material to conduct heat. The physical constant for quantity of heat that passes through unit cube of a substance in unit time when the difference in temperature of two faces is one degree.

Thermoplastic -- A plastic that repeatedly can be softened by heating and hardened by cooling through a temperature range characteristic of the plastic, and when in the softened stage, can be shaped by flow into articles by molding or extrusion.

Thermoset -- A class of polymers that, when cured using heat, chemical, or other means, changes into a substantially infusible and insoluble material.

Tolerance -- The total amount by which a quantity is allowed to vary.

Tolerance Limit -- A lower (upper) confidence limit on a specified percentile of a distribution. For example, the B-basis value is a 95% lower confidence limit on the tenth percentile of a distribution.

Tolerance Limit Factor -- The factor which is multiplied by the estimate of variability in computing the tolerance limit.

Toughness -- A measure of a material's ability to absorb work, or the actual work per unit volume or unit mass of material that is required to rupture it. Toughness is proportional to the area under the load-elongation curve from the origin to the breaking point.

Tow -- An untwisted bundle of continuous filaments. Commonly used in referring to man-made fibers, particularly carbon and graphite fibers, in the composites industry.

Transformation -- A transformation of data values is a change in the units of measurement accomplished by applying a mathematical function to all data values. For example, if the data is given by x, then $y = x + 1$, x , $1/x$, $\log x$, and $\cos x$ are transformations.

Transition, First Order -- A change of state associated with crystallization or melting in a polymer.

Transversely Isotropic -- Descriptive term for a material exhibiting a special case of orthotropy in which properties are identical in two orthotropic dimensions, but not the third; having identical properties in both transverse directions but not the longitudinal direction.

Traveller -- A small piece of the same product (panel, tube, etc.) as the test specimen, used for example to measure moisture content as a result of conditioning.

Twist -- The number of turns about its axis per unit of length in a yarn or other textile strand. It may be expressed as turns per inch (tpi) or turns per centimeter (tpcm).

Twist, Direction of -- The direction of twist in yarns and other textile strands is indicated by the capital letters S and Z. Yarn has S twist if, when held in a vertical position, the visible spirals or helices around its central axis are in the direction of slope of the central portion of the letter S, and Z twist is in the other direction.

Twist multiplier -- The ratio of turns per inch to the square root of the cotton count.

Typical Basis -- A typical property value is a sample mean. Note that the typical value is defined as the simple arithmetic mean which has a statistical connotation of 50% reliability with a 50% confidence.

Unbond -- An area within a bonded interface between two adherends in which the intended bonding action failed to take place. Also used to denote specific areas deliberately prevented from bonding in order to simulate a defective bond, such as in the generation of quality standards specimens. (See **Disbond, Debond**).

Unidirectional Fiber-Reinforced Composite -- Any fiber-reinforced composite with all fibers aligned in a single direction.

Unit Cell -- The term applied to the path of a yarn in a braided fabric representing a unit cell of a repeating geometric pattern. The smallest element representative of the braided structure.

Unstructured Data -- See Volume 1, Section 8.1.4.

Upper Confidence Limit -- See **Confidence Interval**.

Vacuum Bag Molding -- A process in which the lay-up is cured under pressure generated by drawing a vacuum in the space between the lay-up and a flexible sheet placed over it and sealed at the edges.

Variance -- See **Sample Variance**.

Viscosity -- The property of resistance to flow exhibited within the body of a material.

Void - Any pocket of enclosed gas or near-vacuum within a composite.

Warp -- The longitudinally oriented yarn in a woven fabric (see **Fill**); a group of yarns in long lengths and approximately parallel.

Wet Lay-up -- A method of making a reinforced product by applying a liquid resin system while or after the reinforcement is put in place.

Weibull Distribution (Two - Parameter) -- A probability distribution for which the probability that a randomly selected observation from this population lies between a and b ($0 < a < b < \infty$) is given by Equation 1.8(d) where α is called the scale parameter and β is called the shape parameter. (See Volume 1, Section 8.1.4.)

$$\exp\left[-\left(\frac{a}{\alpha}\right)^{\beta}\right] - \exp\left[-\left(\frac{b}{\alpha}\right)^{\beta}\right] \qquad 1.8(d)$$

Wet Lay-up -- A method of making a reinforced product by applying a liquid resin system while the reinforcement is put in place.

Wet Strength -- The strength of an organic matrix composite when the matrix resin is saturated with absorbed moisture. (See **Saturation**).

Wet Winding -- A method of filament winding in which the fiber reinforcement is coated with the resin system as a liquid just prior to wrapping on a mandrel.

Whisker -- A short single crystal fiber or filament. Whisker diameters range from 1 to 25 microns, with aspect ratios between 100 and 15,000.

Winding -- A process in which continuous material is applied under controlled tension to a form in a predetermined geometric relationship to make a structure.

Discussion: A matrix material to bind the fibers together may be added before, during or after winding. Filament winding is the most common type.

Work Life -- The period during which a compound, after mixing with a catalyst, solvent, or other compounding ingredient, remains suitable for its intended use.

Woven Fabric Composite -- A major form of advanced composites in which the fiber constituent consists of woven fabric. A woven fabric composite normally is a laminate comprised of a number of laminae, each of which consists of one layer of fabric embedded in the selected matrix material. Individual fabric laminae are directionally oriented and combined into specific multiaxial laminates for application to specific envelopes of strength and stiffness requirements.

Yarn -- A generic term for strands or bundles of continuous filaments or fibers, usually twisted and suitable for making textile fabric.

Yarn, Plied -- Yarns made by collecting two or more single yarns together. Normally, the yarns are twisted together though sometimes they are collected without twist.

Yield Strength -- The stress at which a material exhibits a specified limiting deviation from the proportionality of stress to strain. (The deviation is expressed in terms of strain such as 0.2 percent for the Offset Method or 0.5 percent for the Total Extension Under Load Method.)

X-Axis -- In composite laminates, an axis in the plane of the laminate which is used as the 0 degree reference for designating the angle of a lamina.

X-Y Plane -- In composite laminates, the reference plane parallel to the plane of the laminate.

Y-Axis -- In composite laminates, the axis in the plane of the laminate which is perpendicular to the x-axis.

Z-Axis -- In composite laminates, the reference axis normal to the plane of the laminate.

REFERENCES

1.6.1(a) *DOD/NASA Advanced Composites Design Guide,* Vol. 4, Section 4.0.5, Air Force Wright Aeronautical Laboratories, Dayton, OH, prepared by Rockwell International Corporation, 1983 (distribution limited).

1.6.1(b) ASTM Guide E1309, "Identification of Composite Materials in Computerized Material Property Databases," *Annual Book of ASTM Standards,* Vol. 15.03, American Society for Testing and Materials, West Conshohocken, PA.

1.7(a) Military Standardization Handbook, *Metallic Materials and Elements for Aerospace Vehicle Structures*, MIL-HDBK-5D, Change Notice 2, May, 1985.

1.7(b) *DOD/NASA Advanced Composites Design Guide,* Air Force Wright Aeronautical Laboratories, Dayton, OH, prepared by Rockwell International Corporation, 1983 (distribution limited).

1.7(c) ASTM Terminology E206, "Definitions of Terms Relating to Fatigue Testing and the Statistical Analysis of Fatigue Data," *Annual Book of ASTM Standards,* Vol. 03.01, American Society for Testing and Materials, West Conshohocken, PA.
(canceled March 27, 1987; replaced by ASTM E1150).

1.7.2(a) ASTM Practice E380, "Metric Practice," Annual Book of ASTM Standards, Vol. 14.01, American Society for Testing and Materials, West Conshohocken, PA.
(canceled April 28, 1997; now sold in book form called "Metric 97").

1.7.2(b) *Engineering Design Handbook: Metric Conversion Guide*, DARCOM P 706-470, July 1976.

1.7.2(c) *The International System of Units (SI)*, NBS Special Publication 330, National Bureau of Standards, 1986 edition.

1.7.2(d) *Units and Systems of Weights and Measures, Their Origin, Development, and Present Status*, NBS Letter Circular LC 1035, National Bureau of Standards, November 1985.

1.7.2(e) *The International System of Units Physical Constants and Conversion Factors*, NASA Special Publication 7012, 1964.

INDEX

CHAPTER 1

PAGE

CHAPTER 2 FIBER PROPERTIES

CHAPTER 3 MATRIX PROPERTIES

CHAPTER 4 CARBON FIBER COMPOSITES

4.1 INTRODUCTION

4.2 CARBON - EPOXY COMPOSITES

4.2.1 T-500 12k/976 unitape data set description

Material Description:

Material: T-500 12k/976

Form: Unidirectional tape, fiber areal weight of 142 g/m^2, typical cured resin content of 28-34%, typical cured ply thickness of 0.0053 inches.

Processing: Autoclave cure; 240°F, 85 psi, 1 hour; 350°F, 100 psi for 2 hours.

General Supplier Information:

Fiber: T-500 fibers are continuous carbon filaments made from PAN precursor, surface treated to improve handling characteristics and structural properties. Filament count is 12,000 filaments/tow. Typical tensile modulus is 35.5 x 10^6 psi. Typical tensile strength is 575,000 psi.

Matrix: 976 is a high flow, modified epoxy resin that meets the NASA outgassing requirements. 10 days out-time at 72°F.

Maximum Short Term Service Temperature: 350°F (dry), 250°F (wet)

Typical applications: General purpose commercial and military structural applications, good hot/wet properties.

4.2.1 T500 12k/976 unidirectional tape*

MATERIAL:	T-500 12k/976 unidirectional tape			C/Ep T-500/976 Summary
PREPREG:	Fiberite Hy-E 3076P unidirectional tape			
FIBER:	Union Carbide Thornel T-500 12k	MATRIX:	Fiberite 976	
T_g(dry):	361°F T_g(wet):	T_g METHOD:		
PROCESSING:	240°F, 1 hour, 85 psi, 350°F, 2 hours, 100 psi			

* DATA WERE SUBMITTED BEFORE THE ESTABLISHMENT OF DATA DOCUMENTATION REQUIREMENTS (JUNE 1989). ALL DOCUMENTATION PRESENTLY REQUIRED WERE NOT SUPPLIED FOR THIS MATERIAL.

Date of fiber manufacture		Date of testing	
Date of resin manufacture		Date of data submittal	6/88
Date of prepreg manufacture	12/83	Date of analysis	1/93
Date of composite manufacture			

LAMINA PROPERTY SUMMARY

	75°F/A		-65°F/A		250°F/A			
Tension, 1-axis	II-I		II-I		II-I			
Tension, 2-axis	II-I		II-I		II-I			
Tension, 3-axis								
Compression, 1-axis								
Compression, 2-axis								
Compression, 3-axis								
Shear, 12-plane								
Shear, 23-plane								
Shear, 31-plane								

Classes of data: F - Fully approved, I - Interim, S - Screening in Strength/Modulus/Poisson's ratio/Strain-to-failure order.

* DATA WERE SUBMITTED BEFORE THE ESTABLISHMENT OF DATA DOCUMENTATION REQUIREMENTS (JUNE 1989). ALL DOCUMENTATION PRESENTLY REQUIRED WERE NOT SUPPLIED FOR THIS MATERIAL.

		Nominal	As Submitted	Test Method
Fiber Density	(g/cm^3)	1.79		
Resin Density	(g/cm^3)	1.28		
Composite Density	(g/cm^3)	1.59	1.57 - 1.61	
Fiber Areal Weight	(g/m^2)	142	142 - 146	
Ply Thickness	(in)	0.0053	0.0050 - 0.0057	

LAMINATE PROPERTY SUMMARY

Classes of data: F - Fully approved, I - Interim, S - Screening in Strength/Modulus/Poisson's ratio/Strain-to-failure order.

* DATA WERE SUBMITTED BEFORE THE ESTABLISHMENT OF DATA DOCUMENTATION REQUIREMENTS (JUNE 1989). ALL DOCUMENTATION PRESENTLY REQUIRED WAS NOT SUPPLIED FOR THIS MATERIAL.

MATERIAL:	T-500 12k/976 unidirectional tape			**Table 4.2.1(a) C/Ep 142-UT T-500/976 Tension, 1-axis $[90]_8$ 75/A, -65/A, 200/A Interim**
RESIN CONTENT:	28-34 wt%	COMP: DENSITY:	1.57-1.61 g/cm^3	
FIBER VOLUME:	59-64 %	VOID CONTENT:	0.3-1.7%	
PLY THICKNESS:	0.0050 - 0.0057 in.			
TEST METHOD:		MODULUS CALCULATION:		
ASTM D3039-76		Chord, 20-40% of ultimate load		
NORMALIZED BY:	Specimen thickness to 60% fiber volume			

		Temperature (°F) Moisture Content (%) Equilibrium at T, RH Source Code	75 ambient 13		-65 ambient 13		250 ambient 13	
			Normalized	Measured	Normalized	Measured	Normalized	Measured
F_1^{tu} (ksi)		Mean	295	298	213	213	273	276
		Minimum	257	270	163	196	236	258
		Maximum	329	328	243	235	302	310
		C.V.(%)	6.41	5.74	9.78	5.02	7.39	6.05
		B-value	(1)		(1)		(1)	
		Distribution	ANOVA		Weibull		Weibull	
		C_1	20.5		221		282	
		C_2	4.64		13.1		15.7	
		No. Specimens	15		15		15	
		No. Batches	3		3		3	
		Approval Class	Interim		Interim		Interim	
E_1^t (Msi)		Mean	21.9	22.0	19.0	19.1	22.2	22.4
		Minimum	20.9	20.5	15.9	17.7	18.6	21.0
		Maximum	24.7	24.0	21.5	21.5	25.1	23.8
		C.V.(%)	4.42	4.15	8.11	5.76	6.91	4.17
		No. Specimens	15		15		15	
		No. Batches	3		3		3	
		Approval Class	Interim		Interim		Interim	
ν_{12}^t		Mean						
		No. Specimens						
		No. Batches						
		Approval Class						
ε_1^{tu} ($\mu\varepsilon$)		Mean		13000		10700		11800
		Minimum		11700		9300		10800
		Maximum		13900		12000		12900
		C.V.(%)		4.98		5.98		5.32
		B-value		(1)		(1)		(1)
		Distribution		ANOVA		Weibull		Weibull
		C_1		706		11000		12100
		C_2		4.75		18.8		21.6
		No. Specimens	15		15		15	
		No. Batches	3		3		3	
		Approval Class	Interim		Interim		Interim	

(1) B-values are presented only for fully approved data.

* DATA WERE SUBMITTED BEFORE THE ESTABLISHMENT OF DATA DOCUMENTATION REQUIREMENTS (JUNE 1989). ALL DOCUMENTATION PRESENTLY REQUIRED WAS NOT SUPPLIED FOR THIS MATERIAL.

MATERIAL:	T-500 12k/976 unidirectional tape			**Table 4.2.1(b)** **C/Ep 142-UT** **T-500/976** **Tension, 2-axis** **$[90]_8$** **75/A, -65/A, 200/A** **Interim**
RESIN CONTENT:	28-34 wt%	COMP: DENSITY:	1.57-1.61 lb/in^3	
FIBER VOLUME:	59-64 %	VOID CONTENT:	0.3-1.7%	
PLY THICKNESS:	0.0050-0.0057 in.			
TEST METHOD:		MODULUS CALCULATION:		
ASTM D3039-76		Chord, 20 - 40 % of ultimate load		
NORMALIZED BY:	Not normalized			

Temperature (°F) Moisture Content (%) Equilibrium at T, RH Source Code		75 ambient 13	-65 ambient 13	250 ambient 13			
	Mean	10.2	10.3	7.90			
	Minimum	9.40	9.40	7.00			
	Maximum	11.3	12.1	8.80			
	C.V.(%)	5.59	6.61	5.35			
	B-value	(1)	(1)	(1)			
F_2^{tu}	Distribution	ANOVA	Lognormal	Weibull			
(ksi)	C_1	0.594	2.33	8.09			
	C_2	3.48	0.0636	19.7			
	No. Specimens	15	15	15			
	No. Batches	3	3	3			
	Approval Class	Interim	Interim	Interim			
	Mean	1.3	1.5	1.2			
	Minimum	1.3	1.4	1.1			
	Maximum	1.7	1.6	1.3			
E_2^t	C.V.(%)	7.8	4.8	7.0			
(Msi)	No. Specimens	15	15	15			
	No. Batches	3	3	3			
	Approval Class	Interim	Interim	Interim			
	Mean						
	No. Specimens						
ν_{21}^t	No. Batches						
	Approval Class						
	Mean	7750	7110	6930			
	Minimum	5800	6200	5900			
	Maximum	8900	8600	8000			
	C.V.(%)	10.3	8.28	8.32			
	B-value	(1)	(1)	(1)			
ε_2^{tu}	Distribution	Weibull	Weibull	Weibull			
($\mu\varepsilon$)	C_1	8080	7390	7180			
	C_2	12.4	11.5	13.7			
	No. Specimens	15	15	15			
	No. Batches	3	3	3			
	Approval Class	Interim	Interim	Interim			

(1) B-values are presented only for fully approved data.

4.2.2 HITEX 33 6k/E7K8 unitape data set description

Material Description:

Material:HITEX 33-6k/E7K8

Form: Unidirectional tape, fiber areal weight of 145 g/m^2, typical cured resin content of 34% typical cured ply thickness of 0.0057 inches.

Processing: Good drape. Autoclave cure; 300-310°F, 55 psi for 2 hours. Low exotherm profile for processing of thick parts.

General Supplier Information:

Fiber: HITEX 33 fibers are continuous carbon filaments made from PAN precursor. Filament count is 6,000 filaments/tow. Typical tensile modulus is 33 x 10^6 psi. Typical tensile strength is 560,000 psi.

Matrix: E7K8 is a medium flow, low exotherm epoxy resin. Good tack; up to 20 days out-time at ambient temperature

Maximum Short Term Service Temperature: 300°F (dry), 190°F (wet)

Typical applications: Primary and secondary structural applications on commercial and military aircraft, jet engine applications such as stationary airfoils and thrust reverser blocker doors.

4.2.2 HITEX 33 6k/E7K8 unidirectional tape*

			C/Ep HITEX 33/E7K8 Summary
MATERIAL:	HITEX 33 6k/E7K8 unidirectional tape		
PREPREG:	U.S. Polymeric HITEX 33 6k/E7K8 unidirectional tape, grade 145		
FIBER:	Hitco HITEX 33 6k, no twist	MATRIX:	U.S. Polymeric E7K8
T_g(dry):	T_g(wet):	T_g METHOD:	
PROCESSING:	Autoclave cure: 300 - 310°F, 120 - 130 min., 55 psi		

* DATA WERE SUBMITTED BEFORE THE ESTABLISHMENT OF DATA DOCUMENTATION REQUIREMENTS (JUNE 1989). ALL DOCUMENTATION PRESENTLY REQUIRED WERE NOT SUPPLIED FOR THIS MATERIAL.

Date of fiber manufacture		Date of testing	
Date of resin manufacture		Date of data submittal	1/83
Date of prepreg manufacture		Date of analysis	1/93
Date of composite manufacture			

LAMINA PROPERTY SUMMARY

	75°F/A		-65°F/A	180°F/A		75°F/W	180°F/W	
Tension, 1-axis	SSSS		SS-S			SSS-	SSS-	
Tension, 2-axis	SS--							
Tension, 3-axis								
Compression, 1-axis	SS-S		SS-S			SS--	SS--	
Compression, 2-axis								
Compression, 3-axis								
Shear, 12-plane	S---			S---		S---	S---	
Shear, 23-plane								
Shear, 31-plane								

Classes of data: F - Fully approved, I - Interim, S - Screening in Strength/Modulus/Poisson's ratio/Strain-to-failure order.

* DATA WERE SUBMITTED BEFORE THE ESTABLISHMENT OF DATA DOCUMENTATION REQUIREMENTS (JUNE 1989). ALL DOCUMENTATION PRESENTLY REQUIRED WERE NOT SUPPLIED FOR THIS MATERIAL.

		Nominal	As Submitted	Test Method
Fiber Density	(g/cm^3)	1.80		
Resin Density	(g/cm^3)	1.27		
Composite Density	(g/cm^3)	1.59	1.56 - 1.61	
Fiber Areal Weight	(g/m^2)	145		
Ply Thickness	(in)	0.0057	0.0053 - 0.0058	

LAMINATE PROPERTY SUMMARY

Classes of data: F - Fully approved, I - Interim, S - Screening in Strength/Modulus/Poisson's ratio/Strain-to-failure order.

* DATA WERE SUBMITTED BEFORE THE ESTABLISHMENT OF DATA DOCUMENTATION REQUIREMENTS (JUNE 1989). ALL DOCUMENTATION PRESENTLY REQUIRED WAS NOT SUPPLIED FOR THIS MATERIAL.

MATERIAL: HITEX 33 6k/E7K8 unidirectional tape	**Table 4.2.2(a)** **C/Ep 145-UT** **HITEX 33/E7K8** **Tension, 1-axis** **$[0]_{10}$** **75/A, -65/A, 75/1.5%** **Screening**
RESIN CONTENT: 34 wt% COMP: DENSITY: 1.58 g/cm^3 FIBER VOLUME: 58 % VOID CONTENT: 0.0% PLY THICKNESS: 0.0057 in.	
TEST METHOD: ASTM D3039-76 MODULUS CALCULATION:	
NORMALIZED BY: Fiber volume fraction to 60%	

Temperature (°F)		75		-65		75	
Moisture Content (%)		ambient		ambient		1.5	
Equilibrium at T, RH						(1)	
Source Code		20		20		20	
		Normalized	Measured	Normalized	Measured	Normalized	Measured
	Mean	313	304	296	288	318	310
	Minimum	292	283	267	259	280	272
	Maximum	339	330	327	319	345	335
	C.V.(%)	4.80	4.84	9.19	9.20	7.63	7.65
	B-value	(2)	(2)	(2)	(2)	(2)	(2)
F_1^{tu}	Distribution	Weibull	Weibull	Normal	Normal	Normal	Normal
(ksi)	C_1	320	311	296	288	318	310
	C_2	22.2	21.9	27.2	26.5	24.3	23.7
	No. Specimens	20		5		5	
	No. Batches	1		1		1	
	Approval Class	Screening		Screening		Screening	
	Mean	18.2	17.7	18.5	18.0	18.5	18.0
	Minimum	17.5	17.0	18.1	17.7	18.3	17.8
	Maximum	19.0	18.5	18.6	18.1	18.7	18.2
E_1^t	C.V.(%)	2.58	2.60	1.06	1.07	0.79	0.79
(Msi)	No. Specimens	18		5		5	
	No. Batches	1		1		1	
	Approval Class	Screening		Screening		Screening	
	Mean		0.310				0.310
	No. Specimens	5				5	
ν_{12}^t	No. Batches	1				1	
	Approval Class	Screening				Screening	
	Mean		15900		16100		
	Minimum		15200		15500		
	Maximum		17100		17000		
	C.V.(%)		4.81		3.61		
	B-value		(2)		(2)		
ε_1^{tu}	Distribution		Normal		Normal		
(με)	C_1		15900		16200		
	C_2		765		582		
	No. Specimens	5		5			
	No. Batches	1		1			
	Approval Class	Screening		Screening			

(1) Conditioned for 14 days at 160°F, 85% RH.
(2) B-values are presented only for fully approved data.

* DATA WERE SUBMITTED BEFORE THE ESTABLISHMENT OF DATA DOCUMENTATION REQUIREMENTS (JUNE 1989). ALL DOCUMENTATION PRESENTLY REQUIRED WAS NOT SUPPLIED FOR THIS MATERIAL.

MATERIAL:	HITEX 33 6k/E7K8 unidirectional tape			**Table 4.2.2(b)** **C/Ep 145-UT** **HITEX 33/E7K8** **Tension, 1-axis** **$[0]_{10}$** **180/1.5%** **Screening**
RESIN CONTENT:	34 wt%	COMP: DENSITY:	1.58 g/cm^3	
FIBER VOLUME:	58 %	VOID CONTENT:	0.0%	
PLY THICKNESS:	0.0057 in.			
TEST METHOD: ASTM D3039-76		MODULUS CALCULATION:		
NORMALIZED BY:	Fiber volume fraction to 60%			

Temperature (°F)		180					
Moisture Content (%)		1.5					
Equilibrium at T, RH		(1)					
Source Code		20					
		Normalized	Measured	Normalized	Measured	Normalized	Measured
	Mean	308	300				
	Minimum	296	288				
	Maximum	318	309				
	C.V.(%)	2.65	2.65				
	B-value	(2)	(2)				
F_1^{tu}	Distribution	Normal	Normal				
(ksi)	C_1	308	300				
	C_2	8.17	7.95				
	No. Specimens	5					
	No. Batches	1					
	Approval Class	Screening					
	Mean	18.7	18.2				
	Minimum	17.8	17.3				
	Maximum	19.5	19.0				
E_1^t	C.V.(%)	3.64	3.65				
(Msi)	No. Specimens	5					
	No. Batches	1					
	Approval Class	Screening					
	Mean		0.300				
	No. Specimens	5					
ν_{12}^t	No. Batches	1					
	Approval Class	Screening					
	Mean						
	Minimum						
	Maximum						
	C.V.(%)						
	B-value						
ε_1^{tu}	Distribution						
(με)	C_1						
	C_2						
	No. Specimens						
	No. Batches						
	Approval Class						

(1) Conditioned for 14 days at 160°F, 85% RH.

(2) B-values are presented only for fully approved data.

* DATA WERE SUBMITTED BEFORE THE ESTABLISHMENT OF DATA DOCUMENTATION REQUIREMENTS (JUNE 1989). ALL DOCUMENTATION PRESENTLY REQUIRED WAS NOT SUPPLIED FOR THIS MATERIAL.

MATERIAL:	HITEX 33 6k/E7K8 unidirectional tape			**Table 4.2.2(c)** **C/Ep 145-UT** **HITEX 33/E7K8** **Tension, 2-axis** **$[90]_{20}$** **75/A** **Screening**
RESIN CONTENT:	34 wt%	COMP: DENSITY:	1.58 g/cm^3	
FIBER VOLUME:	58 %	VOID CONTENT:	0.39%	
PLY THICKNESS:	0.0058 in.			
TEST METHOD: ASTM D3039-76		MODULUS CALCULATION:		
NORMALIZED BY:	Not normalized			

Temperature (°F) Moisture Content (%) Equilibrium at T, RH Source Code		75 ambient 20					
	Mean	6.90					
	Minimum	5.58					
	Maximum	8.07					
	C.V.(%)	11.2					
	B-value	(1)					
F_2^{tu}	Distribution	Weibull					
(ksi)	C_1	7.23					
	C_2	10.9					
	No. Specimens	20					
	No. Batches	1					
	Approval Class	Screening					
	Mean	1.25					
	Minimum	1.23					
	Maximum	1.27					
E_2^t	C.V.(%)	0.977					
(Msi)	No. Specimens	20					
	No. Batches	1					
	Approval Class	Screening					
	Mean						
	No. Specimens						
ν_{21}^t	No. Batches						
	Approval Class						
	Mean						
	Minimum						
	Maximum						
	C.V.(%)						
	B-value						
ε_2^{tu}	Distribution						
(με)	C_1						
	C_2						
	No. Specimens						
	No. Batches						
	Approval Class						

(1) B-values are presented only for fully approved data.

* DATA WERE SUBMITTED BEFORE THE ESTABLISHMENT OF DATA DOCUMENTATION REQUIREMENTS (JUNE 1989). ALL DOCUMENTATION PRESENTLY REQUIRED WAS NOT SUPPLIED FOR THIS MATERIAL.

MATERIAL: HITEX 33 6k/E7K8 unidirectional tape

RESIN CONTENT: 34-35 wt% COMP: DENSITY: 1.57-1.58 g/cm³
FIBER VOLUME: 57-58 % VOID CONTENT: 0.0%
PLY THICKNESS: 0.0057 in.

TEST METHOD: SACMA SRM 1-88 MODULUS CALCULATION:

NORMALIZED BY: Fiber volume fraction to 60%

Table 4.2.2(d)
C/Ep 145-UT
HITEX 33/E7K8
Compression, 1-axis
$[0]_{10}$
75/A, -65/A, 75/1.5%
Screening

Temperature (°F)		75		-65		75	
Moisture Content (%)		ambient		ambient		1.5	
Equilibrium at T, RH						(1)	
Source Code		20		20		20	
		Normalized	Measured	Normalized	Measured	Normalized	Measured
	Mean	209	204	230	224	198	193
	Minimum	168	164	209	204	178	174
	Maximum	234	228	254	248	217	211
	C.V.(%)	9.41	9.41	7.98	8.04	8.13	8.03
	B-value	(2)	(2)	(2)	(2)	(2)	(2)
F_1^{cu}	Distribution	Weibull	Weibull	Normal	Normal	Normal	Normal
(ksi)	C_1	218	212	230	224	198	193
	C_2	13.7	13.7	18.3	17.9	16.1	15.7
	No. Specimens	20		5		5	
	No. Batches	1		1		1	
	Approval Class	Screening		Screening		Screening	
	Mean	17.1	16.2	17.9	16.9	18.0	17.0
	Minimum	16.1	15.2	17.5	16.5	17.5	16.6
	Maximum	17.8	16.8	18.1	17.1	18.8	17.8
E_1^c	C.V.(%)	2.89	2.94	1.23	1.35	3.04	5.59
(Msi)	No. Specimens	20		5		5	
	No. Batches	1		1		1	
	Approval Class	Screening		Screening		Screening	
	Mean						
	No. Specimens						
ν_{12}^c	No. Batches						
	Approval Class						
	Mean		12600		13600		
	Minimum		12000		13600		
	Maximum		13400		13700		
	C.V.(%)		2.92		0.48		
	B-value		(2)		(2)		
ε_1^{cu}	Distribution		Weibull		Normal		
(με)	C_1		12800		13600		
	C_2		35.7		65.7		
	No. Specimens	20		5			
	No. Batches	1		1			
	Approval Class	Screening		Screening			

(1) Conditioned for 14 days at 160°F, 85% RH.
(2) B-values are presented only for fully approved data.

* DATA WERE SUBMITTED BEFORE THE ESTABLISHMENT OF DATA DOCUMENTATION REQUIREMENTS (JUNE 1989). ALL DOCUMENTATION PRESENTLY REQUIRED WAS NOT SUPPLIED FOR THIS MATERIAL.

Table 4.2.2(e)
C/Ep 145-UT
HITEX 33/E7K8
Compression, 1-axis
$[0]_{10}$
180/1.5%
Screening

MATERIAL: HITEX 33 6k/E7K8 unidirectional tape

RESIN CONTENT: 34 -35 wt% COMP: DENSITY: 1.57-1.58 g/cm^3
FIBER VOLUME: 57-58 % VOID CONTENT: 0.0%
PLY THICKNESS: 0.0057 in.

TEST METHOD: SACMA SRM 1-88 MODULUS CALCULATION:

NORMALIZED BY: Fiber volume fraction to 60%

Temperature (°F) Moisture Content (%) Equilibrium at T, RH Source Code		180 1.5 (1) 20					
		Normalized	Measured	Normalized	Measured	Normalized	Measured
F_1^{cu} (ksi)	Mean	136	132				
	Minimum	111	108				
	Maximum	161	157				
	C.V.(%)	13.4	13.6				
	B-value	(2)	(2)				
	Distribution	Normal	Normal				
	C_1	136	132				
	C_2	18.3	17.8				
	No. Specimens	5					
	No. Batches	1					
	Approval Class	Screening					
E_1^c (Msi)	Mean	17.6	16.6				
	Minimum	17.0	16.1				
	Maximum	18.0	17.0				
	C.V.(%)	2.47	2.47				
	No. Specimens	5					
	No. Batches	1					
	Approval Class	Screening					
ν_{12}^c	Mean						
	No. Specimens						
	No. Batches						
	Approval Class						
ε_1^{cu} (με)	Mean						
	Minimum						
	Maximum						
	C.V.(%)						
	B-value						
	Distribution						
	C_1						
	C_2						
	No. Specimens						
	No. Batches						
	Approval Class						

(1) Conditioned for 14 days at 160°F, 85% RH.
(2) B-values are presented only for fully approved data.

* DATA WERE SUBMITTED BEFORE THE ESTABLISHMENT OF DATA DOCUMENTATION REQUIREMENTS (JUNE 1989). ALL DOCUMENTATION PRESENTLY REQUIRED WAS NOT SUPPLIED FOR THIS MATERIAL.

MATERIAL:	HITEX 33 6k/E7K8 unidirectional tape			**Table 4.2.2(f)** **C/Ep 145-UT** **HITEX 33/E7K8** **Shear, 12-plane** **$[(\pm45)_2/45]_s$** **75/A, 180/A, 75/1.5%, 180/1.5%** **Screening**
RESIN CONTENT:	29-30 wt%	COMP: DENSITY:	1.59-1.61 g/cm^3	
FIBER VOLUME:	62-64 %	VOID CONTENT:	0.05-0.91%	
PLY THICKNESS:	0.0053 in.			
TEST METHOD:		MODULUS CALCULATION:		
ASTM D3518-76				
NORMALIZED BY:	Not normalized			

Temperature (°F) Moisture Content (%) Equilibrium at T, RH Source Code		75 ambient 20	180 ambient 20	75 1.5 (1) 20	180 1.5 (1) 20		
	Mean	15.0	13.2	16.3	11.7		
	Minimum	13.5	13.1	15.8	11.5		
	Maximum	15.8	13.3	16.7	11.9		
	C.V.(%)	3.52	0.655	2.20	1.27		
	B-value	(2)	(2)	(2)	(2)		
F_{12}^{su}	Distribution	Weibull	Normal	Normal	Normal		
(ksi)	C_1	15.2	13.2	16.3	11.7		
	C_2	34.8	0.0865	0.357	0.148		
	No. Specimens	20	5	5	5		
	No. Batches	1	1	1	1		
	Approval Class	Screening	Screening	Screening	Screening		
	Mean						
	Minimum						
	Maximum						
	C.V.(%)						
	B-value						
γ_{12}^{su}	Distribution						
(με)	C_1						
	C_2						
	No. Specimens						
	No. Batches						
	Approval Class						
	Mean						
	Minimum						
	Maximum						
G_s^{12}	C.V.(%)						
(Msi)	No. Specimens						
	No. Batches						
	Approval Class						

(1) Conditioned for 14 days at 160°F, 85% RH.
(2) B-values are presented only for fully approved data.

4.2.3 AS4 12k/E7K8 unitape data set description

Material Description:

Material: AS4-12k/E7K8

Form: Unidirectional tape, fiber areal weight of 145 g/m^2, typical cured resin content of 32-37%, typical cured ply thickness of 0.0054 inches.

Processing: Good drape. Autoclave cure; 300-310° F, 85 psi for 2 hours. Low exotherm profile for processing of thick parts.

General Supplier Information:

Fiber: AS4 fibers are continuous carbon filaments made from PAN precursor, surface treated to improve handling characteristics and structural properties. Filament count is 12,000 filaments/tow. Typical tensile modulus is 34 x 10^6 psi. Typical tensile strength is 550,000 psi.

Matrix: E7K8 is a medium flow, low exotherm epoxy resin. Good tack; up to 20 days out-time at ambient temperature.

Maximum Short Term Service Temperature: 300°F (dry), 190°F (wet)

Typical applications: Primary and secondary structural applications commercial and military aircraft, jet engine applications such as stationary airfoils and thrust reverser blocker doors.

4.2.3 AS4 12k/E7K8 unidirectional tape*

MATERIAL:	AS4 12k/E7K8 unidirectional tape		**C/Ep AS4/E7K8 Summary**
PREPREG:	U.S. Polymeric AS4 12k/E7K8 unidirectional tape		
FIBER:	Hercules AS4 12k	MATRIX:	U.S. Polymeric E7K8
T_g(dry):	T_g(wet):	T_g METHOD:	
PROCESSING:	Autoclave cure: 300 - 310°F, 120 - 130 min., 55 psi		

* DATA WERE SUBMITTED BEFORE THE ESTABLISHMENT OF DATA DOCUMENTATION REQUIREMENTS (JUNE 1989). ALL DOCUMENTATION PRESENTLY REQUIRED WERE NOT SUPPLIED FOR THIS MATERIAL.

Date of fiber manufacture		Date of testing	
Date of resin manufacture		Date of data submittal	1/88
Date of prepreg manufacture		Date of analysis	1/93
Date of composite manufacture			

LAMINA PROPERTY SUMMARY

	75°F/A		-65°F/A	180°F/A		75°F/W	180°F/W	
Tension, 1-axis	IISS		SS-S			SSSS	SSSS	
Tension, 2-axis	SS--							
Tension, 3-axis								
Compression, 1-axis	SS-S		SS-S			SS--	SS--	
Compression, 2-axis								
Compression, 3-axis								
Shear, 12-plane	I---			S---		S---	S---	
Shear, 23-plane								
Shear, 31-plane								

Classes of data: F - Fully approved, I - Interim, S - Screening in Strength/Modulus/Poisson's ratio/Strain-to-failure order.

* DATA WERE SUBMITTED BEFORE THE ESTABLISHMENT OF DATA DOCUMENTATION REQUIREMENTS (JUNE 1989). ALL DOCUMENTATION PRESENTLY REQUIRED WERE NOT SUPPLIED FOR THIS MATERIAL.

		Nominal	As Submitted	Test Method
Fiber Density	(g/cm^3)	1.80		
Resin Density	(g/cm^3)	1.28		
Composite Density	(g/cm^3)	1.59	1.52 - 1.59	
Fiber Areal Weight	(g/m^2)	145		
Ply Thickness	(in)	0.0054	0.0054 - 0.0057	

LAMINATE PROPERTY SUMMARY

Classes of data: F - Fully approved, I - Interim, S - Screening in Strength/Modulus/Poisson's ratio/Strain-to-failure order.

* DATA WERE SUBMITTED BEFORE THE ESTABLISHMENT OF DATA DOCUMENTATION REQUIREMENTS (JUNE 1989). ALL DOCUMENTATION PRESENTLY REQUIRED WAS NOT SUPPLIED FOR THIS MATERIAL.

MATERIAL:	AS4 12k/E7K8 unidirectional tape			**Table 4.2.3(a)** **C/Ep 145-UT** **AS4/E7K8** **Tension, 1-axis** **$[0]_{10}$** **75/A, -65/A, 75/0.77%** **Interim, Screening**
RESIN CONTENT:	32-37 wt%	COMP: DENSITY:	1.53-1.59 g/cm^3	
FIBER VOLUME:	53-60 %	VOID CONTENT:	0.64-2.2%	
PLY THICKNESS:	0.0054 in.			
TEST METHOD: ASTM D3039-76		MODULUS CALCULATION: Slope of initial linear portion of load-displacement curve		
NORMALIZED BY:	Fiber volume to 60%			

		75		-65		75	
Temperature (°F)		75		-65		75	
Moisture Content (%)		ambient		ambient		0.77	
Equilibrium at T, RH						(1)	
Source Code		20		20		20	
		Normalized	Measured	Normalized	Measured	Normalized	Measured
F_1^{tu} (ksi)	Mean	303	293	291	273	304	294
	Minimum	253	252	255	239	286	276
	Maximum	345	347	327	306	317	306
	C.V.(%)	8.26	8.94	8.93	8.90	4.16	4.22
	B-value	(2)	(2)	(2)	(2)	(2)	(2)
	Distribution	ANOVA	ANOVA	Normal	Normal	Normal	Normal
	C_1	26.7	32.4	291	273	304	294
	C_2	4.40	7.49	26.0	24.4	12.7	12.2
	No. Specimens	20		5		5	
	No. Batches	2		1		1	
	Approval Class	Interim		Screening		Screening	
E_1^t (Msi)	Mean	19.3	18.7	20.1	18.8	19.6	18.9
	Minimum	18.5	17.4	19.7	18.4	19.0	18.4
	Maximum	21.3	21.4	20.6	19.3	20.1	19.4
	C.V.(%)	3.79	6.10	1.67	1.79	2.04	1.96
	No. Specimens	20		5		5	
	No. Batches	2		1		1	
	Approval Class	Interim		Screening		Screening	
ν_{12}^t	Mean		0.320				0.288
	No. Specimens	5				5	
	No. Batches	1				1	
	Approval Class	Screening				Screening	
ε_1^{tu} (με)	Mean		13900		13500		14600
	Minimum		12500		12000		13700
	Maximum		16000		14800		15000
	C.V.(%)		11.0		8.24		3.83
	B-value		(2)		(2)		(2)
	Distribution		Normal		Normal		Normal
	C_1		13900		13500		14600
	C_2		1530		1110		561
	No. Specimens	5		5		5	
	No. Batches	1		1		1	
	Approval Class	Screening		Screening		Screening	

(1) Conditioned for 14 days at 160°F, 85% RH.
(2) B-values are presented only for fully approved data.

* DATA WERE SUBMITTED BEFORE THE ESTABLISHMENT OF DATA DOCUMENTATION REQUIREMENTS (JUNE 1989). ALL DOCUMENTATION PRESENTLY REQUIRED WAS NOT SUPPLIED FOR THIS MATERIAL.

MATERIAL:	AS4 12k/E7K8 unidirectional tape			**Table 4.2.3(b)** **C/Ep 145-UT** **AS4/E7K8** **Tension, 1-axis** **$[0]_{10}$** **180/0.77%** **Screening**
RESIN CONTENT:	32-37 wt%	COMP: DENSITY:	1.53-1.59 g/cm^3	
FIBER VOLUME:	53-60 %	VOID CONTENT:	0.64-2.2%	
PLY THICKNESS:	0.0054 in.			
TEST METHOD:		MODULUS CALCULATION:		
ASTM D3039-76		Slope of initial linear portion of load-displacement curve		
NORMALIZED BY:	Fiber volume to 60%			

Temperature (°F) Moisture Content (%) Equilibrium at T, RH Source Code		180 0.77 (1) 20					
		Normalized	Measured	Normalized	Measured	Normalized	Measured
	Mean	310	296				
	Minimum	284	274				
	Maximum	326	306				
	C.V.(%)	5.87	4.76				
	B-value	(2)	(2)				
F_1^{tu}	Distribution	Normal	Normal				
(ksi)	C_1	310	296				
	C_2	18.2	13.9				
	No. Specimens	5					
	No. Batches	1					
	Approval Class	Screening					
	Mean	20.1	19.2				
	Minimum	19.1	18.5				
	Maximum	21.8	20.4				
E_1^t	C.V.(%)	5.65	4.01				
(Msi)	No. Specimens	5					
	No. Batches	1					
	Approval Class	Screening					
	Mean		0.288				
	No. Specimens	5					
ν_{12}^t	No. Batches	1					
	Approval Class	Screening					
	Mean		14600				
	Minimum		13900				
	Maximum		15400				
	C.V.(%)		4.21				
	B-value		(2)				
ε_1^{tu}	Distribution		Normal				
(με)	C_1		14600				
	C_2		616				
	No. Specimens	5					
	No. Batches	1					
	Approval Class	Screening					

(1) Conditioned for 14 days at 160°F, 85% RH.
(2) B-values are presented only for fully approved data.

* DATA WERE SUBMITTED BEFORE THE ESTABLISHMENT OF DATA DOCUMENTATION REQUIREMENTS (JUNE 1989). ALL DOCUMENTATION PRESENTLY REQUIRED WAS NOT SUPPLIED FOR THIS MATERIAL.

MATERIAL:	AS4 12k/E7K8 unidirectional tape			**Table 4.2.3(c)** **C/Ep 145-UT** **AS4/E7K8** **Tension, 2-axis** **$[90]_{20}$** **75/A** **Screening**
RESIN CONTENT:	32-38 wt%	COMP: DENSITY:	1.54-1.59 g/cm^3	
FIBER VOLUME:	53-60 %	VOID CONTENT:	0.64-0.75%	
PLY THICKNESS:	0.0057 in.			
TEST METHOD: ASTM D3039-76		MODULUS CALCULATION: Slope of initial linear portion of load-displacement curve		
NORMALIZED BY:	Not normalized			

Temperature (°F) Moisture Content (%) Equilibrium at T, RH Source Code		75 ambient 20					
F_2^{tu} (ksi)	Mean	5.47					
	Minimum	4.10					
	Maximum	7.01					
	C.V.(%)	13.2					
	B-value	(1)					
	Distribution	Weibull					
	C_1	5.79					
	C_2	8.04					
	No. Specimens	20					
	No. Batches	1					
	Approval Class	Screening					
E_2^t (Msi)	Mean	1.23					
	Minimum	1.16					
	Maximum	1.32					
	C.V.(%)	3.76					
	No. Specimens	20					
	No. Batches	1					
	Approval Class	Screening					
ν_{21}^t	Mean						
	No. Specimens						
	No. Batches						
	Approval Class						
ε_2^{tu} (με)	Mean						
	Minimum						
	Maximum						
	C.V.(%)						
	B-value						
	Distribution						
	C_1						
	C_2						
	No. Specimens						
	No. Batches						
	Approval Class						

(1) B-values are presented only for fully approved data.

* DATA WERE SUBMITTED BEFORE THE ESTABLISHMENT OF DATA DOCUMENTATION REQUIREMENTS (JUNE 1989). ALL DOCUMENTATION PRESENTLY REQUIRED WAS NOT SUPPLIED FOR THIS MATERIAL.

MATERIAL:	AS4 12k/E7K8 unidirectional tape			**Table 4.2.3(d)** **C/Ep 145-UT** **AS4/E7K8** **Compression, 1-axis** **$[0]_{10}$** **75/A, -65/A, 75/0.77%** **Screening**
RESIN CONTENT:	35-40 wt%	COMP: DENSITY:	1.52-1.58 g/cm^3	
FIBER VOLUME:	51-57 %	VOID CONTENT:	1.4-2.3%	
PLY THICKNESS:	0.0054 in.			
TEST METHOD:		MODULUS CALCULATION:		
SACMA SRM 1-88		Slope of initial linear portion of load-displacement curve		
NORMALIZED BY:	Fiber volume to 60%			

Temperature (°F) Moisture Content (%) Equilibrium at T, RH Source Code		75 ambient 20		-65 ambient 20		75 0.77 (1) 20	
		Normalized	Measured	Normalized	Measured	Normalized	Measured
	Mean	245	209	276	235	215	182
	Minimum	207	176	251	213	196	166
	Maximum	269	229	299	254	238	202
	C.V.(%)	8.00	7.80	6.57	6.60	7.78	7.75
	B-value	(2)	(2)	(2)	(2)	(2)	(2)
F_1^{cu}	Distribution	Weibull	Weibull	Normal	Normal	Normal	Normal
(ksi)	C_1	254	216	276	235	215	183
	C_2	16.3	16.3	18.1	15.4	16.7	14.2
	No. Specimens	20		5		5	
	No. Batches	1		1		1	
	Approval Class	Screening		Screening		Screening	
	Mean	19.0	17.9	17.6	16.5	18.5	17.4
	Minimum	17.3	16.3	16.6	15.7	17.7	16.7
	Maximum	20.4	19.2	18.0	17.0	19.0	17.9
E_1^c	C.V.(%)	4.58	4.54	3.16	3.14	2.95	2.86
(Msi)	No. Specimens	20		5		5	
	No. Batches	1		1		1	
	Approval Class	Screening		Screening		Screening	
	Mean						
	No. Specimens						
ν_{12}^c	No. Batches						
	Approval Class						
	Mean		11700		14400		
	Minimum		10800		13900		
	Maximum		13100		15100		
	C.V.(%)		4.81		3.89		
	B-value		(2)		(2)		
ε_1^{cu}	Distribution		Normal		Normal		
(με)	C_1		11700		14400		
	C_2		564		559		
	No. Specimens	20		5			
	No. Batches	1		1			
	Approval Class	Screening		Screening			

(1) Conditioned for 14 days at 160°F, 85% RH.

(2) B-values are presented only for fully approved data.

* DATA WERE SUBMITTED BEFORE THE ESTABLISHMENT OF DATA DOCUMENTATION REQUIREMENTS (JUNE 1989). ALL DOCUMENTATION PRESENTLY REQUIRED WAS NOT SUPPLIED FOR THIS MATERIAL.

MATERIAL: AS4 12k/E7K8 unidirectional tape		**Table 4.2.3(e)** **C/Ep 145-UT** **AS4/E7K8** **Compression, 1-axis** **$[0]_{10}$** **180/0.77%** **Screening**
RESIN CONTENT: 35-40 wt% FIBER VOLUME: 51-57 % PLY THICKNESS: 0.0054 in.	COMP: DENSITY: 1.52-1.58 g/cm^3 VOID CONTENT: 1.4-2.3%	
TEST METHOD: SACMA SRM 1-88	MODULUS CALCULATION: Slope of initial linear portion of load-displacement curve	
NORMALIZED BY: Fiber volume to 60%		

		Temperature (°F) Moisture Content (%) Equilibrium at T, RH Source Code	180 0.77 (1) 20					
			Normalized	Measured	Normalized	Measured	Normalized	Measured
F_1^{cu} (ksi)	Mean		150	127				
	Minimum		125	106				
	Maximum		176	150				
	C.V.(%)		14.8	15.0				
	B-value		(2)	(2)				
	Distribution		Normal	Normal				
	C_1		150	127				
	C_2		22.2	18.9				
	No. Specimens		5					
	No. Batches		1					
	Approval Class		Screening					
E_1^c (Msi)	Mean		18.0	17.0				
	Minimum		17.4	16.4				
	Maximum		18.4	17.3				
	C.V.(%)		2.46	2.41				
	No. Specimens		5					
	No. Batches		1					
	Approval Class		Screening					
ν_{12}^c	Mean							
	No. Specimens							
	No. Batches							
	Approval Class							
ε_1^{cu} ($\mu\varepsilon$)	Mean							
	Minimum							
	Maximum							
	C.V.(%)							
	B-value							
	Distribution							
	C_1							
	C_2							
	No. Specimens							
	No. Batches							
	Approval Class							

(1) Conditioned for 14 days at 160°F, 85% RH.

(2) B-values are presented only for fully approved data.

* DATA WERE SUBMITTED BEFORE THE ESTABLISHMENT OF DATA DOCUMENTATION REQUIREMENTS (JUNE 1989). ALL DOCUMENTATION PRESENTLY REQUIRED WAS NOT SUPPLIED FOR THIS MATERIAL.

MATERIAL:	AS4 12k/E7K8 unidirectional tape			**Table 4.2.3(f)** **C/Ep 145-UT** **AS4/E7K8** **Shear, 12-plane** **$[(\pm45)_2/45]_s$** **75/A, 180/A, 75/0.77%, 180/0.77%** **Interim, Screening**
RESIN CONTENT:	33-36 wt%	COMP: DENSITY:	1.54-1.55 g/cm^3	
FIBER VOLUME:	55-57 %	VOID CONTENT:	1.9-2.3%	
PLY THICKNESS:	0.0055 in.			
TEST METHOD: ASTM D3518-76		MODULUS CALCULATION:		
NORMALIZED BY:	Not normalized			

Temperature (°F) Moisture Content (%) Equilibrium at T, RH Source Code		75 ambient 20	180 ambient 20	75 0.77 (1) 20	180 0.77 (1) 20		
F_{12}^{su} (ksi)	Mean	16.5	14.6	15.1	13.4		
	Minimum	13.8	14.2	13.5	13.0		
	Maximum	17.0	14.9	15.8	13.8		
	C.V.(%)	6.41	1.90	6.04	2.44		
	B-value	(2)	(2)	(2)	(2)		
	Distribution	ANOVA	Normal	Normal	Normal		
	C_1	2.46	14.6	15.1	13.4		
	C_2	7.58	0.277	0.905	0.328		
	No. Specimens	20	5	5	5		
	No. Batches	2	1	1	1		
	Approval Class	Interim	Screening	Screening	Screening		
G_{12}^{s} (Msi)	Mean						
	Minimum						
	Maximum						
	C.V.(%)						
	No. Specimens						
	No. Batches						
	Approval Class						
γ_{12}^{su} ($\mu\varepsilon$)	Mean						
	Minimum						
	Maximum						
	C.V.(%)						
	B-value						
	Distribution						
	C_1						
	C_2						
	No. Specimens						
	No. Batches						
	Approval Class						

(1) Conditioned for 14 days at 160°F, 85% RH.
(2) B-values are presented only for fully approved data.

4.2.4 Celion 12k/E7K8 unitape data set description

Material Description:

Material: Celion-12k/E7K8

Form: Unidirectional tape, fiber areal weight of 280 g/m^2, typical cured resin content of 29-33%, typical cured ply thickness of 0.011 inches.

Processing: Good drape. Autoclave cure; 300-310°F, 55 psi for 2 hours. Low exotherm profile for processing of thick parts.

General Supplier Information:

Fiber: Celion fibers are continuous carbon filaments made from PAN precursor. Filament count is 12,000 filaments/tow. Typical tensile modulus is 34 x 10^6 psi. Typical tensile strength is 515,000 psi.

Matrix: E7K8 is a medium flow, low exotherm epoxy resin. Good tack; up to 20 days out-time at ambient temperature.

Maximum Short Term Service Temperature: 300°F (dry), 190°F (wet)

Typical Applications: Primary and secondary structural applications on commercial and military aircraft.

4.2.4 Celion 12k/E7K8 unidirectional tape*

MATERIAL:	Celion 12k/E7K8 unidirectional tape		**C/Ep Celion 12k/E7K8 Summary**
PREPREG:	U.S. Polymeric Celion 12k/E7K8 unidirectional tape, grade 280		
FIBER:	Celanese Celion 12k, no twist	MATRIX:	U.S. Polymeric E7K8
T_g(dry):	T_g(wet):	T_g METHOD:	
PROCESSING:	Autoclave cure: 300 - 310°F, 120 - 130 min., 55 psi		

* DATA WERE SUBMITTED BEFORE THE ESTABLISHMENT OF DATA DOCUMENTATION REQUIREMENTS (JUNE 1989). ALL DOCUMENTATION PRESENTLY REQUIRED WERE NOT SUPPLIED FOR THIS MATERIAL.

Date of fiber manufacture	Date of testing	
Date of resin manufacture	Date of data submittal	1/88
Date of prepreg manufacture	Date of analysis	1/93
Date of composite manufacture		

LAMINA PROPERTY SUMMARY

	75°F/A		-65°F/A	180°F/A		75°F/W	180°F/W	
Tension, 1-axis	SSSS		SS-S			SSS-	SSSS	
Tension, 2-axis	SS--							
Tension, 3-axis								
Compression, 1-axis	SS-S		SS-S			SS--	SS--	
Compression, 2-axis								
Compression, 3-axis								
Shear, 12-plane	S---			S---		S---	S---	
Shear, 23-plane								
Shear, 31-plane								

Classes of data: F - Fully approved, I - Interim, S - Screening in Strength/Modulus/Poisson's ratio/Strain-to-failure order.

* DATA WERE SUBMITTED BEFORE THE ESTABLISHMENT OF DATA DOCUMENTATION REQUIREMENTS (JUNE 1989). ALL DOCUMENTATION PRESENTLY REQUIRED WERE NOT SUPPLIED FOR THIS MATERIAL.

		Nominal	As Submitted	Test Method
Fiber Density	(g/cm^3)	1.8		
Resin Density	(g/cm^3)	1.28		
Composite Density	(g/cm^3)	1.59	1.59 - 1.61	
Fiber Areal Weight	(g/m^2)	280		
Ply Thickness	(in)	0.011	0.010 - 0.011	

LAMINATE PROPERTY SUMMARY

Classes of data: F - Fully approved, I - Interim, S - Screening in Strength/Modulus/Poisson's ratio/Strain-to-failure order.

* DATA WERE SUBMITTED BEFORE THE ESTABLISHMENT OF DATA DOCUMENTATION REQUIREMENTS (JUNE 1989). ALL DOCUMENTATION PRESENTLY REQUIRED WAS NOT SUPPLIED FOR THIS MATERIAL.

Table 4.2.4(a)
C/Ep 280-UT
Celion 12k/E7K8
Tension, 1-axis
$[0]_5$
75/A, -65/A, 75/0.77%
Screening

MATERIAL: Celion 12k/E7K8 unidirectional tape

RESIN CONTENT: 29 wt% | COMP: DENSITY: 1.61 g/cm^3
FIBER VOLUME: 63-64 % | VOID CONTENT: 0.53-1.0%
PLY THICKNESS: 0.011 in.

TEST METHOD: ASTM D3039-76 | MODULUS CALCULATION:

NORMALIZED BY: Fiber volume to 60%

Temperature (°F) Moisture Content (%) Equilibrium at T, RH Source Code		75 ambient 20		-65 ambient 20		75 0.77 (1) 20	
		Normalized	Measured	Normalized	Measured	Normalized	Measured
	Mean	293	309	281	302	300	314
	Minimum	265	285	268	287	292	306
	Maximum	317	332	307	330	315	330
	C.V.(%)	4.52	4.52	5.44	5,44	3.22	3.60
	B-value	(2)	(2)	(2)	(2)	(2)	(2)
F_1^{tu}	Distribution	Weibull	Weibull	Normal	Normal	Normal	Normal
(ksi)	C_1	299	316	281	302	300	314
	C_2	25.6	25.9	15.3	16.4	9.67	10.1
	No. Specimens	20		5		5	
	No. Batches	1		1		1	
	Approval Class	Screening		Screening		Screening	
	Mean	20.0	21.1	19.2	20.6	19.0	19.9
	Minimum	18.7	20.1	18.6	20.0	18.5	19.4
	Maximum	21.9	23.0	20.3	21.8	20.0	21.0
E_1^t	C.V.(%)	4.48	4.25	3.40	3.80	3.22	3.60
(Msi)	No. Specimens	20		5		5	
	No. Batches	1		1		1	
	Approval Class	Screening		Screening		Screening	
	Mean		0.286				0.292
	No. Specimens	5				5	
ν_{12}^t	No. Batches	1				1	
	Approval Class	Screening				Screening	
	Mean		14300		14800		
	Minimum		13500		14200		
	Maximum		14700		15800		
	C.V.(%)		3.34		3.87		
	B-value		(2)		(2)		
ε_1^{tu}	Distribution		Normal		Normal		
($\mu\varepsilon$)	C_1		14300		14800		
	C_2		478		573		
	No. Specimens	5		5			
	No. Batches	1		1			
	Approval Class	Screening		Screening			

(1) Conditioned for 14 days at 160°F, 85% RH.
(2) B-values are presented only for fully approved data.

* DATA WERE SUBMITTED BEFORE THE ESTABLISHMENT OF DATA DOCUMENTATION REQUIREMENTS (JUNE 1989). ALL DOCUMENTATION PRESENTLY REQUIRED WAS NOT SUPPLIED FOR THIS MATERIAL.

MATERIAL:	Celion 12k/E7K8 unidirectional tape			Table 4.2.4(b) C/Ep 280-UT Celion 12k/E7K8 Tension, 1-axis $[0]_s$ 180/0.77% Screening
RESIN CONTENT:	29 wt%	COMP: DENSITY:	1.61 g/cm^3	
FIBER VOLUME:	63-64 %	VOID CONTENT:	0.53-1.0%	
PLY THICKNESS:	0.011 in.			
TEST METHOD: ASTM D3039-76		MODULUS CALCULATION:		
NORMALIZED BY:	Fiber volume to 60%			

		Normalized	Measured	Normalized	Measured	Normalized	Measured
Temperature (°F)		180					
Moisture Content (%)		0.77					
Equilibrium at T, RH		(1)					
Source Code		20					
F_1^{tu} (ksi)	Mean	293	311				
	Minimum	269	286				
	Maximum	316	335				
	C.V.(%)	6.43	7.19				
	B-value	(2)	(2)				
	Distribution	Normal	Normal				
	C_1	293	311				
	C_2	18.9	20.0				
	No. Specimens	5					
	No. Batches	1					
	Approval Class	Screening					
E_1^t (Msi)	Mean	19.8	21.0				
	Minimum	19.4	20.6				
	Maximum	20.1	21.4				
	C.V.(%)	1.61	1.81				
	No. Specimens	5					
	No. Batches	1					
	Approval Class	Screening					
ν_{12}^t	Mean		0.322				
	No. Specimens	5					
	No. Batches	1					
	Approval Class	Screening					
ε_1^{tu} (με)	Mean		13800				
	Minimum		12300				
	Maximum		15400				
	C.V.(%)		10.4				
	B-value		(2)				
	Distribution		Normal				
	C_1		13800				
	C_2		1440				
	No. Specimens	5					
	No. Batches	1					
	Approval Class	Screening					

(1) Conditioned for 14 days at 160°F, 85% RH.
(2) B-values are presented only for fully approved data.

* DATA WERE SUBMITTED BEFORE THE ESTABLISHMENT OF DATA DOCUMENTATION REQUIREMENTS (JUNE 1989). ALL DOCUMENTATION PRESENTLY REQUIRED WAS NOT SUPPLIED FOR THIS MATERIAL.

MATERIAL:	Celion 12k/E7K8 unidirectional tape			**Table 4.2.4(c)** **C/Ep 280-UT** **Celion 12k/E7K8** **Tension, 2-axis** **$[90]_{12}$** **75/A** **Screening**
RESIN CONTENT:	31-33 wt%	COMP: DENSITY:	1.59-1.60 g/cm^3	
FIBER VOLUME:	59-61 %	VOID CONTENT:	0.68-0.74%	
PLY THICKNESS:	0.011 in.			
TEST METHOD: ASTM D3039-76		MODULUS CALCULATION:		
NORMALIZED BY:	Not normalized			

Temperature (°F) Moisture Content (%) Equilibrium at T, RH Source Code		75 ambient 20					
F_2^{tu} (ksi)	Mean	6.00					
	Minimum	5.21					
	Maximum	6.89					
	C.V.(%)	8.79					
	B-value	(1)					
	Distribution	Weibull					
	C_1	6.24					
	C_2	12.6					
	No. Specimens	20					
	No. Batches	1					
	Approval Class	Screening					
E_2^t (Msi)	Mean	1.28					
	Minimum	1.19					
	Maximum	1.36					
	C.V.(%)	4.52					
	No. Specimens	20					
	No. Batches	1					
	Approval Class	Screening					
ν_{21}^t	Mean						
	No. Specimens						
	No. Batches						
	Approval Class						
ε_2^{tu} (με)	Mean						
	Minimum						
	Maximum						
	C.V.(%)						
	B-value						
	Distribution						
	C_1						
	C_2						
	No. Specimens						
	No. Batches						
	Approval Class						

(1) B-values are presented only for fully approved data.

* DATA WERE SUBMITTED BEFORE THE ESTABLISHMENT OF DATA DOCUMENTATION REQUIREMENTS (JUNE 1989). ALL DOCUMENTATION PRESENTLY REQUIRED WAS NOT SUPPLIED FOR THIS MATERIAL.

MATERIAL:	Celion 12k/E7K8 unidirectional tape			**Table 4.2.4(d) C/Ep 280-UT Celion 12k/E7K8 Compression, 1-axis $[0]_s$ 75/A, -65/A, 75/0.77% Screening**
RESIN CONTENT:	29-30 wt%	COMP: DENSITY:	1.60-1.61 g/cm^3	
FIBER VOLUME:	62-64 %	VOID CONTENT:	0.78-0.79%	
PLY THICKNESS:	0.010 in.			
TEST METHOD: SACMA SRM 1-88		MODULUS CALCULATION:		
NORMALIZED BY:	Fiber volume to 60%			

Temperature (°F)		75		-65		75	
Moisture Content (%)		ambient		ambient		0.77	
Equilibrium at T, RH						(1)	
Source Code		20		20		20	
		Normalized	Measured	Normalized	Measured	Normalized	Measured
F_1^{cu} (ksi)	Mean	206	213	221	229	207	214
	Minimum	171	177	198	205	198	205
	Maximum	247	255	267	276	219	227
	C.V.(%)	8.62	8.62	12.2	12.2	5.06	5.06
	B-value	(2)	(2)	(2)	(2)	(2)	(2)
	Distribution	Weibull	Weibull	Normal	Normal	Normal	Normal
	C_1	214	221	221	228	207	214
	C_2	12.1	12.1	27.0	28.0	10.5	10.8
	No. Specimens	20		5		5	
	No. Batches	1		1		1	
	Approval Class	Screening		Screening		Screening	
E_1^c (Msi)	Mean	19.9	21.1	22.9	24.3	21.6	22.3
	Minimum	18.1	19.2	20.8	22.0	20.2	21.0
	Maximum	21.7	22.3	23.8	25.1	22.8	23.6
	C.V.(%)	4.95	5.08	5.28	5.90	5.25	5.86
	No. Specimens	20		5		5	
	No. Batches	1		1		1	
	Approval Class	Screening		Screening		Screening	
ν_{12}^c	Mean						
	No. Specimens						
	No. Batches						
	Approval Class						
ε_1^{cu} ($\mu\varepsilon$)	Mean		11200		9870		
	Minimum		10800		9210		
	Maximum		11800		10600		
	C.V.(%)		3.59		5.32		
	B-value		(2)		(2)		
	Distribution		Normal		Normal		
	C_1		11200		9870		
	C_2		401		526		
	No. Specimens	5		5			
	No. Batches	1		1			
	Approval Class	Screening		Screening			

(1) Conditioned for 14 days at 160°F, 85% RH.
(2) B-values are presented only for fully approved data.

* DATA WERE SUBMITTED BEFORE THE ESTABLISHMENT OF DATA DOCUMENTATION REQUIREMENTS (JUNE 1989). ALL DOCUMENTATION PRESENTLY REQUIRED WAS NOT SUPPLIED FOR THIS MATERIAL.

MATERIAL:	Celion 12k/E7K8 unidirectional tape			**Table 4.2.4(e)** **C/Ep 280-UT** **Celion 12k/E7K8** **Compression, 1-axis** **$[0]_5$** **180/0.77%** **Screening**
RESIN CONTENT:	29-30 wt%	COMP: DENSITY:	1.60-1.61 g/cm^3	
FIBER VOLUME:	62-64 %	VOID CONTENT:	0.78-0.79%	
PLY THICKNESS:	0.010 in.			
TEST METHOD: SACMA SRM 1-88		MODULUS CALCULATION:		
NORMALIZED BY:	Fiber volume to 60%			

Temperature (°F)		180					
Moisture Content (%)		0.77					
Equilibrium at T, RH		(1)					
Source Code		20					
		Normalized	Measured	Normalized	Measured	Normalized	Measured
F_1^{cu} (ksi)	Mean	185	192				
	Minimum	158	164				
	Maximum	220	228				
	C.V.(%)	12.9	12.9				
	B-value	(2)	(2)				
	Distribution	Normal	Normal				
	C_1	185	192				
	C_2	24.0	24.8				
	No. Specimens	5					
	No. Batches	1					
	Approval Class	Screening					
E_1^c (Msi)	Mean	21.1	22.3				
	Minimum	19.5	20.6				
	Maximum	23.1	24.5				
	C.V.(%)	6.80	7.63				
	No. Specimens	5					
	No. Batches	1					
	Approval Class	Screening					
ν_{12}^c	Mean						
	No. Specimens						
	No. Batches						
	Approval Class						
ε_1^{cu} (με)	Mean						
	Minimum						
	Maximum						
	C.V.(%)						
	B-value						
	Distribution						
	C_1						
	C_2						
	No. Specimens						
	No. Batches						
	Approval Class						

(1) Conditioned for 14 days at 160°F, 85% RH.
(2) B-values are presented only for fully approved data.

* DATA WERE SUBMITTED BEFORE THE ESTABLISHMENT OF DATA DOCUMENTATION REQUIREMENTS (JUNE 1989). ALL DOCUMENTATION PRESENTLY REQUIRED WAS NOT SUPPLIED FOR THIS MATERIAL.

Table 4.2.4(f)
C/Ep 280-UT
Celion 12k/E7K8
Shear, 12-plane
$[\pm45/45]_s$
75/A, 180/A, 75/0.77%, 180/077%
Screening

MATERIAL: Celion 12k/E7K8 unidirectional tape

RESIN CONTENT: 30-31 wt% | COMP: DENSITY: 1.60 g/cm^3
FIBER VOLUME: 61-62 % | VOID CONTENT: 0.41-0.61%
PLY THICKNESS: 0.011 in.

TEST METHOD: ASTM D3518-76 | MODULUS CALCULATION:

NORMALIZED BY: Not normalized

Temperature (°F)		75	180	75	180		
Moisture Content (%)		ambient	ambient	0.77	0.77		
Equilibrium at T, RH				(1)	(1)		
Source Code		20	20	20	20		
F_{12}^{su} (ksi)	Mean	9.9	10.0	12.0	10.0		
	Minimum	9.3	8.1	11.3	8.2		
	Maximum	11.1	11.1	12.3	11.4		
	C.V.(%)	4.16	11.7	3.41	11.7		
	B-value	(2)	(2)	(2)	(2)		
	Distribution	Nonpara.	Normal	Normal	Normal		
	C_1	10	10.0	12.0	10.0		
	C_2	1.25	1.17	0.407	1.17		
	No. Specimens	20	5	5	5		
	No. Batches	1	1	1	1		
	Approval Class	Screening	Screening	Screening	Screening		
G_{12}^{s} (Msi)	Mean						
	Minimum						
	Maximum						
	C.V.(%)						
	No. Specimens						
	No. Batches						
	Approval Class						
γ_{12}^{su} (με)	Mean						
	Minimum						
	Maximum						
	C.V.(%)						
	B-value						
	Distribution						
	C_1						
	C_2						
	No. Specimens						
	No. Batches						
	Approval Class						

(1) Conditioned for 14 days at 160°F, 85% RH.
(2) B-values are presented only for fully approved data.

4.2.5 S4 12k/938 unitape data set description

Material Description:

Material: AS4-12k/938

Form: Unidirectional tape, fiber areal weight of 145 g/m^2, typical cured resin content of 35-49%, typical cured ply thickness of 0.0055 inches.

Processing: Autoclave cure; 350°F, 85 psi for 2 hours.

General Supplier Information:

Fiber: AS4 fibers are continuous carbon filaments made from PAN precursor, surface treated to improve handling characteristics and structural properties. Filament count is 12,000 filaments/tow. Typical tensile modulus is 34 x 10^6 psi. Typical tensile strength is 550,000 psi.

Matrix: 938 is an epoxy resin. 10 days out-time at 72°F.

Maximum Short Term Service Temperature: 350°F (dry), 200°F (wet)

Typical applications: Commercial and military structural applications

4.2.5 AS4 12k/938 unidirectional tape*

MATERIAL:	AS4 12k/938 unidirectional tape			**C/Ep AS4/938 Summary**
PREPREG:	Fiberite Hy-E 1338H unidirectional tape, grade 145			
FIBER:	Hercules AS4 12k, unsized, no twist	MATRIX:	Fiberite 938	
T_g(dry):	T_g(wet): 260°F	T_g METHOD:		
PROCESSING:	Autoclave cure: 350 ± 10°F, 120 - 135 min., 100 ± 15 psi			

* DATA WERE SUBMITTED BEFORE THE ESTABLISHMENT OF DATA DOCUMENTATION REQUIREMENTS (JUNE 1989). ALL DOCUMENTATION PRESENTLY REQUIRED WERE NOT SUPPLIED FOR THIS MATERIAL.

Date of fiber manufacture		Date of testing	8/85
Date of resin manufacture		Date of data submittal	4/89
Date of prepreg manufacture	7/85	Date of analysis	1/93
Date of composite manufacture			

LAMINA PROPERTY SUMMARY

	75°F/A		-65°F/A	200°F/A		200°F/W		
Tension, 1-axis	II--		II--	II--				
Tension, 2-axis	II--			II--				
Tension, 3-axis								
Compression, 1-axis	II--					II--		
Compression, 2-axis	S---							
Compression, 3-axis								
Shear, 12-plane	S---			I---				
Shear, 23-plane								
Shear, 31-plane								

Classes of data: F - Fully approved, I - Interim, S - Screening in Strength/Modulus/Poisson's ratio/Strain-to-failure order.

* DATA WERE SUBMITTED BEFORE THE ESTABLISHMENT OF DATA DOCUMENTATION REQUIREMENTS (JUNE 1989). ALL DOCUMENTATION PRESENTLY REQUIRED WERE NOT SUPPLIED FOR THIS MATERIAL.

		Nominal	As Submitted	Test Method
Fiber Density	(g/cm^3)	1.80	1.77 - 1.79	
Resin Density	(g/cm^3)	1.30	1.30	
Composite Density	(g/cm^3)	1.60	1.55 - 1.58	
Fiber Areal Weight	(g/m^2)	145	144 - 146	
Ply Thickness	(in)	0.0055	0.0048 - 0.0065	

LAMINATE PROPERTY SUMMARY

Classes of data: F - Fully approved, I - Interim, S - Screening in Strength/Modulus/Poisson's ratio/Strain-to-failure order.

* DATA WERE SUBMITTED BEFORE THE ESTABLISHMENT OF DATA DOCUMENTATION REQUIREMENTS (JUNE 1989). ALL DOCUMENTATION PRESENTLY REQUIRED WAS NOT SUPPLIED FOR THIS MATERIAL.

Table 4.2.5(a)
C/Ep 145-UT
AS4/938
Tension, 1-axis
$[0]_8$
75/A, -65/A, 200/A
Interim

MATERIAL: AS4 12k/938 unidirectional tape

RESIN CONTENT: 35-41 wt% COMP: DENSITY: 1.55-1.57 g/cm^3
FIBER VOLUME: 52-57 % VOID CONTENT: 0.0-<1.0%
PLY THICKNESS: 0.0042-0.0052 in.

TEST METHOD: MODULUS CALCULATION:
ASTM D3039-76 (1)

NORMALIZED BY: Specimen thickness and batch fiber volume to 60%

Temperature (°F) Moisture Content (%) Equilibrium at T, RH Source Code		75 ambient 12		-65 ambient 12		200 ambient 12	
		Normalized	Measured	Normalized	Measured	Normalized	Measured
	Mean	314	272	296	238	321	274
	Minimum	270	230	198	174	263	229
	Maximum	351	330	363	287	356	322
	C.V.(%)	7.45	8.79	14.4	11.0	7.79	8.10
	B-value	(2)	(2)	(2)	(2)	(2)	(2)
F_1^{tu}	Distribution	Weibull	ANOVA	ANOVA	ANOVA	ANOVA	Weibull
(ksi)	C_1	324	26.3	49.1	249	26.9	284
	C_2	16.5	4.12	4.64	11.1	3.78	13.3
	No. Specimens	22		22		20	
	No. Batches	3		3		3	
	Approval Class	Interim		Interim		Interim	
	Mean	22.4	19.4	19.5	19.0	20.4	20.8
	Minimum	18.8	17.1	18.5	16.9	18.4	18.4
	Maximum	26.9	21.0	21.5	22.0	24.0	22.4
E_1^t	C.V.(%)	9.88	4.66	4.07	5.13	7.23	6.06
(Msi)	No. Specimens	22		22		20	
	No. Batches	3		3		3	
	Approval Class	Interim		Interim		Interim	
	Mean						
	No. Specimens						
ν_{12}^t	No. Batches						
	Approval Class						
	Mean						
	Minimum						
	Maximum						
	C.V.(%)						
	B-value						
ε_1^{tu}	Distribution						
(με)	C_1						
	C_2						
	No. Specimens						
	No. Batches						
	Approval Class						

(1) Gage length 2.0 inches.
(2) B-values are presented only for fully approved data.

* DATA WERE SUBMITTED BEFORE THE ESTABLISHMENT OF DATA DOCUMENTATION REQUIREMENTS (JUNE 1989). ALL DOCUMENTATION PRESENTLY REQUIRED WAS NOT SUPPLIED FOR THIS MATERIAL.

MATERIAL: AS4 12k/938 unidirectional tape RESIN CONTENT: 35-40 wt% COMP:DENSITY: 1.56-1.58 g/cm^3 FIBER VOLUME: 52-58 % VOID CONTENT: 0.0-<1.0% PLY THICKNESS: 0.0053-0.0063 in. TEST METHOD: MODULUS CALCULATION: ASTM D3039-76 (1) NORMALIZED BY: Not normalized	**Table 4.2.5(b)** **C/Ep 145-UT** **AS4/938** **Tension, 2-axis** **$[90]_{16}$** **75/A, 200/A** **Interim**

Temperature (°F)		75.0	200				
Moisture Content (%)		ambient	ambient				
Equilibrium at T, RH							
Source Code		12	12				
F_2^{tu} (ksi)	Mean	8.96	8.84				
	Minimum	6.50	6.85				
	Maximum	12.0	10.3				
	C.V.(%)	15.2	12.2				
	B-value	(2)	(2)				
	Distribution	Weibull	ANOVA				
	C_1	9.54	1.18				
	C_2	7.10	3.96				
	No. Specimens	19	17				
	No. Batches	3	3				
	Approval Class	Interim	Interim				
E_2^t (Msi)	Mean	1.29	1.23				
	Minimum	0.970	1.05				
	Maximum	1.72	1.40				
	C.V.(%)	7.89	7.81				
	No. Specimens	19	17				
	No. Batches	3	3				
	Approval Class	Interim	Interim				
ν_{21}^t	Mean						
	No. Specimens						
	No. Batches						
	Approval Class						
ε_2^{tu} (με)	Mean						
	Minimum						
	Maximum						
	C.V.(%)						
	B-value						
	Distribution						
	C_1						
	C_2						
	No. Specimens						
	No. Batches						
	Approval Class						

(1) Gage length 2.0 inches.

(2) B-values are presented only for fully approved data.

* DATA WERE SUBMITTED BEFORE THE ESTABLISHMENT OF DATA DOCUMENTATION REQUIREMENTS (JUNE 1989). ALL DOCUMENTATION PRESENTLY REQUIRED WAS NOT SUPPLIED FOR THIS MATERIAL.

MATERIAL:	AS4 12k/938 unidirectional tape			**Table 4.2.5(c)** **C/Ep 145-UT** **AS4/938** **Compression, 1-axis** **$[0]_8$** **75/A, 200/W** **Interim, Screening**
RESIN CONTENT:	33-38 wt%	COMP: DENSITY:	1.55-1.58 g/cm^3	
FIBER VOLUME:	54-60 %	VOID CONTENT:	0.0-<1.0%	
PLY THICKNESS:	0.0048-0.0060 in.			
TEST METHOD: SACMA SRM 1-88		MODULUS CALCULATION:		
NORMALIZED BY:	Specimen thickness and batch fiber volume to 60%			

		75		200			
Temperature (°F) Moisture Content (%) Equilibrium at T, RH Source Code		75 ambient 12		200 (1) 12			
		Normalized	Measured	Normalized	Measured	Normalized	Measured
F_1^{cu} (ksi)	Mean	228	211	190	168		
	Minimum	186	172	158	138		
	Maximum	265	251	223	194		
	C.V.(%)	9.31	10.2	8.96	9.29		
	B-value	(2)	(2)	(2)	(2)		
	Distribution	Weibull	ANOVA	ANOVA	ANOVA		
	C_1	224	22.4	19.0	17.6		
	C_2	12.5	3.31	4.40	4.57		
	No. Specimens	25		24			
	No. Batches	3		3			
	Approval Class	Interim		Interim			
E_1^c (Msi)	Mean	18.2	18.4	19.1	18.4		
	Minimum	15.7	15.9	16.9	16.6		
	Maximum	21.0	22.5	24.0	21.0		
	C.V.(%)	9.13	12.4	12.8	9.10		
	No. Specimens	15		13			
	No. Batches	2		2			
	Approval Class	Interim		Screening			
ν_{12}^c	Mean						
	No. Specimens						
	No. Batches						
	Approval Class						
ε_1^{cu} (με)	Mean						
	Minimum						
	Maximum						
	C.V.(%)						
	B-value						
	Distribution						
	C_1						
	C_2						
	No. Specimens						
	No. Batches						
	Approval Class						

(1) Specimens conditioned at 140°F and 95% relative humidity for one month.

(2) B-values are presented only for fully approved data.

* DATA WERE SUBMITTED BEFORE THE ESTABLISHMENT OF DATA DOCUMENTATION REQUIREMENTS (JUNE 1989). ALL DOCUMENTATION PRESENTLY REQUIRED WAS NOT SUPPLIED FOR THIS MATERIAL.

Table 4.2.5(d)
C/Ep 145-UT
AS4/938
Compression, 2-axis
$[90]_8$
75/A
Screening

MATERIAL: AS4 12k/938 unidirectional tape

RESIN CONTENT: 36 wt% | COMP: DENSITY: 1.56 g/cm^3
FIBER VOLUME: 56 % | VOID CONTENT: 0.0%
PLY THICKNESS: 0.0058 in.

TEST METHOD: SACMA SRM 1-88

MODULUS CALCULATION:

NORMALIZED BY: Not normalized

Temperature (°F)		75.0					
Moisture Content (%)		ambient					
Equilibrium at T, RH							
Source Code		12					
	Mean	30.4					
	Minimum	26.2					
	Maximum	39.7					
	C.V.(%)	16.4					
	B-value	(1)					
F_2^{cu}	Distribution	Nonpara.					
(ksi)	C_1	6					
	C_2	2.14					
	No. Specimens	10					
	No. Batches	1					
	Approval Class	Screening					
	Mean						
	Minimum						
	Maximum						
E_2^c	C.V.(%)						
(Msi)	No. Specimens						
	No. Batches						
	Approval Class						
	Mean						
	No. Specimens						
ν_{21}^c	No. Batches						
	Approval Class						
	Mean						
	Minimum						
	Maximum						
	C.V.(%)						
	B-value						
ε_2^{cu}	Distribution						
(με)	C_1						
	C_2						
	No. Specimens						
	No. Batches						
	Approval Class						

(1) B-values are presented only for fully approved data.

* DATA WERE SUBMITTED BEFORE THE ESTABLISHMENT OF DATA DOCUMENTATION REQUIREMENTS (JUNE 1989). ALL DOCUMENTATION PRESENTLY REQUIRED WAS NOT SUPPLIED FOR THIS MATERIAL.

Table 4.2.5(e)
C/Ep 145-UT
AS4/938
Shear, 12-plane
$[\pm45]_{2S}$
75/A, 200/A
Screening, Interim

MATERIAL: AS4 12k/938 unidirectional tape

RESIN CONTENT: 35-37 wt% | COMP: DENSITY: 1.56-1.58 g/cm^3
FIBER VOLUME: 54-57 % | VOID CONTENT: 0.0-<1.0%
PLY THICKNESS: 0.0051-0.0063 in.

TEST METHOD: ASTM D3518-76 | MODULUS CALCULATION:

NORMALIZED BY: Not normalized

Temperature (°F) Moisture Content (%) Equilibrium at T, RH Source Code		75.0 ambient 12	200 ambient 12					
	Mean	13.0	13.9					
	Minimum	10.8	11.9					
	Maximum	13.9	16.0					
	C.V.(%)	6.36	7.63					
	B-value	(1)	(1)					
F_{12}^{su}	Distribution	Weibull	ANOVA					
(ksi)	C_1	13.4	1.26					
	C_2	25.4	4.96					
	No. Specimens	13	18					
	No. Batches	3	3					
	Approval Class	Screening	Interim					
	Mean							
	Minimum							
	Maximum							
G_{12}^{s}	C.V.(%)							
(Msi)	No. Specimens							
	No. Batches							
	Approval Class							
	Mean							
	Minimum							
	Maximum							
	C.V.(%)							
	B-value							
γ_{12}^{su}	Distribution							
($\mu\varepsilon$)	C_1							
	C_2							
	No. Specimens							
	No. Batches							
	Approval Class							

(1) B-values are presented only for fully approved data.

4.2.6 T-300 3k/934 plain weave data set description

Material Description:

Material: T-300 3k/934

Form: Plain weave fabric, fiber areal weight of 196 g/m^2, typical cured resin content of 34%, typical cured ply thickness of 0.0078 inches.

Processing: Autoclave cure; 355°F, 85-100 psi for 2 hours.

General Supplier Information:

Fiber: T-300 fibers are continuous, no twist carbon filaments made from PAN precursor, surface treated to improve handling characteristics and structural properties. Filament count is 3,000 filaments/tow. Typical tensile modulus is 33 x 10^6. Typical tensile strength is 530,000 psi.

Matrix: 934 is a high flow, epoxy resin with good hot/wet properties and meets NASA outgassing requirements.

Maximum Short Term Service Temperature: 350°F (dry), 200°F (wet)

Typical applications: Aircraft primary and secondary structure, critical space structure.

4.2.6 T300 3k/934 plain weave*

MATERIAL:	T-300 3k/934 plain weave			**C/Ep T-300/934 Summary**
PREPREG:	Fiberite HMF-322/34 plain weave			
FIBER:	Toray T-300 3k	MATRIX:	Fiberite 934	
T_g(dry):	410°F T_g(wet):	T_g METHOD:	DSC	
PROCESSING:	Autoclave cure: 355 ± 10°F, 120 - 130 min., 85-100 psig			

* DATA WERE SUBMITTED BEFORE THE ESTABLISHMENT OF DATA DOCUMENTATION REQUIREMENTS (JUNE 1989). ALL DOCUMENTATION PRESENTLY REQUIRED WERE NOT SUPPLIED FOR THIS MATERIAL.

Date of fiber manufacture		Date of testing	
Date of resin manufacture		Date of data submittal	6/88
Date of prepreg manufacture	2/84	Date of analysis	1/93
Date of composite manufacture			

LAMINA PROPERTY SUMMARY

	75°F/A		-65°F/A	250°F/A		160°F/W	250°F/W	
Tension, 1-axis	IS-I		IS-I	SS-S		II--	II--	
Tension, 2-axis	II-I		II-I	SS-S		II--	II--	
Tension, 3-axis								
Compression, 1-axis	II--		II--	SI--		I---	I---	
Compression, 2-axis	II--		II--	SI--		I---	I---	
Compression, 3-axis								
Shear, 12-plane								
Shear, 23-plane								
SB Strength, 31-plane	S---		S---	S---				

Classes of data: F - Fully approved, I - Interim, S - Screening in Strength/Modulus/Poisson's ratio/Strain-to-failure order.

* DATA WERE SUBMITTED BEFORE THE ESTABLISHMENT OF DATA DOCUMENTATION REQUIREMENTS (JUNE 1989). ALL DOCUMENTATION PRESENTLY REQUIRED WERE NOT SUPPLIED FOR THIS MATERIAL.

		Nominal	As Submitted	Test Method
Fiber Density	(g/cm^3)		1.73 - 1.74	
Resin Density	(g/cm^3)	1.30		
Composite Density	(g/cm^3)	1.55	1.54 - 1.57	
Fiber Areal Weight	(g/m^2)	194	1.92 - 2.00	
Ply Thickness	(in)		0.0073 - 0.0084	

LAMINATE PROPERTY SUMMARY

Classes of data: F - Fully approved, I - Interim, S - Screening in Strength/Modulus/Poisson's ratio/Strain-to-failure order.

* DATA WERE SUBMITTED BEFORE THE ESTABLISHMENT OF DATA DOCUMENTATION REQUIREMENTS (JUNE 1989). ALL DOCUMENTATION PRESENTLY REQUIRED WAS NOT SUPPLIED FOR THIS MATERIAL.

MATERIAL:	T-300 3k/934 plain weave			**Table 4.2.6(a)** **C/Ep 194-PW** **T-300/934** **Tension, 1-axis** **$[0_f]_{12}$** **75/A, -65/A, 250/A** **Interim, Screening**
RESIN CONTENT:	33-35 wt%	COMP: DENSITY:	1.54-1.57 g/cm^3	
FIBER VOLUME:	58-60 %	VOID CONTENT:	<0.5-1.2%	
PLY THICKNESS:	0.0074-0.0082 in.			
TEST METHOD:		MODULUS CALCULATION:		
ASTM D3039-76 (2)		Chord between 20 and 40% of typical ultimate load		
NORMALIZED BY:	Specimen thickness and batch fiber volume to 57%			

		Temperature (°F) 75		-65		250	
	Moisture Content (%)	ambient		ambient		ambient	
	Equilibrium at T, RH						
	Source Code	12		12		12	
		Normalized	Measured	Normalized	Measured	Normalized	Measured
F_1^{tu} (ksi)	Mean	91	94	83	85	109	113
	Minimum	82	85	78	79	104	109
	Maximum	99	100	87	90	114	118
	C.V.(%)	4.1	4.0	3.2	3.3	3.54	3.42
	B-value	(1)	(1)	(1)	(1)	(1)	(1)
	Distribution	Weibull	Weibull	Weibull	Weibull	Normal	Normal
	C_1	93.0	96	83.7	86	86.0	113
	C_2	28.2	31	35.8	36	2.86	3.87
	No. Specimens	20		20		5	
	No. Batches	4		4		1	
	Approval Class	Interim		Interim		Screening	
E_1^t (Msi)	Mean	9.1	9.4	10.	10.	9.3	9.7
	Minimum	8.4	8.7	8.6	9.0	9.1	9.4
	Maximum	9.5	9.9	12	12	10.0	10.7
	C.V.(%)	3.3	3.6	11	10.	4.6	5.6
	No. Specimens	20		20		5	
	No. Batches	4		4		1	
	Approval Class	Interim		Interim		Screening	
ν_{12}^t	Mean						
	No. Specimens						
	No. Batches						
	Approval Class						
ε_1^{tu} (με)	Mean		9780		8990		11300
	Minimum		8880		7990		10900
	Maximum		11200		9800		11800
	C.V.(%)		5.61		6.07		3.11
	B-value		(1)		(1)		(1)
	Distribution		ANOVA		ANOVA		Normal
	C_1		577		592		11300
	C_2		3.12		3.61		351
	No. Specimens	20		20		5	
	No. Batches	4		4		1	
	Approval Class	Interim		Interim		Screening	

(1) B-values are presented only for fully approved data.

(2) Width 0.5 inch, speed of testing 0.05 in./in./min, gage length below recommendation

* DATA WERE SUBMITTED BEFORE THE ESTABLISHMENT OF DATA DOCUMENTATION REQUIREMENTS (JUNE 1989). ALL DOCUMENTATION PRESENTLY REQUIRED WAS NOT SUPPLIED FOR THIS MATERIAL.

MATERIAL:	T-300 3k/934 plain weave			**Table 4.2.6(b)** **C/Ep 194-PW** **T-300/934** **Tension, 1-axis** **$[0_f]_{12}$** **160/W, 250/W** **Interim**
RESIN CONTENT:	33-35 wt%	COMP: DENSITY:	1.54-1.57 g/cm^3	
FIBER VOLUME:	58-60 %	VOID CONTENT:	<0.5-1.2%	
PLY THICKNESS:	0.0074-0.0082 in.			
TEST METHOD:		MODULUS CALCULATION:		
ASTM D3039-76 (2)		Chord between 20 and 40% of typical ultimate load		
NORMALIZED BY:	Specimen thickness and batch fiber volume to 57%			

		160		250			
Temperature (°F) Moisture Content (%) Equilibrium at T, RH Source Code		160 (1) 12		250 (1) 12			
		Normalized	Measured	Normalized	Measured	Normalized	Measured
F_1^{tu} (ksi)	Mean	96	98	79	82		
	Minimum	84	88	61	66		
	Maximum	104	106	95	97		
	C.V.(%)	5.7	5.11	14	11		
	B-value	(2)	(2)	(2)	(2)		
	Distribution	ANOVA	Weibull	ANOVA	Weibull		
	C_1	6.0	101	12	86		
	C_2	4.8	24	5.3	11		
	No. Specimens	15		15			
	No. Batches	3		3			
	Approval Class	Interim		Interim			
E_1^t (Msi)	Mean	9.8	10.0	9.4	9.7		
	Minimum	8.1	8.6	6.8	7.1		
	Maximum	11.0	11.7	12.0	13.0		
	C.V.(%)	8.7	8.7	17.	18		
	No. Specimens	15		15			
	No. Batches	3		3			
	Approval Class	Interim		Interim			
ν_{12}^t	Mean						
	No. Specimens						
	No. Batches						
	Approval Class						
ε_1^{tu} (με)	Mean						
	Minimum						
	Maximum						
	C.V.(%)						
	B-value						
	Distribution						
	C_1						
	C_2						
	No. Specimens						
	No. Batches						
	Approval Class						

(1) Immersed in water at 160°F for 14 days.
(2) B-values are presented only for fully approved data.
(3) Width 0.5 inch, speed of testing 0.05 in./in./min, gage length below recommendation.

* DATA WERE SUBMITTED BEFORE THE ESTABLISHMENT OF DATA DOCUMENTATION REQUIREMENTS (JUNE 1989). ALL DOCUMENTATION PRESENTLY REQUIRED WAS NOT SUPPLIED FOR THIS MATERIAL.

Table 4.2.6(c)
C/Ep 194-PW
T-300/934
Tension, 2-axis
$[90_f]_{12}$
75/A, -65/A, 250/A
Interim, Screening

MATERIAL: T-300 3k/934 plain weave

RESIN CONTENT: 33-35 wt% COMP: DENSITY: 1.54-1.57 g/cm^3
FIBER VOLUME: 58-60 % VOID CONTENT: <0.5-1.2%
PLY THICKNESS: 0.0074-0.0082 in.

TEST METHOD: ASTM D3039-76

MODULUS CALCULATION: Chord between 20 and 40% of typical ultimate load

NORMALIZED BY: Specimen thickness and batch fiber volume to 57%

		Temperature (°F) / Moisture Content (%) / Equilibrium at T, RH / Source Code: 75 / ambient / / 12		-65 / ambient / / 12		250 / ambient / / 12	
		Normalized	Measured	Normalized	Measured	Normalized	Measured
F_2^{tu} (ksi)	Mean	88	91	80.	82	94	98
	Minimum	80.	82	70.	72	90.	94
	Maximum	97	99	91	95	97	101
	C.V.(%)	5.7	5.5	6.2	6.5	2.6	2.7
	B-value	(1)	(1)	(1)	(1)	(1)	(1)
	Distribution	ANOVA	ANOVA	ANOVA	ANOVA	Normal	Normal
	C_1	5.4	5.4	5.2	5.7	93.7	97.8
	C_2	3.5	3.4	3.3	3.4	2.47	2.59
	No. Specimens	20		20		5	
	No. Batches	4		4		1	
	Approval Class	Interim		Interim		Screening	
E_2^t (Msi)	Mean	9.0	9.3	9.1	9.5	8.1	8.5
	Minimum	8.3	8.7	8.1	8.3	8.0	8.3
	Maximum	9.9	10.3	10.8	11.1	8.2	8.6
	C.V.(%)	5.0	4.8	9.3	9.2	1.1	1.5
	No. Specimens	20		20		5	
	No. Batches	4		4		1	
	Approval Class	Interim		Interim		Screening	
ν_{21}^t	Mean						
	No. Specimens						
	No. Batches						
	Approval Class						
ε_2^{tu} (με)	Mean		9630		9100		11400
	Minimum		8680		7750		10400
	Maximum		11100		10700		12400
	C.V.(%)		6.18		7.44		8.59
	B-value		(1)		(1)		(1)
	Distribution		ANOVA		ANOVA		Normal
	C_1		616		710		11400
	C_2		2.82		3.08		981
	No. Specimens	20		20		5	
	No. Batches	4		4		1	
	Approval Class	Interim		Interim		Screening	

(1) B-values are presented only for fully approved data.

* DATA WERE SUBMITTED BEFORE THE ESTABLISHMENT OF DATA DOCUMENTATION REQUIREMENTS (JUNE 1989). ALL DOCUMENTATION PRESENTLY REQUIRED WAS NOT SUPPLIED FOR THIS MATERIAL.

Table 4.2.6(d)
C/Ep 194-PW
T-300/934
Tension, 2-axis
$[90_f]_{12}$
160/W, 250/W
Interim

MATERIAL: T-300 3k/934 plain weave

RESIN CONTENT: 33-35 wt% COMP: DENSITY: 1.54-1.57 g/cm^3
FIBER VOLUME: 58-60 % VOID CONTENT: <0.5-1.2%
PLY THICKNESS: 0.0074-0.0082 in.

TEST METHOD: ASTM D3039-76

MODULUS CALCULATION: Chord between 20 and 40% of typical ultimate load

NORMALIZED BY: Specimen thickness and batch fiber volume to 57%

		160		250			
Temperature (°F)		160		250			
Moisture Content (%)		(1)		(1)			
Equilibrium at T, RH							
Source Code		12		12			
		Normalized	Measured	Normalized	Measured	Normalized	Measured
F_2^{tu} (ksi)	Mean	97	100	81	83		
	Minimum	90.	92	73	75		
	Maximum	111	113	89	91		
	C.V.(%)	6.8	6.3	5.1	4.8		
	B-value	(2)	(2)	(2)	(2)		
	Distribution	ANOVA	ANOVA	ANOVA	ANOVA		
	C_1	7.3	6.8	4.4	4.2		
	C_2	4.8	4.5	4.5	4.2		
	No. Specimens	15		15			
	No. Batches	3		3			
	Approval Class	Interim		Interim			
E_2^t (Msi)	Mean	10.	10.	9.9	10.		
	Minimum	8.0	8.2	8.2	8.5		
	Maximum	11.8	12.1	11.9	12.1		
	C.V.(%)	11	11	11	11		
	No. Specimens	15		15			
	No. Batches	3		3			
	Approval Class	Interim		Interim			
ν_{21}^t	Mean						
	No. Specimens						
	No. Batches						
	Approval Class						
ε_2^{tu} ($\mu\varepsilon$)	Mean						
	Minimum						
	Maximum						
	C.V.(%)						
	B-value						
	Distribution						
	C_1						
	C_2						
	No. Specimens						
	No. Batches						
	Approval Class						

(1) Immersed in water at 160°F for 14 days.
(2) B-values are presented only for fully approved data.

* DATA WERE SUBMITTED BEFORE THE ESTABLISHMENT OF DATA DOCUMENTATION REQUIREMENTS (JUNE 1989). ALL DOCUMENTATION PRESENTLY REQUIRED WAS NOT SUPPLIED FOR THIS MATERIAL.

MATERIAL: T-300 3k/934 plain weave						
RESIN CONTENT: 33-35 wt%	COMP: DENSITY: 1.54-1.57 g/cm^3					**Table 4.2.6(e)** **C/Ep 194-PW** **T-300/934** **Compression, 1-axis** **$[0_f]_{12}$** **75/A, -65/A, 250/A** **Interim, Screening**
FIBER VOLUME: 58-60 %	VOID CONTENT: <0.5-1.2%					
PLY THICKNESS: 0.0074-0.0082 in.						
TEST METHOD: SACMA SRM 1-88	MODULUS CALCULATION: Chord between 20 and 40% of typical ultimate load					
NORMALIZED BY: Specimen thickness and batch fiber volume to 57%						
Temperature (°F)	75		-65		250	
Moisture Content (%)	ambient		ambient		ambient	
Equilibrium at T, RH						
Source Code	12		12		12	
	Normalized	Measured	Normalized	Measured	Normalized	Measured
F_1^{cu} (ksi) Mean	95	98	104	108	100.	105
Minimum	83	87	87	90.	94	98
Maximum	120	125	133	139	107	111
C.V.(%)	10.	10.	13	14	5.6	5.1
B-value	(1)	(1)	(1)	(1)	(1)	(1)
Distribution	ANOVA	ANOVA	ANOVA	ANOVA	Normal	Normal
C_1	10.	11	15	16	100.	105
C_2	3.9	3.9	3.7	3.8	5.64	5.4
No. Specimens	20		20		5	
No. Batches	4		4		1	
Approval Class	Interim		Interim		Screening	
E_1^c (Msi) Mean	8.4	8.8	8.2	8.6	8.4	8.9
Minimum	7.7	8.0	7.4	7.8	7.9	8.1
Maximum	9.0	9.4	8.9	9.7	10.0	10.1
C.V.(%)	5.1	5.3	5.1	5.7	6.3	6.4
No. Specimens	20		20		19	
No. Batches	4		4		4	
Approval Class	Interim		Interim		Interim	
ν_{12}^c Mean						
No. Specimens						
No. Batches						
Approval Class						
ε_1^{cu} (με) Mean						
Minimum						
Maximum						
C.V.(%)						
B-value						
Distribution						
C_1						
C_2						
No. Specimens						
No. Batches						
Approval Class						

(1) B-values are presented only for fully approved data.

(2) Tab thickness of 0.112 - 0.120 inch is larger than 0.070 inch nominal thickness per method.

(3) Specimen thickness of 0.09 - 0.10 inch is less than nominal 0.12 inch thickness per method.

* DATA WERE SUBMITTED BEFORE THE ESTABLISHMENT OF DATA DOCUMENTATION REQUIREMENTS (JUNE 1989). ALL DOCUMENTATION PRESENTLY REQUIRED WAS NOT SUPPLIED FOR THIS MATERIAL.

Table 4.2.6(f)
C/Ep 194-PW
T-300/934
Compression, 1-axis
$[0_f]_{12}$
160/W, 250/W
Interim

MATERIAL: T-300 3k/934 plain weave

RESIN CONTENT: 33-35 wt% | COMP: DENSITY: 1.54-1.57 g/cm^3
FIBER VOLUME: 58-60 % | VOID CONTENT: <0.5-1.2%
PLY THICKNESS: 0.0074-0.0082 in.

TEST METHOD: SACMA SRM 1-88

MODULUS CALCULATION: Chord between 20 and 40% of typical ultimate load

NORMALIZED BY: Specimen thickness and batch fiber volume to 57%

		160		250			
Temperature (°F)		160		250			
Moisture Content (%)		(1)		(1)			
Equilibrium at T, RH							
Source Code		12		12			
		Normalized	Measured	Normalized	Measured	Normalized	Measured
F_1^{cu} (3) (ksi)	Mean	74	76	44	46		
	Minimum	67	68	40	41		
	Maximum	81	84	49	51		
	C.V.(%)	6.9	5.6	6.2	6.2		
	B-value	(2)	(2)	(2)	(2)		
	Distribution	ANOVA	ANOVA	Weibull	Weibull		
	C_1	5.6	6.2	45.4	46.8		
	C_2	4.9	5.0	17.4	16.9		
	No. Specimens	15		15			
	No. Batches	3		3			
	Approval Class	Interim		Interim			
E_1^c (Msi)	Mean						
	Minimum						
	Maximum						
	C.V.(%)						
	No. Specimens						
	No. Batches						
	Approval Class						
ν_{12}^c	Mean						
	No. Specimens						
	No. Batches						
	Approval Class						
ε_1^{cu} (με)	Mean						
	Minimum						
	Maximum						
	C.V.(%)						
	B-value						
	Distribution						
	C_1						
	C_2						
	No. Specimens						
	No. Batches						
	Approval Class						

(1) Immersed in water at 160°F for 14 days.
(2) B-values are presented only for fully approved data.
(3) Tab thickness of 0.112 - 0.120 inch is larger than 0.070 inch nominal thickness per method.

* DATA WERE SUBMITTED BEFORE THE ESTABLISHMENT OF DATA DOCUMENTATION REQUIREMENTS (JUNE 1989). ALL DOCUMENTATION PRESENTLY REQUIRED WAS NOT SUPPLIED FOR THIS MATERIAL.

Table 4.2.6(g)
C/Ep 194-PW
T-300/934
Compression, 2-axis
$[90_f]_{12}$
75/A, -65/A, 250/A
Interim, Screening

MATERIAL: T-300 3k/934 plain weave

RESIN CONTENT:	33-35 wt%	COMP: DENSITY:	1.54-1.57 g/cm^3
FIBER VOLUME:	58-60 %	VOID CONTENT:	<0.5-1.2%
PLY THICKNESS:	0.0074-0.0082 in.		

TEST METHOD: SACMA SRM 1-88

MODULUS CALCULATION: Chord between 20 and 40% of typical ultimate load

NORMALIZED BY: Specimen thickness and batch fiber volume to 57%

Temperature (°F) Moisture Content (%) Equilibrium at T, RH Source Code		75 ambient 12		-65 ambient 12		250 ambient 12	
		Normalized	Measured	Normalized	Measured	Normalized	Measured
F_2^{cu} (2) (ksi)	Mean	90.	93	103	106	82	85
	Minimum	81	85	94	98	77	81
	Maximum	100.	104	116	121	84	88
	C.V.(%)	5.9	6.0	6.2	6.1	3.4	3.4
	B-value	(1)	(1)	(1)	(1)	(1)	(1)
	Distribution	ANOVA	ANOVA	Normal	Normal	Normal	Normal
	C_1	5.6	5.9	103	106	81.7	85.3
	C_2	3.2	3.2	6.18	6.4	2.74	2.86
	No. Specimens	20		20		5	
	No. Batches	4		4		1	
	Approval Class	Interim		Interim		Screening	
E_2^c (3) (Msi)	Mean	8.3	8.6	8.4	8.8	8.8	9.0
	Minimum	7.4	7.7	7.5	7.7	7.9	8.1
	Maximum	9.3	9.5	9.0	9.4	10.2	10.6
	C.V.(%)	7.0	6.6	5.1	5.5	8.4	8.9
	No. Specimens	20		20		20	
	No. Batches	4		4		4	
	Approval Class	Interim		Interim		Interim	
ν_{21}^c	Mean						
	No. Specimens						
	No. Batches						
	Approval Class						
ε_2^{cu} (με)	Mean						
	Minimum						
	Maximum						
	C.V.(%)						
	B-value						
	Distribution						
	C_1						
	C_2						
	No. Specimens						
	No. Batches						
	Approval Class						

(1) B-values are presented only for fully approved data.
(2) Tab thickness of 0.112-0.120 inch is larger than 0.070 inch nominal thickness per method.
(3) Specimen thickness of 0.09-0.10 inch is less than nominal 0.120 inch thickness per method.

* DATA WERE SUBMITTED BEFORE THE ESTABLISHMENT OF DATA DOCUMENTATION REQUIREMENTS (JUNE 1989). ALL DOCUMENTATION PRESENTLY REQUIRED WAS NOT SUPPLIED FOR THIS MATERIAL.

MATERIAL:	T-300 3k/934 plain weave			**Table 4.2.6(h)**
RESIN CONTENT:	33-35 wt%	COMP: DENSITY:	1.54-1.57 g/cm^3	**C/Ep 194-PW**
FIBER VOLUME:	58-60 %	VOID CONTENT:	<0.5-1.2%	**T-300/934**
PLY THICKNESS:	0.0074-0.0082 in.			**Compression, 2-axis**
				$[90_f]_{12}$
TEST METHOD:		MODULUS CALCULATION:		**160/W, 250/W**
SACMA SRM 1-88		Chord between 20 and 40% of typical ultimate load		**Interim**
NORMALIZED BY:	Specimen thickness and batch fiber volume to 57%			

Temperature (°F) Moisture Content (%) Equilibrium at T, RH Source Code		160 wet (1) 12		250 wet (1) 12			
		Normalized	Measured	Normalized	Measured	Normalized	Measured
F_2^{cu} (3) (ksi)	Mean	75	77	46	47		
	Minimum	63	66	38	39		
	Maximum	81	83	59	60		
	C.V.(%)	7.2	6.5	11	11		
	B-value	(2)	(2)	(2)	(2)		
	Distribution	ANOVA	ANOVA	ANOVA	ANOVA		
	C_1	6.0	5.4	5.9	5.8		
	C_2	5.0	4.7	5.1	5.0		
	No. Specimens	15		15			
	No. Batches	3		3			
	Approval Class	Interim		Interim			
E_2^c (Msi)	Mean						
	Minimum						
	Maximum						
	C.V.(%)						
	No. Specimens						
	No. Batches						
	Approval Class						
ν_{21}^c	Mean						
	No. Specimens						
	No. Batches						
	Approval Class						
ε_2^{cu} (με)	Mean						
	Minimum						
	Maximum						
	C.V.(%)						
	B-value						
	Distribution						
	C_1						
	C_2						
	No. Specimens						
	No. Batches						
	Approval Class						

(1) Immersed in water at 160°F for 14 days.
(2) B-values are presented only for fully approved data.
(3) Tab thickness of 0.112-0.120 inch is larger than 0.070 nominal thickness per method.

* DATA WERE SUBMITTED BEFORE THE ESTABLISHMENT OF DATA DOCUMENTATION REQUIREMENTS (JUNE 1989). ALL DOCUMENTATION PRESENTLY REQUIRED WAS NOT SUPPLIED FOR THIS MATERIAL.

Table 4.2.6(i)
C/Ep 194-PW
T-300/934
SBS, 31-plane
$[0_f]_{12}$
75/A, -65/A, 250/A
Screening

MATERIAL: T-300 3k/934 plain weave

RESIN CONTENT: 33-35 wt% COMP: DENSITY: 1.54-1.57 g/cm^3
FIBER VOLUME: 58-60 % VOID CONTENT: <0.5-1.2%
PLY THICKNESS: 0.0074-0.0082 in.

TEST METHOD: ASTM D-2344-68 (1)

MODULUS CALCULATION: Chord between 20 and 40% of typical ultimate load

NORMALIZED BY: Not normalized

		75	-65	250		
Temperature (°F)		75	-65	250		
Moisture Content (%)		ambient	ambient	ambient		
Equilibrium at T, RH						
Source Code		12	12	12		
	Mean	12.0	11.9	9.2		
	Minimum	10.5	10.0	9.1		
	Maximum	13.4	13.9	9.5		
	C.V.(%)	6.89	8.38	2.1		
	B-value	(2)	(2)	(2)		
	Distribution	ANOVA	ANOVA	Normal		
F_{31}^{sbs}	C_1	1.07	0.901	9.2		
(ksi)	C_2	3.41	3.71	0.20		
	No. Specimens	20	20	5		
	No. Batches	4	4	1		
	Approval Class	Screening	Screening	Screening		

(1) Length-to-thickness ratio is approximately 11.
(2) B-values are presented only for fully approved data.

4.2.7 Celion 12k/938 unitape data set description

Material Description:

Material: Celion-12k/938

Form: Unidirectional tape, fiber areal weight of 145 g/m^2, typical cured resin content of 28-40%, typical cured ply thickness of 0.0040-0.0073 inches.

Processing: Autoclave cure; 355°F, 85-100 psi for 2 hours.

General Supplier Information:

Fiber: Celion fibers are continuous carbon filaments made from PAN precursor. Filament count is 12,000 filaments/tow. Typical tensile modulus is 34 x 10^6 psi. Typical tensile strength is 515,000 psi.

Matrix: 938 is an epoxy resin. 10 days out-time at 72°F.

Maximum Short Term Service Temperature: 350°F (dry), 200°F (wet)

Typical applications: Commercial and military structural applications.

4.2.7 Celion 12k/938 unidirectional tape*

				C/Ep Celion 12k/938 Summary
MATERIAL:	Celion 12k/938 unidirectional tape			
PREPREG:	Fiberite Hy-E 1638N			
FIBER:	Celanese Celion 12k, EP06, no twist	MATRIX:	Fiberite 938	
T_g(dry):	T_g(wet):	T_g METHOD:		
PROCESSING:	Autoclave cure: 355 ± 10°F, 120 - 130 min., 85 - 100 psig			

* DATA WERE SUBMITTED BEFORE THE ESTABLISHMENT OF DATA DOCUMENTATION REQUIREMENTS (JUNE 1989). ALL DOCUMENTATION PRESENTLY REQUIRED WERE NOT SUPPLIED FOR THIS MATERIAL.

Date of fiber manufacture	5/85	Date of testing	7/85
Date of resin manufacture		Date of data submittal	6/88
Date of prepreg manufacture		Date of analysis	1/93
Date of composite manufacture			

LAMINA PROPERTY SUMMARY

	75°F/A		-67°F/A	250°F/A		180°F/W		
Tension, 1-axis	IIII		SSSS	IISI		IISI		
Tension, 2-axis	II-I		II-I	SS-S		II-I		
Tension, 3-axis								
Compression, 1-axis	II--		II--	II--		II--		
Compression, 2-axis	II--		II--	SI--		I---		
Compression, 3-axis								
Shear, 12-plane	I---		S---	S---		I---		
Shear, 23-plane								
SB Strength, 31-plane	I---							

Classes of data: F - Fully approved, I - Interim, S - Screening in Strength/Modulus/Poisson's ratio/Strain-to-failure order.

* DATA WERE SUBMITTED BEFORE THE ESTABLISHMENT OF DATA DOCUMENTATION REQUIREMENTS (JUNE 1989). ALL DOCUMENTATION PRESENTLY REQUIRED WERE NOT SUPPLIED FOR THIS MATERIAL.

		Nominal	As Submitted	Test Method
Fiber Density	(g/cm^3)	1.78		
Resin Density	(g/cm^3)	1.30		
Composite Density	(g/cm^3)		1.54 - 1.61	
Fiber Areal Weight	(g/m^2)	145	144 - 147	
Ply Thickness	(in)		0.0040 - 0.0073	

LAMINATE PROPERTY SUMMARY

Classes of data: F - Fully approved, I - Interim, S - Screening in Strength/Modulus/Poisson's ratio/Strain-to-failure order.

* DATA WERE SUBMITTED BEFORE THE ESTABLISHMENT OF DATA DOCUMENTATION REQUIREMENTS (JUNE 1989). ALL DOCUMENTATION PRESENTLY REQUIRED WAS NOT SUPPLIED FOR THIS MATERIAL.

MATERIAL:	Celion 12k/938 unidirectional tape			**Table 4.2.7(a)** **C/Ep 145-UT** **Celion 12k/938** **Tension, 1-axis** **$[0]_7$** **75/A, -67/A, 250/A** **Interim, Screening**
RESIN CONTENT:	28-36 wt%	COMP: DENSITY:	1.55-1.61 g/cm^3	
FIBER VOLUME:	56-65 %	VOID CONTENT:	<1.1%	
PLY THICKNESS:	0.0040-0.0063 in.			
TEST METHOD: ASTM D3039-76		MODULUS CALCULATION: Secant at 25% of typical ultimate load		
NORMALIZED BY:	Fiber volume 60%			

		75		-67		250	
Temperature (°F) Moisture Content (%) Equilibrium at T, RH Source Code		75 ambient 12		-67 ambient 12		250 ambient 12	
		Normalized	Measured	Normalized	Measured	Normalized	Measured
F_1^{tu} (ksi)	Mean	273	271	262	278	309	319
	Minimum	223	207	235	254	295	306
	Maximum	324	319	290	303	328	337
	C.V.(%)	7.56	9.76	7.67	6.25	3.00	2.82
	B-value	(1)	(1)	(1)	(1)	(1)	(1)
	Distribution	ANOVA	ANOVA	ANOVA	ANOVA	Weibull	Weibull
	C_1	21.0	29.3	25.1	20.9	314	323
	C_2	2.42	4.36	18.0	16.2	34.5	36.1
	No. Specimens	102		10		15	
	No. Batches	3		2		3	
	Approval Class	Interim		Screening		Interim	
E_1^t (Msi)	Mean	19.7	19.5	19.0	20.2	20.1	20.7
	Minimum	16.9	16.5	17.3	18.1	16.9	17.9
	Maximum	23.1	21.8	20.3	22.0	23.4	23.4
	C.V.(%)	5.22	5.59	4.94	5.94	9.12	7.49
	No. Specimens	102		10		15	
	No. Batches	3		2		3	
	Approval Class	Interim		Screening		Interim	
ν_{12}^t (2)	Mean		0.317		0.279		0.280
	No. Specimens	102		10		10	
	No. Batches	3		2		2	
	Approval Class	Interim		Screening		Interim	
ε_1^{tu} ($\mu\varepsilon$)	Mean		13100		12800		14800
	Minimum		10600		11500		12900
	Maximum		14800		14000		16100
	C.V.(%)		6.95		6.72		5.81
	B-value		(1)		(1)		(1)
	Distribution		ANOVA		ANOVA		Weibull
	C_1		946		1060		15100
	C_2		3.14		17.2		21.4
	No. Specimens	102		10		15	
	No. Batches	3		2		3	
	Approval Class	Interim		Screening		Interim	

(1) B-values are presented only for fully approved data.

(2) Poisson's ratio measured at 25% of typical ultimate load.

* DATA WERE SUBMITTED BEFORE THE ESTABLISHMENT OF DATA DOCUMENTATION REQUIREMENTS (JUNE 1989). ALL DOCUMENTATION PRESENTLY REQUIRED WAS NOT SUPPLIED FOR THIS MATERIAL.

Table 4.2.7(b)
C/Ep 145-UT
Celion 12k/938
Tension, 1-axis
$[0]_7$
180/1.1%
Interim, Screening

MATERIAL: Celion 12k/938 unidirectional tape

RESIN CONTENT: 28-36 wt% | COMP: DENSITY: 1.55-1.59 g/cm^3
FIBER VOLUME: 56-64 % | VOID CONTENT: <1.4%
PLY THICKNESS: 0.0044-0.0063 in.

TEST METHOD: ASTM D3039-76

MODULUS CALCULATION: Secant at 25% of typical ultimate load

NORMALIZED BY: Fiber volume fraction to 60%

		Normalized	Measured	Normalized	Measured	Normalized	Measured
Temperature (°F)		180					
Moisture Content (%)		1.1					
Equilibrium at T, RH		(1)					
Source Code		12					
F_1^{tu} (ksi)	Mean	277	282				
	Minimum	236	219				
	Maximum	307	328				
	C.V.(%)	8.89	14.3				
	B-value	(3)	(3)				
	Distribution	ANOVA	ANOVA				
	C_1	27.7	46.7				
	C_2	5.36	5.89				
	No. Specimens	15					
	No. Batches	3					
	Approval Class	Interim					
E_1^t (Msi)	Mean	18.9	19.2				
	Minimum	17.7	16.4				
	Maximum	20.5	21.9				
	C.V.(%)	4.81	9.74				
	No. Specimens	15					
	No. Batches	3					
	Approval Class	Interim					
ν_{12}^t (2)	Mean		0.345				
	No. Specimens	14					
	No. Batches	3					
	Approval Class	Screening					
ε_1^{tu} (με)	Mean		14000				
	Minimum		11800				
	Maximum		15700				
	C.V.(%)		8.13				
	B-value		(3)				
	Distribution		ANOVA				
	C_1		1180				
	C_2		3.36				
	No. Specimens	15					
	No. Batches	3					
	Approval Class	Interim					

(1) Conditioned at 160°F, 88% R.H. until weight gain was between 1.0 and 1.2%.
(2) Poisson's ratio measured at 25% of typical ultimate load.
(3) B-values are presented only for fully approved data.

* DATA WERE SUBMITTED BEFORE THE ESTABLISHMENT OF DATA DOCUMENTATION REQUIREMENTS (JUNE 1989). ALL DOCUMENTATION PRESENTLY REQUIRED WAS NOT SUPPLIED FOR THIS MATERIAL.

MATERIAL: Celion 12k/938 unidirectional tape	**Table 4.2.7(c)** **C/Ep 145-UT** **Celion 12k/938** **Tension, 2-axis** **$[90]_{20}$** **75/A, -67/A, 250/A, 180/1.1%** **Interim, Screening**
RESIN CONTENT: 32-37 wt%	COMP: DENSITY: 1.55-1.58 g/cm^3
FIBER VOLUME: 55-60 %	VOID CONTENT: <1.3%
PLY THICKNESS: 0.0053-0.0064 in.	
TEST METHOD: ASTM D3039-76	MODULUS CALCULATION: Secant at 25% of typical ultimate load
NORMALIZED BY: Not normalized	

Temperature (°F)		75	-67	250	180		
Moisture Content (%)		ambient	ambient	ambient	1.1		
Equilibrium at T, RH					(1)		
Source Code		12	12	12	12		
F_2^{tu} (ksi)	Mean	9.6	9.5	8.8	5.8		
	Minimum	7.5	8.5	7.1	5.0		
	Maximum	13.9	10.4	10.7	6.6		
	C.V.(%)	13	6.6	11	8.4		
	B-value	(2)	(2)	(2)	(2)		
	Distribution	ANOVA	Weibull	Weibull	ANOVA		
	C_1	1.3	9.8	9.2	0.54		
	C_2	2.7	18	10	5.1		
	No. Specimens	101	15	10	15		
	No. Batches	3	3	2	3		
	Approval Class	Interim	Interim	Screening	Interim		
E_2^t (Msi)	Mean	1.35	1.35	1.22	1.19		
	Minimum	1.14	1.25	0.94	1.03		
	Maximum	1.82	1.51	1.52	1.36		
	C.V.(%)	9.29	4.96	12.5	8.65		
	No. Specimens	101	15	10	15		
	No. Batches	3	3	2	3		
	Approval Class	Interim	Interim	Screening	Interim		
ν_{21}^t	Mean						
	No. Specimens						
	No. Batches						
	Approval Class						
ε_2^{tu} (με)	Mean	7200	6700	7600	4900		
	Minimum	1300	5500	6900	4200		
	Maximum	9500	7900	9300	5800		
	C.V.(%)	15	9.2	9.5	8.6		
	B-value	(2)	(2)	(2)	(2)		
	Distribution	Nonpara.	Weibull	Normal	Weibull		
	C_1	5	7000	7600	5100		
	C_2		12	720	12		
	No. Specimens	97	15	10	15		
	No. Batches	3	3	2	3		
	Approval Class	Interim	Interim	Screening	Interim		

(1) Conditioned at 160°F, 88% R.H. until weight gain was between 1.0 and 1.2%.

(2) B-values are presented only for fully approved data.

* DATA WERE SUBMITTED BEFORE THE ESTABLISHMENT OF DATA DOCUMENTATION REQUIREMENTS (JUNE 1989). ALL DOCUMENTATION PRESENTLY REQUIRED WAS NOT SUPPLIED FOR THIS MATERIAL.

Table 4.2.7(d)
C/Ep 145-UT
Celion 12k/938
Compression, 1-axis
$[0]_7$
75/A, -67/A, 250/A
Interim

MATERIAL: Celion 12k/938 unidirectional tape

RESIN CONTENT: 26-35 wt% COMP: DENSITY: 1.56-1.61 g/cm³
FIBER VOLUME: 57-67 % VOID CONTENT: <1.5%
PLY THICKNESS: 0.0046-0.0073 in.

TEST METHOD: SACMA SRM 1-88

MODULUS CALCULATION: Chord modulus between 20% and 40% of typical ultimate load

NORMALIZED BY: Fiber volume fraction to 60%

		75		-67		250	
Temperature (°F) Moisture Content (%) Equilibrium at T, RH Source Code		75 ambient 12		-67 ambient 12		250 ambient 12	
		Normalized	Measured	Normalized	Measured	Normalized	Measured
F_1^{cu} (ksi)	Mean	201	198	240	240	195	201
	Minimum	166	172	204	216	180	179
	Maximum	255	246	286	276	214	229
	C.V.(%)	9.88	8.99	11.3	8.25	5.48	7.26
	B-value	(1)	(1)	(1)	(1)	(1)	(1)
	Distribution	ANOVA	ANOVA	ANOVA	ANOVA	ANOVA	ANOVA
	C_1	21.4	18.7	31.1	21.9	11.9	16.7
	C_2	3.93	3.35	5.59	4.97	5.07	5.59
	No. Specimens	102		15		15	
	No. Batches	3		3		3	
	Approval Class	Interim		Interim		Interim	
E_1^c (Msi)	Mean	17.2	18.2	18.8	19.1	18.1	18.1
	Minimum	14.7	15.0	16.6	16.6	17.1	16.3
	Maximum	21.0	21.5	21.7	22.5	19.1	20.3
	C.V.(%)	6.87	7.64	7.14	9.74	3.73	7.07
	No. Specimens	97		15		15	
	No. Batches	3		3		3	
	Approval Class	Interim		Interim		Interim	
ν_{12}^c	Mean						
	No. Specimens						
	No. Batches						
	Approval Class						
ε_1^{cu} (με)	Mean						
	Minimum						
	Maximum						
	C.V.(%)						
	B-value						
	Distribution						
	C_1						
	C_2						
	No. Specimens						
	No. Batches						
	Approval Class						

(1) B-values are presented only for fully approved data.

* DATA WERE SUBMITTED BEFORE THE ESTABLISHMENT OF DATA DOCUMENTATION REQUIREMENTS (JUNE 1989). ALL DOCUMENTATION PRESENTLY REQUIRED WAS NOT SUPPLIED FOR THIS MATERIAL.

Table 4.2.7(e)
C/Ep 145-UT
Celion 12k/938
Compression, 1-axis
$[0]_7$
180/1.1%
Interim

MATERIAL: Celion 12k/938 unidirectional tape

RESIN CONTENT: 28-34 wt% COMP: DENSITY: 1.58-1.60 g/cm^3
FIBER VOLUME: 58-65 % VOID CONTENT: <1.0%
PLY THICKNESS: 0.0044-0.0073 in.

TEST METHOD: SACMA SRM 1-88

MODULUS CALCULATION: Chord modulus between 20% and 40% of typical ultimate load

NORMALIZED BY: Fiber volume fraction to 60%

Temperature (°F)		180					
Moisture Content (%)		1.1					
Equilibrium at T, RH		(1)					
Source Code		12					
		Normalized	Measured	Normalized	Measured	Normalized	Measured
	Mean	185	188				
	Minimum	157	160				
	Maximum	206	217				
	C.V.(%)	7.40	7.55				
	B-value	(2)	(2)				
F_1^{cu}	Distribution	Weibull	Weibull				
(ksi)	C_1	191	194				
	C_2	16.3	14.4				
	No. Specimens	15					
	No. Batches	3					
	Approval Class	Interim					
	Mean	18.2	19.2				
	Minimum	15.7	15.8				
	Maximum	22.3	23.7				
E_1^c	C.V.(%)	8.88	10.5				
(Msi)	No. Specimens	15					
	No. Batches	3					
	Approval Class	Interim					
	Mean						
	No. Specimens						
ν_{12}^c	No. Batches						
	Approval Class						
	Mean						
	Minimum						
	Maximum						
	C.V.(%)						
	B-value						
ε_1^{cu}	Distribution						
(με)	C_1						
	C_2						
	No. Specimens						
	No. Batches						
	Approval Class						

(1) Conditioned at 160°F, 88% R.H. until weight gain was between 1.0 and 1.2%.
(2) B-values are presented only for fully approved data.

* DATA WERE SUBMITTED BEFORE THE ESTABLISHMENT OF DATA DOCUMENTATION REQUIREMENTS (JUNE 1989). ALL DOCUMENTATION PRESENTLY REQUIRED WAS NOT SUPPLIED FOR THIS MATERIAL.

Table 4.2.7(f)
C/Ep 145-UT
Celion 12k/938
Shear, 12-plane
$[\pm 45]_{2S}$
75/A, -65/A, 250/A, 180/1.1%
Interim, Screening

MATERIAL: Celion 12k/938 unidirectional tape

RESIN CONTENT: 28-34 wt% COMP: DENSITY: 1.57-1.61 g/cm^3
FIBER VOLUME: 58-65 % VOID CONTENT: <1.4%
PLY THICKNESS: 0.0044-0.0064 in.

TEST METHOD: ASTM D3518-76 MODULUS CALCULATION:

NORMALIZED BY: Not normalized

Temperature (°F) Moisture Content (%) Equilibrium at T, RH Source Code		75 ambient 12	-67 ambient 12	250 ambient 12	180 1.1 (1) 12	
	Mean	14	16	14	14	
	Minimum	11	14	13	13	
	Maximum	16	18	15	14	
	C.V.(%)	7.3	10.	6.1	3.6	
	B-value	(2)	(2)	(2)	(2)	
	Distribution	ANOVA	ANOVA	Weibull	ANOVA	
F_{12}^{su}	C_1	1.1	1.8	14	0.53	
(ksi)	C_2	4.4	5.8	19	4.6	
	No. Specimens	102	14	14	15	
	No. Batches	3	3	3	3	
	Approval Class	Interim	Screening	Screening	Interim	
	Mean					
	Minimum					
	Maximum					
G_{12}^{s}	C.V.(%)					
(Msi)	No. Specimens					
	No. Batches					
	Approval Class					
	Mean					
	Minimum					
	Maximum					
	C.V.(%)					
	B-value					
γ_{12}^{su}	Distribution					
($\mu\varepsilon$)	C_1					
	C_2					
	No. Specimens					
	No. Batches					
	Approval Class					

(1) Conditioned at 160°F, 88% R.H. until weight gain was between 1.0 and 1.2%.
(2) B-values are presented only for fully approved data.

* DATA WERE SUBMITTED BEFORE THE ESTABLISHMENT OF DATA DOCUMENTATION REQUIREMENTS (JUNE 1989). ALL DOCUMENTATION PRESENTLY REQUIRED WAS NOT SUPPLIED FOR THIS MATERIAL.

MATERIAL:	Celion 12k/938 unidirectional tape			**Table 4.2.7(g)** **C/Ep 145-UT** **Celion 12k/938** **Shear, 31-plane** **$[0]_{14}$** **75/A** **Screening**
RESIN CONTENT:	31-40 wt%	COMP: DENSITY:	1.54-1.59 g/cm^3	
FIBER VOLUME:	52-62 %	VOID CONTENT:	<1.0%	
PLY THICKNESS:	0.0051-0.0064 in.			
TEST METHOD: ASTM D2344-68		MODULUS CALCULATION:		
NORMALIZED BY:	Not normalized			

Temperature (°F) Moisture Content (%) Equilibrium at T, RH Source Code		75 ambient 12				
F_{31}^{sbs} (ksi)	Mean	18.3				
	Minimum	16.6				
	Maximum	19.7				
	C.V.(%)	3.29				
	B-value	(1)				
	Distribution	ANOVA				
	C_1	0.619				
	C_2	2.76				
	No. Specimens	102				
	No. Batches	3				
	Approval Class	Screening				

(1) B-values are presented only for fully approved data.

4.2.8 AS4 12k/3502 unitape data set description

Material Description:

Material: AS4-12k/3502

Form: Unidirectional tape, fiber areal weight of 150 g/m^2, typical cured resin content of 32-45%, typical cured ply thickness of 0.0052 inches.

Processing: Good drape. Autoclave cure; 275° F, 85 psi for 45 minutes; 350°F, 85 psi, hold for 2 hours. Post cure at 400°F to develop optimum 350°F properties.

General Supplier Information:

Fiber: AS4 fibers are continuous high strength, high strain, standard, modulus carbon filaments made from PAN precursor. The fibers are surface treated to improve handling characteristics and structural properties. Filament count is 12,000 filaments/tow. Typical tensile modulus is 34 x 10^{6}psi. Typical tensile strength is 550,000 psi.

Matrix: 3502 is an epoxy resin. Good tack; up to 10 days out-time at ambient temperature.

Maximum Short Term Service Temperature: 350°F (dry), 180°F (wet)

Typical applications: Primary and secondary structural applications on commercial and military aircraft.

4.2.8 AS4 12k/3502 unidirectional tape*

MATERIAL:	AS4 12k/3502 unidirectional tape			**C/Ep AS4/3502 Summary**
PREPREG:	Hercules AS4/3502 unidirectional tape			
FIBER:	Hercules AS4 12k, surface-treated, 0 twist	MATRIX:	Hercules 3502	
T_g(dry):	407°F T_g(wet):	T_g METHOD:	TMA	
PROCESSING:	Autoclave cure: 280 ± 5°F, 90 min, 85+15-0 psi; 350°F, 120 min.			

* Additional data set found on p. 70.

Date of fiber manufacture	4/83 - 6/83	Date of testing	11/83 - 7/84
Date of resin manufacture	6/83	Date of data submittal	12/93, 5/94
Date of prepreg manufacture	6/83 - 7/83	Date of analysis	8/94
Date of composite manufacture	8/83 - 5/84		

LAMINA PROPERTY SUMMARY

	75°F/A		-65°F/A		180°F/W	250°F/W		
Tension, 1-axis	FF--		FF--		FF--	FF--		
Tension, 2-axis	FF--		FF--		FF--	FF--		
Tension, 3-axis								
Compression, 1-axis	FF--		II--		FF--	FF--		
Compression, 2-axis	FF--		II--		FF--	FF--		
Compression, 3-axis								
Shear, 12-plane	FF--		II--		FF--	II--		
Shear, 23-plane								
Shear, 31-plane								

Classes of data: F - Fully approved, I - Interim, S - Screening in Strength/Modulus/Poisson's ratio/Strain-to-failure order.

	Nominal	As Submitted	Test Method
Fiber Density (g/cm^3)	1.79	1.77 - 1.80	
Resin Density (g/cm^3)	1.26	1.24 - 1.29	
Composite Density (g/cm^3)	1.57	1.56 - 1.59	
Fiber Areal Weight (g/m^2)	147	146 - 150	
Ply Thickness (in)	0.0055	0.0049 - 0.0061	

LAMINATE PROPERTY SUMMARY

Classes of data: F - Fully approved, I - Interim, S - Screening in Strength/Modulus/Poisson's ratio/Strain-to-failure order.

MATERIAL:	AS4 12k/3502 unidirectional tape			**Table 4.2.8(a)** **C/Ep 147-UT** **AS4/3502** **Tension, 1-axis** **$[0]_8$** **75/A, -65/A, 180/W** **Fully Approved**
RESIN CONTENT: FIBER VOLUME: PLY THICKNESS:	30-33 wt% 59-61 % 0.0049-0.0061 in.	COMP: DENSITY: VOID CONTENT:	1.56-1.59 g/cm³ 0.0-1.0%	
TEST METHOD: ASTM D3039-76		MODULUS CALCULATION: Linear portion of curve		
NORMALIZED BY:	Ply thickness to 0.0055 in. (59%)			

Temperature (°F) Moisture Content (%) Equilibrium at T, RH Source Code		75 ambient 49		-65 ambient 49		180 1.1 - 1.3 (1) 49	
		Normalized	Measured	Normalized	Measured	Normalized	Measured
	Mean	258	253	231	227	261	255
	Minimum	191	186	162	151	140	135
	Maximum	317	302	285	280	317	315
	C.V.(%)	9.83	9.13	13.4	13.6	14.8	15.2
	B-value	205		173		200	
F_1^{tu}	Distribution	Weibull		Weibull		Weibull	
(ksi)	C_1	269		244		276	
	C_2	11.2		8.82		9.39	
	No. Specimens	36		38		40	
	No. Batches	5		5		5	
	Approval Class	Fully Approved		Fully Approved		Fully Approved	
	Mean	19.3		19.2		19.7	
	Minimum	15.6		16.8		15.1	
	Maximum	21.0		23.2		23.3	
E_1^t	C.V.(%)	5.74		6.31		6.87	
(Msi)	No. Specimens	36		38		40	
	No. Batches	5		5		5	
	Approval Class	Fully Approved		Fully Approved		Fully Approved	
	Mean						
	No. Specimens						
ν_{12}^t	No. Batches						
	Approval Class						
	Mean						
	Minimum						
	Maximum						
	C.V.(%)						
	B-value						
ε_1^{tu}	Distribution						
(με)	C_1						
	C_2						
	No. Specimens						
	No. Batches						
	Approval Class						

(1) Conditioned at 160°F, 95-100% relative humidity until the moisture content was between 1.1 and 1.3%.

MATERIAL:	AS4 12k/3502 unidirectional tape		
RESIN CONTENT:	30-33 wt%	COMP: DENSITY:	1.56-1.59 g/cm³
FIBER VOLUME:	59-61 %	VOID CONTENT:	0.0-1.0%
PLY THICKNESS:	0.0055-0.0059 in.		
TEST METHOD: ASTM D3039-76		MODULUS CALCULATION: Linear portion of curve	
NORMALIZED BY:	Ply thickness to 0.0055 in. (59%)		

Table 4.2.8(b)
C/Ep 147-UT
AS4/3502
Tension, 1-axis
$[0]_8$
250/W
Fully Approved

Temperature (°F)		250					
Moisture Content (%)		1.1 - 1.3					
Equilibrium at T, RH		(1)					
Source Code		49					
		Normalized	Measured	Normalized	Measured	Normalized	Measured
F_1^{tu} (ksi)	Mean	256	250				
	Minimum	200	195				
	Maximum	301	305				
	C.V.(%)	9.39	10.2				
	B-value	191					
	Distribution	ANOVA					
	C_1	25.0					
	C_2	2.61					
	No. Specimens	30					
	No. Batches	5					
	Approval Class	Fully Approved					
E_1^t (Msi)	Mean	20.1					
	Minimum	17.8					
	Maximum	23.9					
	C.V.(%)	7.32					
	No. Specimens	30					
	No. Batches	5					
	Approval Class	Fully Approved					
ν_{12}^t	Mean						
	No. Specimens						
	No. Batches						
	Approval Class						
ε_1^{tu} (με)	Mean						
	Minimum						
	Maximum						
	C.V.(%)						
	B-value						
	Distribution						
	C_1						
	C_2						
	No. Specimens						
	No. Batches						
	Approval Class						

(1) Conditioned at 160°F, 95-100% relative humidity until the moisture content was between 1.1 and 1.3%.

MATERIAL:	AS4 12k/3502 unidirectional tape			**Table 4.2.8(c)** **C/Ep 147-UT** **AS4/3502** **Tension, 2-axis** **$[90]_{24}$** **75/A, -65/A, 180/W, 250/W** **Fully Approved**
RESIN CONTENT:	31-33 wt%	COMP: DENSITY:	1.56-1.59 g/cm^3	
FIBER VOLUME:	59-60 %	VOID CONTENT:	0.0-1.0%	
PLY THICKNESS:	0.0052-0.0059 in.			
TEST METHOD: ASTM D3039-76		MODULUS CALCULATION: Linear portion of curve		
NORMALIZED BY:	Not normalized			

Temperature (°F)		75	-65	180	250	
Moisture Content (%)		ambient	ambient	1.1 - 1.3	1.1 - 1.3	
Equilibrium at T, RH				(1)	(1)	
Source Code		49	49	49	49	
F_2^t (ksi)	Mean	7.76	6.65	4.39	2.68	
	Minimum	6.26	2.48	3.52	2.13	
	Maximum	10.2	8.93	5.20	3.40	
	C.V.(%)	10.7	18.0	8.44	12.3	
	B-value	6.28	4.57	3.46	1.65	
	Distribution	Normal	Weibull	ANOVA	ANOVA	
	C_1	7.76	7.09	0.380	0.348	
	C_2	0.832	7.20	2.43	2.94	
	No. Specimens	30	30	30	30	
	No. Batches	5	5	5	5	
	Approval Class	Fully Approved	Fully Approved	Fully Approved	Fully Approved	
E_2^t (Msi)	Mean	1.35	1.44	1.21	0.958	
	Minimum	1.28	1.32	1.14	0.912	
	Maximum	1.49	1.58	1.35	1.06	
	C.V.(%)	4.26	4.16	4.02	3.61	
	No. Specimens	30	30	30	30	
	No. Batches	5	5	5	5	
	Approval Class	Fully Approved	Fully Approved	Fully Approved	Fully Approved	
ν_{21}^t	Mean					
	No. Specimens					
	No. Batches					
	Approval Class					
ε_2^t (με)	Mean					
	Minimum					
	Maximum					
	C.V.(%)					
	B-value					
	Distribution					
	C_1					
	C_2					
	No. Specimens					
	No. Batches					
	Approval Class					

(1) Conditioned at 160°F, 95-100% relative humidity until the moisture content was between 1.1 and 1.3%.

MATERIAL:	AS4 12k/3502 unidirectional tape			**Table 4.2.8(d)** **C/Ep 147-UT** **AS4/3502** **Compression, 1-axis** **$[0]_{19}$** **75/A, -65/A, 180/W** **Fully Approved, Interim**
RESIN CONTENT:	33-37 wt%	COMP: DENSITY:	1.56-1.57 g/cm^3	
FIBER VOLUME:	55-59 %	VOID CONTENT:	0.0%	
PLY THICKNESS:	0.0054-0.0060 in.			
TEST METHOD: ASTM D3410A-75		MODULUS CALCULATION: Linear portion of curve		
NORMALIZED BY:	Ply thickness to 0.0055 in. (59%)			

Temperature (°F)		75		-65		180	
Moisture Content (%)		ambient		ambient		1.1 - 1.3	
Equilibrium at T, RH						(1)	
Source Code		49		49		49	
		Normalized	Measured	Normalized	Measured	Normalized	Measured
F_1^{cu} (ksi)	Mean	204	198	233	225	176	171
	Minimum	168	160	207	201	146	137
	Maximum	226	221	252	251	200	197
	C.V.(%)	6.45	6.62	5.63	6.23	6.31	8.09
	B-value	171		(2)		145	
	Distribution	ANOVA		Weibull		ANOVA	
	C_1	13.5		238		11.5	
	C_2	2.44		23.0		2.65	
	No. Specimens	30		15		30	
	No. Batches	5		5		5	
	Approval Class	Fully Approved		Interim		Fully Approved	
E_1^c (Msi)	Mean	18.0		18.8		18.6	
	Minimum	16.9		17.1		17.5	
	Maximum	19.4		20.5		20.0	
	C.V.(%)	3.19		5.43		3.36	
	No. Specimens	30		16		30	
	No. Batches	5		5		5	
	Approval Class	Fully Approved		Interim		Fully Approved	
ν_{12}^c	Mean						
	No. Specimens						
	No. Batches						
	Approval Class						
ε_1^{cu} (με)	Mean						
	Minimum						
	Maximum						
	C.V.(%)						
	B-value						
	Distribution						
	C_1						
	C_2						
	No. Specimens						
	No. Batches						
	Approval Class						

(1) Conditioned at 160°F, 95-100% relative humidity until the moisture content was between 1.1 and 1.3%.
(2) B-values are presented only for fully approved data.

MATERIAL:	AS4 12k/3502 unidirectional tape			**Table 4.2.8(e)** **C/Ep 147-UT** **AS4/3502** **Compression, 1-axis** **$[0]_{19}$** **250/W** **Fully Approved**
RESIN CONTENT:	33-37 wt%	COMP: DENSITY:	1.56-1.57 g/cm³	
FIBER VOLUME:	55-59 %	VOID CONTENT:	0.0%	
PLY THICKNESS:	0.0054-0.0060 in.			
TEST METHOD: ASTM D3410A-75		MODULUS CALCULATION: Linear portion of curve		
NORMALIZED BY:	Ply thickness to 0.0055 in. (59%)			

Temperature (°F)		250					
Moisture Content (%)		1.1 - 1.3					
Equilibrium at T, RH		(1)					
Source Code		49					
		Normalized	Measured	Normalized	Measured	Normalized	Measured
	Mean	147	143				
	Minimum	118	110				
	Maximum	170	163				
	C.V.(%)	9.42	9.40				
	B-value	119					
F_1^{cu}	Distribution	Weibull					
(ksi)	C_1	153					
	C_2	12.5					
	No. Specimens	30					
	No. Batches	5					
	Approval Class	Fully Approved					
	Mean	18.7					
	Minimum	17.3					
	Maximum	20.6					
E_1^c	C.V.(%)	3.99					
(Msi)	No. Specimens	30					
	No. Batches	5					
	Approval Class	Fully Approved					
	Mean						
	No. Specimens						
ν_{12}^c	No. Batches						
	Approval Class						
	Mean						
	Minimum						
	Maximum						
	C.V.(%)						
	B-value						
ε_1^{cu}	Distribution						
(με)	C_1						
	C_2						
	No. Specimens						
	No. Batches						
	Approval Class						

(1) Conditioned at 160°F, 95-100% relative humidity until the moisture content was between 1.1 and 1.3%.

MATERIAL:	AS4 12k/3502 unidirectional tape			Table 4.2.8(f) C/Ep 147-UT AS4/3502 Compression, 2-axis $[90]_{24}$ 75/A, -65/A, 180/W, 250/W Fully Approved, Interim
RESIN CONTENT:	31-33 wt%	COMP: DENSITY:	1.56-1.59 g/cm^3	
FIBER VOLUME:	59-60 %	VOID CONTENT:	0.0-1.0%	
PLY THICKNESS:	0.0054-0.0058 in.			
TEST METHOD:		MODULUS CALCULATION:		
ASTM D695M (1) (4)		Linear portion of curve		
NORMALIZED BY:	Not normalized			

Temperature (°F)		75	-65	180	250	
Moisture Content (%)		ambient	ambient	1.1 - 1.3	1.1 - 1.3	
Equilibrium at T, RH				(2)	(2)	
Source Code		49	49	49	49	
	Mean	34.6	49.8	24.7	18.4	
	Minimum	27.5	42.5	23.0	17.0	
	Maximum	40.4	57.2	26.7	19.9	
	C.V.(%)	9.53	10.4	3.23	4.99	
	B-value	26.6	(3)	22.3	15.3	
	Distribution	ANOVA	Weibull	ANOVA	ANOVA	
F_2^{cu}	C_1	3.37	52.1	0.836	0.990	
(ksi)	C_2	2.38	11.3	2.80	3.18	
	No. Specimens	30	15	30	30	
	No. Batches	5	5	5	5	
	Approval Class	Fully Approved	Interim	Fully Approved	Fully Approved	
	Mean	1.41	1.68	1.24	1.09	
	Minimum	1.29	1.57	1.14	0.973	
	Maximum	1.60	1.95	1.41	1.41	
E_2^c	C.V.(%)	4.86	6.07	4.90	9.44	
(Msi)	No. Specimens	30	15	30	30	
	No. Batches	5	5	5	5	
	Approval Class	Fully Approved	Interim	Fully Approved	Interim	
	Mean					
ν_{21}^c	No. Specimens					
	No. Batches					
	Approval Class					
	Mean					
	Minimum					
	Maximum					
	C.V.(%)					
	B-value					
ε_2^{cu}	Distribution					
(με)	C_1					
	C_2					
	No. Specimens					
	No. Batches					
	Approval Class					

(1) Tabbed specimen - length 3.12 inch, width 0.50 inch, gage length 0.50 inch.
(2) Conditioned at 160°F, 95-100% relative humidity until the moisture content was between 1.1 and 1.3%.
(3) B-values are presented only for fully approved data.
(4) The test method, ASTM D695M-96, was withdrawn on July 10, 1996.

MATERIAL:	AS4 12k/3502 unidirectional tape			**Table 4.2.8(g)** **C/Ep 147-UT** **AS4/3502** **Shear, 12-plane** **$[\pm45]_{4S}$** **75/A, -65/A, 180/W, 250/W** **Fully Approved, Interim**
RESIN CONTENT:	31-33 wt%	COMP: DENSITY:	1.56-1.59 g/cm^3	
FIBER VOLUME:	59-60 %	VOID CONTENT:	0.0-1.0%	
PLY THICKNESS:	0.0053-0.0059 in.			
TEST METHOD: ASTM D3518-76		MODULUS CALCULATION: Linear portion of curve		
NORMALIZED BY:	Not normalized			

Temperature (°F)		75	-65	180	250	
Moisture Content (%)		ambient	ambient	1.1 - 1.3	1.1 - 1.3	
Equilibrium at T, RH				(1)	(1)	
Source Code		49	49	49	49	
F_{12}^{su} (ksi)	Mean	14.8	15.3	13.5	11.5	
	Minimum	13.7	13.3	12.5	10.5	
	Maximum	15.8	16.2	14.1	12.4	
	C.V.(%)	3.18	4.58	3.39	4.27	
	B-value	13.4	(2)	11.8	10.3	
	Distribution	ANOVA	ANOVA	ANOVA	ANOVA	
	C_1	0.503	0.706	0.502	0.503	
	C_2	2.91	2.04	3.24	2.32	
	No. Specimens	36	23	37	42	
	No. Batches	5	5	5	5	
	Approval Class	Fully Approved	Interim	Fully Approved	Fully Approved	
G_{12}^{s} (Msi)	Mean	0.543	0.769	0.217	0.141	
	Minimum	0.496	0.738	0.169	0.103	
	Maximum	0.593	0.863	0.260	0.205	
	C.V.(%)	5.16	3.69	9.25	17.9	
	No. Specimens	33	23	33	41	
	No. Batches	5	5	5	5	
	Approval Class	Fully Approved	Interim	Fully Approved	Fully Approved	
γ_{12}^{su} (με)	Mean					
	Minimum					
	Maximum					
	C.V.(%)					
	B-value					
	Distribution					
	C_1					
	C_2					
	No. Specimens					
	No. Batches					
	Approval Class					

(1) Conditioned at 160°F, 95-100% relative humidity until the moisture content was between 1.1 and 1.3%.
(2) B-values are presented only for fully approved data.

MATERIAL:	AS4 12k/3502 unidirectional tape			C/Ep AS4/3502 Summary
PREPREG:	Hercules AS4/3502 unidirectional tape			
FIBER:	Hercules AS4 12k, surface-treated	MATRIX:	Hercules 3502	
T_g(dry):	460°F T_g(wet):	T_g METHOD:	TMA	
PROCESSING:	Autoclave cure: 275°F, 45 min., 350°F, 2 hours, 85 psig; Postcure: 400°F, 4 hours			

* ALL DOCUMENTATION PRESENTLY REQUIRED WERE NOT SUPPLIED FOR THIS MATERIAL.

Date of fiber manufacture	12/80 - 2/82	Date of testing	
Date of resin manufacture		Date of data submittal	6/90
Date of prepreg manufacture	12/80 - 2/82	Date of analysis	1/93
Date of composite manufacture			

LAMINA PROPERTY SUMMARY

	75°F/A		-65°F/A	265°F/A		75°F/W	265°F/W	
Tension, 1-axis	IIII			IIII			IIII	
Tension, 2-axis	II-I					II-I	II-I	
Tension, 3-axis								
Compression, 1-axis			II-I	II-I			II-I	
Compression, 2-axis								
Compression, 3-axis								
Shear, 12-plane								
Shear, 23-plane								
Shear, 31-plane								

Classes of data: F - Fully approved, I - Interim, S - Screening in Strength/Modulus/Poisson's ratio/Strain-to-failure order.

* ALL DOCUMENTATION PRESENTLY REQUIRED WERE NOT SUPPLIED FOR THIS MATERIAL.

		Nominal	As Submitted	Test Method
Fiber Density	(g/cm^3)	1.79	1.78 - 1.81	
Resin Density	(g/cm^3)	1.26		
Composite Density	(g/cm^3)	1.58		
Fiber Areal Weight	(g/m^2)			
Ply Thickness	(in)		0.0047 - 0.0062	

LAMINATE PROPERTY SUMMARY

Classes of data: F - Fully approved, I - Interim, S - Screening in Strength/Modulus/Poisson's ratio/Strain-to-failure order.

* ALL DOCUMENTATION PRESENTLY REQUIRED WERE NOT SUPPLIED FOR THIS MATERIAL.

MATERIAL:	AS4 12k/3502 unidirectional tape			**Table 4.2.8(h)** **C/Ep 147-UT** **AS4/3502** **Tension, 1-axis** **$[0]_6$** **75/A, 265/A, 265/W** **Interim**
RESIN CONTENT:	25-29 wt%	COMP: DENSITY:	1.59-1.62 g/cm^3	
FIBER VOLUME:	63-68 %	VOID CONTENT:		
PLY THICKNESS:	0.0055-0.0058 in.			
TEST METHOD: ASTM D3039-76		MODULUS CALCULATION:		
NORMALIZED BY:	Specimen thickness and batch fiber volume to 60%			

		75		265		265	
Temperature (°F) Moisture Content (%) Equilibrium at T, RH Source Code		75 ambient (1) 26		265 ambient (1) 26		265 wet (2) 26	
		Normalized	Measured	Normalized	Measured	Normalized	Measured
F_1^{tu} (ksi)	Mean	253	275	269	292	251	273
	Minimum	212	226	148	165	183	196
	Maximum	294	323	314	358	287	315
	C.V.(%)	8.35	9.49	15.2	16.5	9.09	10.4
	B-value	(3)	(3)	(3)	(3)	(3)	(3)
	Distribution	ANOVA	ANOVA	ANOVA	ANOVA	ANOVA	ANOVA
	C_1	21.5	27.2	24.0	30.2	24.0	30.2
	C_2	2.20	2.60	2.83	3.01	2.83	3.01
	No. Specimens	30		20		25	
	No. Batches	5		4		5	
	Approval Class	Interim		Interim		Interim	
E_1^t (Msi)	Mean	18.7	20.4	18.4	20.0	19.0	20.6
	Minimum	17.3	18.9	17.4	19.1	18.0	19.2
	Maximum	20.2	22.2	19.7	20.8	19.7	22.1
	C.V.(%)	3.88	3.37	3.52	2.59	3.53	3.22
	No. Specimens	29		20		25	
	No. Batches	5		4		5	
	Approval Class	Interim		Interim		Interim	
ν_{12}^t	Mean		0.340		0.356		0.280
	No. Specimens	30		20		25	
	No. Batches	5		4		5	
	Approval Class	Interim		Interim		Interim	
ε_1^{tu} ($\mu\varepsilon$)	Mean		12400		13900		12400
	Minimum		10200		10400		9220
	Maximum		14400		15700		13900
	C.V.(%)		8.65		12.0		8.95
	B-value		(3)		(3)		(3)
	Distribution		ANOVA		ANOVA		ANOVA
	C_1		1120		1850		1170
	C_2		2.62		3.92		2.87
	No. Specimens	30		20		25	
	No. Batches	5		4		5	
	Approval Class	Interim		Interim		Interim	

(1) Conditioned at 180°F, ambient relative humidity for 2 days.
(2) Conditioned at 180°F, 75% relative humidity for 10 days.
(3) B-values are presented only for fully approved data.

* ALL DOCUMENTATION PRESENTLY REQUIRED WAS NOT SUPPLIED FOR THIS MATERIAL.

Table 4.2.8(i)
C/Ep 147-UT
AS4/3502
Tension, 2-axis
$[90]_{15}$
75/A, 75/W, 265/W
Interim

MATERIAL: AS4 12k/3502 unidirectional tape

RESIN CONTENT: 25-29 wt%
COMP: DENSITY: 1.59-1.62 g/cm^3
FIBER VOLUME: 63-68 %
VOID CONTENT:
PLY THICKNESS: 0.055-0.0059 in.

TEST METHOD: ASTM D3039-76

MODULUS CALCULATION:

NORMALIZED BY: Not normalized

Temperature (°F)		75	75	265			
Moisture Content (%)		ambient	wet	wet			
Equilibrium at T, RH		(1)	(2)	(2)			
Source Code		26	26	26			
	Mean	8.04	3.27	3.29			
	Minimum	5.93	2.54	2.62			
	Maximum	10.6	4.15	4.15			
	C.V.(%)	13.5	16.3	13.0			
	B-value	(3)	(3)	(3)			
F_2^{tu}	Distribution	ANOVA	ANOVA	ANOVA			
(ksi)	C_1	1.11	0.560	0.452			
	C_2	2.36	3.79	3.16			
	No. Specimens	30	15	20			
	No. Batches	5	3	4			
	Approval Class	Interim	Interim	Interim			
	Mean	1.50	1.04	1.04			
	Minimum	1.43	0.95	0.95			
	Maximum	1.58	1.10	1.10			
E_2^t	C.V.(%)	2.76	5.1	4.3			
(Msi)	No. Specimens	30	15	20			
	No. Batches	5	3	4			
	Approval Class	Interim	Interim	Interim			
	Mean						
	No. Specimens						
ν_{21}^t	No. Batches						
	Approval Class						
	Mean	5500	3320	3440			
	Minimum	4000	2750	2840			
	Maximum	7390	4200	4200			
	C.V.(%)	13.7	13.3	12.1			
	B-value	(3)	(3)	(3)			
ε_2^{tu}	Distribution	Weibull	ANOVA	ANOVA			
($\mu\varepsilon$)	C_1	5820	506	456			
	C_2	7.67	5.66	3.79			
	No. Specimens	30	15	20			
	No. Batches	5	3	4			
	Approval Class	Interim	Interim	Interim			

(1) Conditioned at 180°F, ambient relative humidity for 2 days.
(2) Conditioned at 180°F, 75% relative humidity for 63 days.
(3) B-values are presented only for fully approved data.

* ALL DOCUMENTATION PRESENTLY REQUIRED WAS NOT SUPPLIED FOR THIS MATERIAL.

MATERIAL:	AS4 12k/3502 unidirectional tape			**Table 4.2.8(j) C/Ep 147-UT AS4/3502 Compression, 1-axis $[0]_6$ -65/A, 265/A, 265/W Interim**
RESIN CONTENT:	25-29 wt%	COMP: DENSITY:	1.59-1.62 g/cm³	
FIBER VOLUME:	63-68 %	VOID CONTENT:		
PLY THICKNESS:	0.0047-0.0062 in.			
TEST METHOD: ASTM D3410C		MODULUS CALCULATION:		
NORMALIZED BY:	Specimen thickness and batch fiber volume to 60%			

Temperature (°F) Moisture Content (%) Equilibrium at T, RH Source Code		-65 ambient (1) 26		265 ambient (1) 26		265 wet (2) 26	
		Normalized	Measured	Normalized	Measured	Normalized	Measured
F_1^{cu} (ksi)	Mean	226	253	228	249	176	192
	Minimum	173	206	142	150	139	146
	Maximum	307	325	275	292	208	228
	C.V.(%)	16.8	14.1	15.0	15.1	11.5	13.3
	B-value	(3)	(3)	(3)	(3)	(3)	(3)
	Distribution	Weibull	Weibull	Weibull	Weibull	Weibull	Weibull
	C_1	242	269	241	264	184	203
	C_2	6.23	7.45	8.66	9.19	10.6	9.32
	No. Specimens	15		15		15	
	No. Batches	3		3		3	
	Approval Class	Interim		Interim		Interim	
E_1^c (Msi)	Mean	19.3	21.1	21.2	23.2	19.6	21.4
	Minimum	17.1	19.3	17.1	19.3	18.5	20.5
	Maximum	21.8	23.7	23.1	26.3	20.6	22.5
	C.V.(%)	6.63	7.30	9.53	9.70	3.85	3.70
	No. Specimens	15		15		15	
	No. Batches	3		3		3	
	Approval Class	Interim		Interim		Interim	
ν_{12}^c	Mean						
	No. Specimens						
	No. Batches						
	Approval Class						
ε_1^{cu} (με)	Mean		16200		13400		10500
	Minimum		11100		7370		7770
	Maximum		21200		16000		12800
	C.V.(%)		17.4		16.2		14.1
	B-value		(3)		(3)		(3)
	Distribution		Weibull		Weibull		Weibull
	C_1		17400		14200		11100
	C_2		6.39		8.53		8.71
	No. Specimens	15		15		15	
	No. Batches	3		3		3	
	Approval Class	Interim		Interim		Interim	

(1) Conditioned at 180°F, ambient relative humidity for 2 days.
(2) Conditioned at 150°F, 98% relative humidity for 14 days.
(3) B-values are presented only for fully approved data.

4.2.9 Celion 3000/E7K8 plain weave fabric data set description

Material Description:

Material: Celion 3000/E7K8

Form: Plain weave fabric, areal weight of 195 g/m^2, typical cured resin content of 37-44%, typical cured ply thickness of 0.0075-0.0084 inches.

Processing: Good drape. Autoclave cure; 310°F, 85 psi for 2 hours. Low exotherm profile for processing of thick parts.

General Supplier Information:

Fiber: Celion 3000 fibers are continuous carbon filaments made from PAN precursor. Filament count is 3000 filaments/tow. Typical tensile modulus is 34 x 10^6 psi. Typical tensile strength is 515,000 psi.

Matrix: E7K8 is a medium flow, low exotherm epoxy resin. Good tack; up to 20 days out-time at ambient temperature.

Maximum Short Term Service Temperature: 300°F (dry), 190°F (wet)

Typical applications: Primary and secondary structural applications on commercial and military aircraft, jet engine applications such as stationary airfoils and thrust reverser blocker doors.

4.2.9 Celion 3000/E7K8 plain weave*

MATERIAL:	Celion 3000/E7K8 plain weave		**C/Ep Celion 3000/E7K8 Summary**
PREPREG:	U.S. Polymeric Celion 3000/E7K8 Plain Weave, Grade 195		
FIBER:	Celanese Celion 3000	MATRIX:	U.S. Polymeric E7K8
T_g(dry):	T_g(wet):	T_g METHOD:	
PROCESSING:	Autoclave: 310°F, 2 hours, 85 psig		

* DATA WERE SUBMITTED BEFORE THE ESTABLISHMENT OF DATA DOCUMENTATION REQUIREMENTS (JUNE 1989). ALL DOCUMENTATION PRESENTLY REQUIRED WERE NOT SUPPLIED FOR THIS MATERIAL.

Date of fiber manufacture		Date of testing	
Date of resin manufacture		Date of data submittal	1/88
Date of prepreg manufacture	2/86 - 3/86	Date of analysis	1/93
Date of composite manufacture			

LAMINA PROPERTY SUMMARY

	75°F/A		-65°F/A	180°F/A		75°F/W	180°F/W	
Tension, 1-axis	SS-S		SS--			SSSS	SSS-	
Tension, 2-axis	SS-S		SS-S			SS-S	SS-S	
Tension, 3-axis								
Compression, 1-axis	SS-S		SS-S	SS-S		SS-S	SS-S	
Compression, 2-axis	SS-S		SS--	SS--		SS-S	SS--	
Compression, 3-axis								
Shear, 12-plane								
Shear, 23-plane								
SB Strength, 31-plane	S---		S---	S---		S---	S---	

Classes of data: F - Fully approved, I - Interim, S - Screening in Strength/Modulus/Poisson's ratio/Strain-to-failure order.

* DATA WERE SUBMITTED BEFORE THE ESTABLISHMENT OF DATA DOCUMENTATION REQUIREMENTS (JUNE 1989). ALL DOCUMENTATION PRESENTLY REQUIRED WERE NOT SUPPLIED FOR THIS MATERIAL.

	Nominal	As Submitted	Test Method
Fiber Density (g/cm^3)	1.8		
Resin Density (g/cm^3)	1.28		
Composite Density (g/cm^3)	1.54	1.37 - 1.55	
Fiber Areal Weight (g/m^2)	195		
Ply Thickness (in)	0.0075	0.0078 - 0.011	

LAMINATE PROPERTY SUMMARY

Classes of data: F - Fully approved, I - Interim, S - Screening in Strength/Modulus/Poisson's ratio/Strain-to-failure order.

* DATA WERE SUBMITTED BEFORE THE ESTABLISHMENT OF DATA DOCUMENTATION REQUIREMENTS (JUNE 1989). ALL DOCUMENTATION PRESENTLY REQUIRED WAS NOT SUPPLIED FOR THIS MATERIAL.

MATERIAL:	Celion 3000/E7K8 Plain Weave			**Table 4.2.9(a)**
				C/Ep 195-PW
RESIN CONTENT:	37-38 wt%	COMP: DENSITY:	1.55 g/cm³	**Celion, 3000/E7K8**
FIBER VOLUME:	55-56 %	VOID CONTENT:	0.0%	**Tension, 1-axis**
PLY THICKNESS:	0.0078-0.0085 in.			**$[0_f]_{10}$**
				75/A, -65/A
TEST METHOD: ASTM D3039-76		MODULUS CALCULATION:		**Screening**
NORMALIZED BY:	Specimen thickness and batch fiber volume to 57%			

Temperature (°F)		75		-65			
Moisture Content (%)		ambient		ambient			
Equilibrium at T, RH							
Source Code		20		20			
		Normalized	Measured	Normalized	Measured	Normalized	Measured
F_1^{tu} (ksi)	Mean	132	128	110	106		
	Minimum	120	115	101	98.4		
	Maximum	143	140	118	113		
	C.V.(%)	4.7	5.8	6.2	5.4		
	B-value	(1)	(1)	(1)	(1)		
	Distribution	Weibull	Weibull	Normal	Normal		
	C_1	135	132	110	106		
	C_2	25.7	21.4	6.88	5.74		
	No. Specimens	20		5			
	No. Batches	1		1			
	Approval Class	Screening		Screening			
E_1^t (Msi)	Mean	9.67	9.38	9.98	9.66		
	Minimum	9.49	8.85	9.82	9.46		
	Maximum	9.98	9.74	10.0	9.90		
	C.V.(%)	1.2	2.5	1.0	1.8		
	No. Specimens	20		5			
	No. Batches	1		1			
	Approval Class	Screening		Screening			
ν_{12}^t	Mean		0.0580				
	No. Specimens	5					
	No. Batches	1					
	Approval Class	Screening					
ε_1^{tu} (με)	Mean		13700		11000		
	Minimum		12300		10200		
	Maximum		14800		11600		
	C.V.(%)		4.5		5.4		
	B-value		(1)		(1)		
	Distribution		Weibull		Normal		
	C_1		14000		11000		
	C_2		26.8		592		
	No. Specimens	20		5			
	No. Batches	1		1			
	Approval Class	Screening		Screening			

(1) B-values are presented only for fully approved data.

* DATA WERE SUBMITTED BEFORE THE ESTABLISHMENT OF DATA DOCUMENTATION REQUIREMENTS (JUNE 1989). ALL DOCUMENTATION PRESENTLY REQUIRED WAS NOT SUPPLIED FOR THIS MATERIAL.

MATERIAL:	Celion 3000/E7K8 Plain Weave			**Table 4.2.9(b)** **C/Ep 195-PW** **Celion, 3000/E7K8** **Tension, 1-axis** **$[0_f]_{10}$** **75/W, 180/W** **Screening**
RESIN CONTENT:	37 wt%	COMP: DENSITY:	1.55 g/cm^3	
FIBER VOLUME:	55 %	VOID CONTENT:	0.0%	
PLY THICKNESS:	0.0078-0.0081 in.			
TEST METHOD: ASTM D3039-76		MODULUS CALCULATION:		
NORMALIZED BY:	Specimen thickness and batch fiber volume to 57%			

Temperature (°F) Moisture Content (%) Equilibrium at T, RH Source Code		75 wet (1) 20		180 wet (1) 20			
		Normalized	Measured	Normalized	Measured	Normalized	Measured
	Mean	125	122	123	120		
	Minimum	111	105	114	112		
	Maximum	130	129	131	127		
	C.V.(%)	6.3	8.1	6.5	6.3		
	B-value	(2)	(2)	(2)	(2)		
F_1^{tu}	Distribution	Normal	Normal	Normal	Normal		
(ksi)	C_1	125	122	123	120		
	C_2	7.93	9.93	7.99	7.52		
	No. Specimens	5		5			
	No. Batches	1		1			
	Approval Class	Screening		Screening			
	Mean	9.23	9.01	9.55	9.33		
	Minimum	8.93	8.81	9.37	9.15		
	Maximum	9.53	9.20	9.84	9.63		
E_1^t	C.V.(%)	2.5	1.7	1.9	2.0		
(Msi)	No. Specimens	5		5			
	No. Batches	1		1			
	Approval Class	Screening		Screening			
	Mean		0.0620		0.0560		
	No. Specimens	5		5			
ν_{12}^t	No. Batches	1		1			
	Approval Class	Screening		Screening			
	Mean		13700		12800		
	Minimum		12100		11200		
	Maximum		14300		14100		
	C.V.(%)		6.9		9.6		
	B-value		(2)		(2)		
ε_1^{tu}	Distribution		Normal		Normal		
(με)	C_1		13700		12800		
	C_2		939		1230		
	No. Specimens	5		5			
	No. Batches	1		1			
	Approval Class	Screening		Screening			

(1) Conditioned at 160°F, 85% relative humidity for 7 days.

(2) B-values are presented only for fully approved data.

* DATA WERE SUBMITTED BEFORE THE ESTABLISHMENT OF DATA DOCUMENTATION REQUIREMENTS (JUNE 1989). ALL DOCUMENTATION PRESENTLY REQUIRED WAS NOT SUPPLIED FOR THIS MATERIAL.

MATERIAL:	Celion 3000/E7K8 Plain Weave		
RESIN CONTENT:	39-44 wt%	COMP: DENSITY:	1.55 g/cm^3
FIBER VOLUME:	51-54 %	VOID CONTENT:	0.04-0.5%
PLY THICKNESS:	0.0079-0.0084 in.		
TEST METHOD: ASTM D3039-76		MODULUS CALCULATION:	
NORMALIZED BY:	Specimen thickness and batch fiber volume to 57%		

Table 4.2.9(c)
C/Ep 195-PW
Celion, 3000/E7K8
Tension, 1-axis
$[0_f]_{12}$
75/A, -65/A
Screening

Temperature (°F) Moisture Content (%) Equilibrium at T, RH Source Code		75 ambient 20		-65 ambient 20			
		Normalized	Measured	Normalized	Measured	Normalized	Measured
F_1^{tu} (ksi)	Mean	132	122	122	115		
	Minimum	106	100	117	111		
	Maximum	147	136	126	123		
	C.V.(%)	7.5	7.5	2.8	4.3		
	B-value	(1)	(1)	(1)	(1)		
	Distribution	Weibull	Weibull	Normal	Normal		
	C_1	136	126	122	116		
	C_2	16.4	17.3	3.44	4.97		
	No. Specimens	20		5			
	No. Batches	1		1			
	Approval Class	Screening		Screening			
E_1^t (Msi)	Mean	9.96	9.21	9.29	8.82		
	Minimum	9.30	8.74	8.95	8.51		
	Maximum	9.98	9.78	9.66	9.41		
	C.V.(%)	1.2	2.5	2.8	4.0		
	No. Specimens	20		5			
	No. Batches	1		1			
	Approval Class	Screening		Screening			
ν_{12}^t	Mean						
	No. Specimens						
	No. Batches						
	Approval Class						
ε_1^{tu} (με)	Mean		14100				
	Minimum		13600				
	Maximum		14600				
	C.V.(%)		2.6				
	B-value		(1)				
	Distribution		Normal				
	C_1		14100				
	C_2		371				
	No. Specimens	5					
	No. Batches	1					
	Approval Class	Screening					

(1) B-values are presented only for fully approved data.

* DATA WERE SUBMITTED BEFORE THE ESTABLISHMENT OF DATA DOCUMENTATION REQUIREMENTS (JUNE 1989). ALL DOCUMENTATION PRESENTLY REQUIRED WAS NOT SUPPLIED FOR THIS MATERIAL.

Table 4.2.9(d)
C/Ep 195-PW
Celion, 3000/E7K8
Tension, 1-axis
$[0_f]_{12}$
75/W, 180/W
Screening

MATERIAL: Celion 3000/E7K8 Plain Weave

RESIN CONTENT: 42 wt% COMP: DENSITY: 1.55 g/cm³
FIBER VOLUME: 51 % VOID CONTENT: 0.48%
PLY THICKNESS: 0.0081-0.0083 in.

TEST METHOD: MODULUS CALCULATION:

ASTM D3039-76

NORMALIZED BY: Specimen thickness and batch fiber volume to 57%

Temperature (°F) Moisture Content (%) Equilibrium at T, RH Source Code		75 wet (1) 20		180 wet (1) 20			
		Normalized	Measured	Normalized	Measured	Normalized	Measured
F_1^{tu} (ksi)	Mean	145	129	148	133		
	Minimum	143	125	139	124		
	Maximum	148	131	154	142		
	C.V.(%)	1.6	1.8	4.0	5.6		
	B-value	(2)	(2)	(2)	(2)		
	Distribution	Normal	Normal	Normal	Normal		
	C_1	145	129	148	133		
	C_2	2.23	2.37	5.94	7.50		
	No. Specimens	5		5			
	No. Batches	1		1			
	Approval Class	Screening		Screening			
E_1^t (Msi)	Mean	10.6	9.42	10.3	9.21		
	Minimum	10.1	8.79	10.1	8.91		
	Maximum	11.4	10.0	10.5	9.53		
	C.V.(%)	4.9	5.0	1.3	2.7		
	No. Specimens	5		5			
	No. Batches	1		1			
	Approval Class	Screening		Screening			
ν_{12}^t	Mean		0.0560		0.0560		
	No. Specimens	5		5			
	No. Batches	1		1			
	Approval Class	Screening		Screening			
ε_1^{tu} (με)	Mean		13400				
	Minimum		12300				
	Maximum		14300				
	C.V.(%)		5.30				
	B-value		(2)				
	Distribution		Normal				
	C_1		13400				
	C_2		713				
	No. Specimens	5					
	No. Batches	1					
	Approval Class	Screening					

(1) Conditioned at 160°F, 85% relative humidity for 7 days.
(2) B-values are presented only for fully approved data.

* DATA WERE SUBMITTED BEFORE THE ESTABLISHMENT OF DATA DOCUMENTATION REQUIREMENTS (JUNE 1989). ALL DOCUMENTATION PRESENTLY REQUIRED WAS NOT SUPPLIED FOR THIS MATERIAL.

MATERIAL: Celion 3000/E7K8 Plain Weave	**Table 4.2.9(e)** **C/Ep 195-PW** **Celion, 3000/E7K8** **Tension, 2-axis** **$[0_f]_{10}$** **75/A, -65/A** **Screening**
RESIN CONTENT: 36 wt% COMP: DENSITY: 1.55 g/cm^3 FIBER VOLUME: 56 % VOID CONTENT: 0.0% PLY THICKNESS: 0.0078-0.0084 in.	
TEST METHOD: ASTM D3039-76	MODULUS CALCULATION:
NORMALIZED BY: Specimen thickness and batch fiber volume to 57%	

Temperature (°F) Moisture Content (%) Equilibrium at T, RH Source Code		75 ambient 20		-65 ambient 20			
		Normalized	Measured	Normalized	Measured	Normalized	Measured
F_2^{tu} (ksi)	Mean	128	127	113	111		
	Minimum	120	115	101	100		
	Maximum	137	134	125	122		
	C.V.(%)	3.6	3.7	9.1	8.9		
	B-value	(1)	(1)	(1)	(1)		
	Distribution	Normal	Normal	Normal	Normal		
	C_1	128	127	113	111		
	C_2	4.64	4.69	10.3	9.89		
	No. Specimens	20		5			
	No. Batches	1		1			
	Approval Class	Screening		Screening			
E_2^t (Msi)	Mean	9.50	9.37	9.51	9.34		
	Minimum	9.36	9.04	9.29	9.20		
	Maximum	9.69	9.71	9.65	9.68		
	C.V.(%)	0.98	1.8	1.6	2.1		
	No. Specimens	20		5			
	No. Batches	1		1			
	Approval Class	Screening		Screening			
ν_{21}^t	Mean						
	No. Specimens						
	No. Batches						
	Approval Class						
ε_2^{tu} (με)	Mean		13400		11700		
	Minimum		12600		10700		
	Maximum		14200		12700		
	C.V.(%)		3.5		7.7		
	B-value		(1)		(1)		
	Distribution		Weibull		Normal		
	C_1		13600		11700		
	C_2		32.5		902		
	No. Specimens	20		5			
	No. Batches	1		1			
	Approval Class	Screening		Screening			

(1) B-values are presented only for fully approved data.

* DATA WERE SUBMITTED BEFORE THE ESTABLISHMENT OF DATA DOCUMENTATION REQUIREMENTS (JUNE 1989). ALL DOCUMENTATION PRESENTLY REQUIRED WAS NOT SUPPLIED FOR THIS MATERIAL.

MATERIAL:	Celion 3000/E7K8 Plain Weave		
RESIN CONTENT:	36 wt%	COMP: DENSITY:	1.55 g/cm^3
FIBER VOLUME:	56 %	VOID CONTENT:	0.0%
PLY THICKNESS:	0.0078-0.0084 in.		
TEST METHOD: ASTM D3039-76		MODULUS CALCULATION:	
NORMALIZED BY:	Specimen thickness and batch fiber volume to 57%		

Table 4.2.9(f)
C/Ep 195-PW
Celion, 3000/E7K8
Tension, 2-axis
$[90_f]_{10}$
75/W, 180/W
Screening

Temperature (°F)		75		180			
Moisture Content (%)		wet		wet			
Equilibrium at T, RH		(1)		(1)			
Source Code		20		20			
		Normalized	Measured	Normalized	Measured	Normalized	Measured
	Mean	119	117	130	128		
	Minimum	105	104	129	125		
	Maximum	130	126	132	131		
	C.V.(%)	7.8	7.3	0.89	1.8		
	B-value	(2)	(2)	(2)	(2)		
F_2^{tu}	Distribution	Normal	Normal	Normal	Normal		
(ksi)	C_1	119	117	130	128		
	C_2	9.35	8.51	1.16	2.35		
	No. Specimens	5		5			
	No. Batches	1		1			
	Approval Class	Screening		Screening			
	Mean	9.08	8.92	9.35	9.18		
	Minimum	8.98	8.73	9.26	8.96		
	Maximum	9.21	9.14	9.48	9.38		
E_2^t	C.V.(%)	1.2	1.6	1.2	1.8		
(Msi)	No. Specimens	5		5			
	No. Batches	1		1			
	Approval Class	Screening		Screening			
	Mean						
	No. Specimens						
ν_{21}^t	No. Batches						
	Approval Class						
	Mean		13100		14200		
	Minimum		11400		13700		
	Maximum		14400		14800		
	C.V.(%)		8.7		3.5		
	B-value		(2)		(2)		
ε_2^{tu}	Distribution		Normal		Normal		
($\mu\varepsilon$)	C_1		13100		14200		
	C_2		1135		490		
	No. Specimens	5		5			
	No. Batches	1		1			
	Approval Class	Screening		Screening			

(1) Conditioned at 160°F, 85% relative humidity for 7 days.
(2) B-values are presented only for fully approved data.

* DATA WERE SUBMITTED BEFORE THE ESTABLISHMENT OF DATA DOCUMENTATION REQUIREMENTS (JUNE 1989). ALL DOCUMENTATION PRESENTLY REQUIRED WAS NOT SUPPLIED FOR THIS MATERIAL.

Table 4.2.9(g)
C/Ep 195-PW
Celion, 3000/E7K8
Compression, 1-axis
$[0_f]_{10}$
75/A, -65/A, 180/A
Screening

MATERIAL: Celion 3000/E7K8 Plain Weave

RESIN CONTENT: 36-40 wt% COMP: DENSITY: 1.55 g/cm^3
FIBER VOLUME: 53-55 % VOID CONTENT: 0.0-0.75%
PLY THICKNESS: 0.0079-0.0084 in.

TEST METHOD: SACMA SRM 1-88

MODULUS CALCULATION:

NORMALIZED BY: Specimen thickness and batch fiber volume to 57%

		75		-65		180	
Temperature (°F) / Moisture Content (%) / Equilibrium at T, RH / Source Code		75 ambient, 20		-65 ambient, 20		180 ambient, 20	
		Normalized	Measured	Normalized	Measured	Normalized	Measured
	Mean	104	101	121	118	97.4	94.5
	Minimum	90.5	87.7	113	111	87.5	85.1
	Maximum	122	120	132	126	105	100
	C.V.(%)	8.3	8.7	5.9	4.7	7.2	7.1
	B-value	(1)	(1)	(1)	(1)	(1)	(1)
F_1^{cu}	Distribution	Weibull	Weibull	Normal	Normal	Normal	Normal
(ksi)	C_1	108	105	121	118	97.4	94.5
	C_2	13.0	12.1	7.19	5.58	7.00	6.72
	No. Specimens	20		5		5	
	No. Batches	1		1		1	
	Approval Class	Screening		Screening		Screening	
	Mean	9.88	9.02	9.83	9.33	9.45	9.16
	Minimum	9.56	8.65	9.75	9.20	9.14	8.89
	Maximum	10.3	9.29	9.95	9.48	9.66	9.37
E_1^c	C.V.(%)	2.3	2.0	1.0	1.1	2.3	2.0
(Msi)	No. Specimens	20		5		5	
	No. Batches	1		1		1	
	Approval Class	Screening		Screening		Screening	
	Mean						
	No. Specimens						
ν_{12}^c	No. Batches						
	Approval Class						
	Mean		10900		12200		10400
	Minimum		10500		12000		10200
	Maximum		11200		12300		10800
	C.V.(%)		2.2		1.0		2.3
	B-value		(1)		(1)		(1)
ε_1^{cu}	Distribution		Weibull		Normal		Normal
(με)	C_1		11000		12200		10400
	C_2		54.2		122		239
	No. Specimens	20		5		5	
	No. Batches	1		1		1	
	Approval Class	Screening		Screening		Screening	

(1) B-values are presented only for fully approved data.

* DATA WERE SUBMITTED BEFORE THE ESTABLISHMENT OF DATA DOCUMENTATION REQUIREMENTS (JUNE 1989). ALL DOCUMENTATION PRESENTLY REQUIRED WAS NOT SUPPLIED FOR THIS MATERIAL.

MATERIAL:	Celion 3000/E7K8 Plain Weave			**Table 4.2.9(h) C/Ep 195-PW Celion, 3000/E7K8 Compression, 1-axis $[0_f]_{10}$ 75/W, 180/W Screening**
RESIN CONTENT:	36-37 wt%	COMP: DENSITY:	1.55 g/cm^3	
FIBER VOLUME:	54-56 %	VOID CONTENT:	0.0-0.70%	
PLY THICKNESS:	0.0073-0.0086 in.			
TEST METHOD: SACMA SRM 1-88		MODULUS CALCULATION:		
NORMALIZED BY:	Specimen thickness and batch fiber volume to 57%			

Temperature (°F) Moisture Content (%) Equilibrium at T, RH Source Code		75 wet (1) 20		180 wet (1) 20			
		Normalized	Measured	Normalized	Measured	Normalized	Measured
	Mean	94.9	92.6	78.9	77.6		
	Minimum	89.7	88.2	72.7	70.5		
	Maximum	102	98.8	83.2	82.3		
	C.V.(%)	5.5	4.9	5.7	6.0		
	B-value	(2)	(2)	(2)	(2)		
F_1^{cu}	Distribution	Normal	Normal	Normal	Normal		
(ksi)	C_1	94.9	92.6	78.9	77.6		
	C_2	5.47	4.57	4.53	4.65		
	No. Specimens	5		5			
	No. Batches	1		1			
	Approval Class	Screening		Screening			
	Mean	9.39	8.92	8.97	8.52		
	Minimum	8.80	8.12	8.45	8.18		
	Maximum	10.2	9.79	9.54	8.80		
E_1^c	C.V.(%)	6.3	6.8	4.4	3.5		
(Msi)	No. Specimens	5		5			
	No. Batches	1		1			
	Approval Class	Screening		Screening			
	Mean						
	No. Specimens						
ν_{12}^c	No. Batches						
	Approval Class						
	Mean		9800		8130		
	Minimum		8970		7620		
	Maximum		10400		8600		
	C.V.(%)		6.0		4.4		
	B-value		(2)		(2)		
ε_1^{cu}	Distribution		Normal		Normal		
(με)	C_1		9800		8130		
	C_2		590		356		
	No. Specimens	5		5			
	No. Batches	1		1			
	Approval Class	Screening		Screening			

(1) Conditioned at 160°F, 85% relative humidity for 7 days.
(2) B-values are presented only for fully approved data.

* DATA WERE SUBMITTED BEFORE THE ESTABLISHMENT OF DATA DOCUMENTATION REQUIREMENTS (JUNE 1989). ALL DOCUMENTATION PRESENTLY REQUIRED WAS NOT SUPPLIED FOR THIS MATERIAL.

Table 4.2.9(i)
C/Ep 195-PW
Celion, 3000/E7K8
Compression, 1-axis
$[0_f]_{12}$
75/A, -65/A, 180/A
Screening

MATERIAL: Celion 3000/E7K8 Plain Weave

RESIN CONTENT: 38-40 wt% | COMP: DENSITY: 1.55 g/cm^3
FIBER VOLUME: 52-54 % | VOID CONTENT: 0.0%
PLY THICKNESS: 0.0078-0.0084 in.

TEST METHOD: SACMA SRM 1-88 | MODULUS CALCULATION:

NORMALIZED BY: Specimen thickness and batch fiber volume to 57%

		75		-65		180	
Temperature (°F) / Moisture Content (%) / Equilibrium at T, RH / Source Code		75 / ambient / 20		-65 / ambient / 20		180 / ambient / 20	
		Normalized	Measured	Normalized	Measured	Normalized	Measured
	Mean	114	107	133	122	103	97.6
	Minimum	86.4	84.4	127	116	96.0	89.2
	Maximum	128	121	139	129	114	107
	C.V.(%)	9.5	9.1	3.9	4.6	6.8	7.2
	B-value	(1)	(1)	(1)	(1)	(1)	(1)
F_1^{cu}	Distribution	Weibull	Weibull	Normal	Normal	Normal	Normal
(ksi)	C_1	118	111	133	122	103	97.6
	C_2	13.8	14.0	5.22	5.60	6.99	7.04
	No. Specimens	20		5		5	
	No. Batches	1		1		1	
	Approval Class	Screening		Screening		Screening	
	Mean	8.22	7.80	8.45	7.71	8.40	7.67
	Minimum	8.07	7.51	8.27	7.43	8.20	7.58
	Maximum	8.50	8.05	8.73	8.09	8.54	7.84
E_1^c	C.V.(%)	1.6	2.2	2.3	3.4	1.5	1.4
(Msi)	No. Specimens	20		5		5	
	No. Batches	1		1		1	
	Approval Class	Screening		Screening		Screening	
	Mean						
	No. Specimens						
ν_{12}^c	No. Batches						
	Approval Class						
	Mean		13500				
	Minimum		13000				
	Maximum		13700				
	C.V.(%)		1.6				
	B-value		(1)				
ε_1^{cu}	Distribution		Nonpara.				
(με)	C_1		10				
	C_2		1.25				
	No. Specimens	20					
	No. Batches	1					
	Approval Class	Screening					

(1) B-values are presented only for fully approved data.

* DATA WERE SUBMITTED BEFORE THE ESTABLISHMENT OF DATA DOCUMENTATION REQUIREMENTS (JUNE 1989). ALL DOCUMENTATION PRESENTLY REQUIRED WAS NOT SUPPLIED FOR THIS MATERIAL.

Table 4.2.9(j)
C/Ep 195-PW
Celion, 3000/E7K8
Compression, 1-axis
$[0_f]_{12}$
75/W, 180/W
Screening

MATERIAL: Celion 3000/E7K8 Plain Weave

RESIN CONTENT: 38-40 wt% COMP: DENSITY: 1.55 g/cm^3
FIBER VOLUME: 52-54 % VOID CONTENT: 0.0-0.04%
PLY THICKNESS: 0.0080-0.0084 in.

TEST METHOD: SACMA SRM 1-88

MODULUS CALCULATION:

NORMALIZED BY: Specimen thickness and batch fiber volume to 57%

		75		180			
Temperature (°F)		75		180			
Moisture Content (%)		wet		wet			
Equilibrium at T, RH		(1)		(1)			
Source Code		20		20			
		Normalized	Measured	Normalized	Measured	Normalized	Measured
F_1^{cu} (ksi)	Mean	96.1	90.7	80.2	75.7		
	Minimum	83.9	78.4	74.4	72.2		
	Maximum	107	101	83.3	79.9		
	C.V.(%)	9.3	9.4	4.7	4.4		
	B-value	(2)	(2)	(2)	(2)		
	Distribution	Normal	Normal	Normal	Normal		
	C_1	96.1	90.7	80.2	75.7		
	C_2	8.91	8.55	3.73	3.31		
	No. Specimens	5		5			
	No. Batches	1		1			
	Approval Class	Screening		Screening			
E_1^c (Msi)	Mean	9.08	8.30	9.36	8.54		
	Minimum	8.84	7.91	9.14	8.20		
	Maximum	9.17	8.62	9.57	8.84		
	C.V.(%)	1.5	3.2	2.0	2.9		
	No. Specimens	5		5			
	No. Batches	1		1			
	Approval Class	Screening		Screening			
ν_{12}^c	Mean						
	No. Specimens						
	No. Batches						
	Approval Class						
ε_1^{cu} (με)	Mean		10700				
	Minimum		10600				
	Maximum		11000				
	C.V.(%)		1.5				
	B-value		(2)				
	Distribution		Normal				
	C_1		10700				
	C_2		164				
	No. Specimens	5					
	No. Batches	1					
	Approval Class	Screening					

(1) Conditioned at 160°F, 85% relative humidity for 7 days.
(2) B-values are presented only for fully approved data.

* DATA WERE SUBMITTED BEFORE THE ESTABLISHMENT OF DATA DOCUMENTATION REQUIREMENTS (JUNE 1989). ALL DOCUMENTATION PRESENTLY REQUIRED WAS NOT SUPPLIED FOR THIS MATERIAL.

Table 4.2.9(k)
C/Ep 195-PW
Celion 3000/E7K8
SBS, 31-plane
$[0_f]_{14}$
75/A, -65/A, 180/A, 75/W, 180/W
Screening

MATERIAL: Celion 3000/E7K8 Plain Weave

RESIN CONTENT: 36-39 wt% COMP: DENSITY: 1.55 g/cm^3
FIBER VOLUME: 54-56 % VOID CONTENT: 0.0-0.75%
PLY THICKNESS: 0.0079-0.0081 in.

TEST METHOD: ASTM D2344-68 MODULUS CALCULATION:

NORMALIZED BY: Not normalized

Temperature (°F) Moisture Content (%) Equilibrium at T, RH Source Code		75 ambient 20	-65 ambient 20	180 ambient 20	75 wet (1) 20	180 wet (1) 20
	Mean	10.3	11.6	9.70	9.81	6.92
	Minimum	9.43	10.7	9.34	9.24	6.60
	Maximum	11.4	13.6	9.94	10.4	7.22
	C.V.(%)	5.7	10.8	3.0	7.0	3.4
	B-value	(2)	(2)	(2)	(2)	(2)
	Distribution	Normal	Normal	Normal	Normal	Normal
F_{31}^{sbs}	C_1	10.3	11.6	9.70	9.81	6.92
(ksi)	C_2	0.446	1.25	0.293	0.505	0.237
	No. Specimens	20	5	5	5	5
	No. Batches	1	1	1	1	1
	Approval Class	Screening	Screening	Screening	Screening	Screening

(1) Conditioned at 160°F, 85% relative humidity for 7 days.
(2) B-values are presented only for fully approved data.

* DATA WERE SUBMITTED BEFORE THE ESTABLISHMENT OF DATA DOCUMENTATION REQUIREMENTS (JUNE 1989). ALL DOCUMENTATION PRESENTLY REQUIRED WAS NOT SUPPLIED FOR THIS MATERIAL.

MATERIAL:	Celion 3000/E7K8 Plain Weave			**Table 4.2.9(l)** **C/Ep 195-PW** **Celion 3000/E7K8** **SBS, 31-plane** **$[0_f]_{12}$** **75/A, -65/A, 180/A, 75/W, 180/W** **Screening**
RESIN CONTENT:	39 wt%	COMP: DENSITY:	1.55 g/cm^3	
FIBER VOLUME:	54 %	VOID CONTENT:	0.29%	
PLY THICKNESS:	0.0080 in.			
TEST METHOD: ASTM D2344-68		MODULUS CALCULATION:		
NORMALIZED BY:	Not normalized			

Temperature (°F) Moisture Content (%) Equilibrium at T, RH Source Code		75 ambient 20	-65 ambient 20	180 ambient 20	75 wet (1) 20	180 wet (1) 20
	Mean	9.76	10.2	9.72	9.72	8.72
	Minimum	9.00	9.54	8.76	8.76	8.35
	Maximum	10.7	10.5	10.3	10.3	9.00
	C.V.(%)	4.8	3.9	6.1	6.1	2.8
	B-value	(2)	(2)	(2)	(2)	(2)
	Distribution	Normal	Normal	Normal	Normal	Normal
F_{31}^{sbs}	C_1	9.76	10.2	9.72	9.72	8.72
(ksi)	C_2	0.470	0.395	0.591	0.591	0.247
	No. Specimens	20	5	5	5	5
	No. Batches	1	1	1	1	1
	Approval Class	Screening	Screening	Screening	Screening	Screening

(1) Conditioned at 160°F, 85% relative humidity for 7 days.
(2) B-values are presented only for fully approved data.

4.2.10 HITEX 33 6k/E7K8 plain weave fabric data set description

Material Description:

Material: HITEX 33-6k/E7K8

Form: Plain weave fabric, areal weight of 195 g/m^2, typical cured resin content of 37-41%, typical cured ply thickness of 0.0085 inches.

Processing: Good drape. Autoclave cure; 310°F, 85 psi for 2 hours. Low exotherm profile for processing of thick parts.

General Supplier Information:

Fiber: HITEX 33 fibers are continuous carbon filaments made from PAN precursor. Filament count is 6000 filaments/tow. Typical tensile modulus is 33 x 10^6 psi. Typical tensile strength is 560,000 psi.

Matrix: E7K8 is a medium flow, low exotherm epoxy resin. Good tack; up to 20 days out-time at ambient temperature.

Maximum Short Term Service Temperature: 300°F (dry), 190°F (wet)

Typical applications: Primary and secondary structural applications on commercial and military aircraft, jet engine applications such as stationary airfoils and thrust reverser blocker doors.

4.2.10 HITEX 33 6k/E7K8 plain weave*

C/Ep
HITEX 33/E7K8
Summary

MATERIAL:	HITEX 33 6k/E7K8 plain weave		
PREPREG:	U.S. Polymeric Hitex 33 6k/E7K8 Plain weave		
FIBER:	Hitco HITEX 33 6k G'	MATRIX:	U.S. Polymeric E7K8
T_g(dry):	T_g(wet):	T_g METHOD:	
PROCESSING:	Autoclave: 310°F, 2 hours, 85 psig		

* DATA WERE SUBMITTED BEFORE THE ESTABLISHMENT OF DATA DOCUMENTATION REQUIREMENTS (JUNE 1989). ALL DOCUMENTATION PRESENTLY REQUIRED WERE NOT SUPPLIED FOR THIS MATERIAL.

Date of fiber manufacture		Date of testing	
Date of resin manufacture		Date of data submittal	1/88
Date of prepreg manufacture		Date of analysis	1/93
Date of composite manufacture			

LAMINA PROPERTY SUMMARY

	75°F/A		-65°F/A	180°F/A		75°F/W	180°F/W	
Tension, 1-axis								
Tension, 2-axis	SSSS		SS-S			SSSS	SSSS	
Tension, 3-axis								
Compression, 1-axis	SS-S		SS--	SS--		SS-S	SS--	
Compression, 2-axis	SS-S		SS--	SS--		SS-S	SS--	
Compression, 3-axis								
Shear, 12-plane								
Shear, 23-plane								
SB Strength, 31-plane	S---		S---			S---	S---	

Classes of data: F - Fully approved, I - Interim, S - Screening in Strength/Modulus/Poisson's ratio/Strain-to-failure order.

* DATA WERE SUBMITTED BEFORE THE ESTABLISHMENT OF DATA DOCUMENTATION REQUIREMENTS (JUNE 1989). ALL DOCUMENTATION PRESENTLY REQUIRED WERE NOT SUPPLIED FOR THIS MATERIAL.

		Nominal	As Submitted	Test Method
Fiber Density	(g/cm^3)	1.77		
Resin Density	(g/cm^3)	1.27		
Composite Density	(g/cm^3)	1.56		
Fiber Areal Weight	(g/m^2)	195		
Ply Thickness	(in)	0.0085	0.0077 - 0.0099	

LAMINATE PROPERTY SUMMARY

Classes of data: F - Fully approved, I - Interim, S - Screening in Strength/Modulus/Poisson's ratio/Strain-to-failure order.

* DATA WERE SUBMITTED BEFORE THE ESTABLISHMENT OF DATA DOCUMENTATION REQUIREMENTS (JUNE 1989). ALL DOCUMENTATION PRESENTLY REQUIRED WAS NOT SUPPLIED FOR THIS MATERIAL.

MATERIAL:	HITEX 33 6k/E7K8 plain weave			**Table 4.2.10(a)** **C/Ep 195-PW** **HITEX 33/E7K8** **Tension, 2-axis** **$[90_f]_{12}$** **75/A, -65/A, 75/W** **Screening**
RESIN CONTENT:	37-41 wt%	COMP: DENSITY:	1.53-1.55 g/cm^3	
FIBER VOLUME:	51-55 %	VOID CONTENT:	0.0%	
PLY THICKNESS:	0.0087-0.0098 in.			
TEST METHOD: ASTM D3039-76		MODULUS CALCULATION:		
NORMALIZED BY:	Specimen thickness and batch fiber volume to 57%			

		75		-65		75	
Temperature (°F) Moisture Content (%) Equilibrium at T, RH Source Code		75 ambient 20		-65 ambient 20		75 wet (1) 20	
		Normalized	Measured	Normalized	Measured	Normalized	Measured
F_2^{tu} (ksi)	Mean	131	124	126	122	134	119
	Minimum	120	103	122	111	130	114
	Maximum	139	136	131	131	137	125
	C.V.(%)	4.3	6.8	3.1	6.7	2.8	3.8
	B-value	(2)	(2)	(2)	(2)	(2)	(2)
	Distribution	Weibull	Weibull	Normal	Normal	Normal	Normal
	C_1	134	128	126	122	134	120
	C_2	28.2	17.8	3.88	8.16	3.69	4.55
	No. Specimens	20		5		5	
	No. Batches	1		1		1	
	Approval Class	Screening		Screening		Screening	
E_2^t (Msi)	Mean	8.65	8.14	8.10	7.82	9.61	8.55
	Minimum	8.01	7.52	7.73	7.54	9.26	8.20
	Maximum	9.65	8.62	8.29	8.26	9.94	9.13
	C.V.(%)	6.2	3.1	2.7	3.4	2.8	4.1
	No. Specimens	20		5		5	
	No. Batches	1		1		1	
	Approval Class	Screening		Screening		Screening	
ν_{21}^t	Mean		0.0460				0.0540
	No. Specimens	5				5	
	No. Batches	1				1	
	Approval Class	Screening				Screening	
ε_2^{tu} (με)	Mean		14300		15600		10500
	Minimum		13700		14600		9930
	Maximum		14900		16500		10800
	C.V.(%)		3.8		4.4		3.2
	B-value		(2)		(2)		(2)
	Distribution		Normal		Normal		Normal
	C_1		14300		15600		10500
	C_2		541		687		335
	No. Specimens	5		5		5	
	No. Batches	1		1		1	
	Approval Class	Screening		Screening		Screening	

(1) Conditioned at 160°F, 85% relative humidity for 14 days.
(2) B-values are presented only for fully approved data.

* DATA WERE SUBMITTED BEFORE THE ESTABLISHMENT OF DATA DOCUMENTATION REQUIREMENTS (JUNE 1989). ALL DOCUMENTATION PRESENTLY REQUIRED WAS NOT SUPPLIED FOR THIS MATERIAL.

Table 4.2.10(b)
C/Ep 195-PW
HITEX 33/E7K8
Tension, 2-axis
$[90_f]_{12}$
180/W
Screening

MATERIAL: HITEX 33 6k/E7K8 plain weave

RESIN CONTENT: 41 wt% COMP: DENSITY: 1.53 g/cm^3
FIBER VOLUME: 51 % VOID CONTENT: 0.0%
PLY THICKNESS: 0.0089-0.0094 in.

TEST METHOD: ASTM D3039-76 MODULUS CALCULATION:

NORMALIZED BY: Specimen thickness and batch fiber volume to 57%

		Normalized	Measured	Normalized	Measured	Normalized	Measured
Temperature (°F)		180					
Moisture Content (%)		wet					
Equilibrium at T, RH		(1)					
Source Code		20					
F_2^{tu} (ksi)	Mean	138	122				
	Minimum	120	107				
	Maximum	155	135				
	C.V.(%)	10.2	9.1				
	B-value	(2)	(2)				
	Distribution	Normal	Normal				
	C_1	138	123				
	C_2	14.1	11.1				
	No. Specimens	5					
	No. Batches	1					
	Approval Class	Screening					
E_2^t (Msi)	Mean	9.91	8.80				
	Minimum	9.11	8.23				
	Maximum	10.7	9.23				
	C.V.(%)	7.2	5.3				
	No. Specimens	5					
	No. Batches	1					
	Approval Class	Screening					
ν_{21}^t	Mean		0.0700				
	No. Specimens	5					
	No. Batches	1					
	Approval Class	Screening					
ε_2^{tu} (με)	Mean		10400				
	Minimum		9840				
	Maximum		10800				
	C.V.(%)		3.6				
	B-value		(2)				
	Distribution		Normal				
	C_1		10400				
	C_2		372				
	No. Specimens	5					
	No. Batches	1					
	Approval Class	Screening					

(1) Conditioned at 160°F, 85% relative humidity for 14 days.
(2) B-values are presented only for fully approved data.

* DATA WERE SUBMITTED BEFORE THE ESTABLISHMENT OF DATA DOCUMENTATION REQUIREMENTS (JUNE 1989). ALL DOCUMENTATION PRESENTLY REQUIRED WAS NOT SUPPLIED FOR THIS MATERIAL.

MATERIAL:	HITEX 33 6k/E7K8 plain weave			**Table 4.2.10(c)** **C/Ep 195-PW** **HITEX 33/E7K8** **Compression, 1-axis** **$[0_f]_{12}$** **75/A, -65/A, 180/A** **Screening**
RESIN CONTENT:	45 wt%	COMP: DENSITY:	1.51 g/cm^3	
FIBER VOLUME:	47 %	VOID CONTENT:	0.0%	
PLY THICKNESS:	0.0079-0.0099 in.			
TEST METHOD: SACMA SRM 1-88		MODULUS CALCULATION:		
NORMALIZED BY:	Specimen thickness and batch fiber volume to 57%			

		75		-65		180	
Temperature (°F) Moisture Content (%) Equilibrium at T, RH Source Code		75 ambient 20		-65 ambient 20		180 ambient 20	
		Normalized	Measured	Normalized	Measured	Normalized	Measured
F_1^{cu} (ksi)	Mean	136	112	155	128	130	107
	Minimum	111	98.4	147	118	118	94.9
	Maximum	158	128	164	139	139	117
	C.V.(%)	8.4	7.5	5.5	7.5	6.3	7.8
	B-value	(1)	(1)	(1)	(1)	(1)	(1)
	Distribution	Weibull	Weibull	Normal	Normal	Normal	Normal
	C_1	141	116	155	128	130	107
	C_2	13.3	14.5	8.51	9.57	8.21	8.22
	No. Specimens	20		5		5	
	No. Batches	1		1		1	
	Approval Class	Screening		Screening		Screening	
E_1^c (Msi)	Mean	9.11	7.53	10.1	8.30	9.37	7.75
	Minimum	8.64	6.83	9.72	7.74	9.15	7.38
	Maximum	9.63	8.17	10.8	8.76	9.66	8.66
	C.V.(%)	3.0	5.2	4.0	5.1	2.4	7.1
	No. Specimens	20		5		5	
	No. Batches	1		1		1	
	Approval Class	Screening		Screening		Screening	
ν_{12}^c	Mean						
	No. Specimens						
	No. Batches						
	Approval Class						
ε_1^{cu} (με)	Mean		14400				
	Minimum		13700				
	Maximum		15200				
	C.V.(%)		3.1				
	B-value		(1)				
	Distribution		Weibull				
	C_1		14600				
	C_2		34.7				
	No. Specimens	20					
	No. Batches	1					
	Approval Class	Screening					

(1) B-values are presented only for fully approved data.

* DATA WERE SUBMITTED BEFORE THE ESTABLISHMENT OF DATA DOCUMENTATION REQUIREMENTS (JUNE 1989). ALL DOCUMENTATION PRESENTLY REQUIRED WAS NOT SUPPLIED FOR THIS MATERIAL.

MATERIAL:	HITEX 33 6k/E7K8 plain weave			**Table 4.2.10(d)** **C/Ep 195-PW** **HITEX 33/E7K8** **Compression, 1-axis** **$[0_f]_{12}$** **75/W, 180/W** **Screening**
RESIN CONTENT:	45 wt%	COMP: DENSITY:	1.51 g/cm^3	
FIBER VOLUME:	47 %	VOID CONTENT:	0.0%	
PLY THICKNESS:	0.0081-0.0098 in.			
TEST METHOD: SACMA SRM 1-88		MODULUS CALCULATION:		
NORMALIZED BY:	Specimen thickness and batch fiber volume to 57%			

		75		180			
Temperature (°F)		75		180			
Moisture Content (%)		wet		wet			
Equilibrium at T, RH		(1)		(1)			
Source Code		20		20			
		Normalized	Measured	Normalized	Measured	Normalized	Measured
F_1^{cu} (ksi)	Mean	133	110	68.5	56.4		
	Minimum	130	100	54.2	46.7		
	Maximum	139	116	75.8	62.2		
	C.V.(%)	2.8	5.8	13.6	12.0		
	B-value	(2)	(2)	(2)	(2)		
	Distribution	Normal	Normal	Normal	Normal		
	C_1	133	110	68.5	56.4		
	C_2	3.71	6.36	9.31	6.79		
	No. Specimens	5		5			
	No. Batches	1		1			
	Approval Class	Screening		Screening			
E_1^c (Msi)	Mean	8.78	7.24	9.43	7.78		
	Minimum	8.41	7.04	9.32	7.69		
	Maximum	9.07	7.51	9.64	7.89		
	C.V.(%)	3.2	2.5	1.4	9.5		
	No. Specimens	5		5			
	No. Batches	1		1			
	Approval Class	Screening		Screening			
ν_{12}^c	Mean						
	No. Specimens						
	No. Batches						
	Approval Class						
ε_1^{cu} (με)	Mean		14600				
	Minimum		14000				
	Maximum		15400				
	C.V.(%)		3.6				
	B-value		(2)				
	Distribution		Normal				
	C_1		14600				
	C_2		525				
	No. Specimens	5					
	No. Batches	1					
	Approval Class	Screening					

(1) Conditioned at 160°F, 85% relative humidity for 14 days.

(2) B-values are presented only for fully approved data.

* DATA WERE SUBMITTED BEFORE THE ESTABLISHMENT OF DATA DOCUMENTATION REQUIREMENTS (JUNE 1989). ALL DOCUMENTATION PRESENTLY REQUIRED WAS NOT SUPPLIED FOR THIS MATERIAL.

MATERIAL:	HITEX 33 6k/E7K8 plain weave			**Table 4.2.10(e) C/Ep 195-PW HITEX 33/E7K8 Compression, 2-axis $[90_f]_6$ 75/A, -65/A, 180/A Screening**
RESIN CONTENT:	39-41 wt%	COMP: DENSITY:	1.53 g/cm³	
FIBER VOLUME:	51-52 %	VOID CONTENT:	0.0%	
PLY THICKNESS:	0.0083-0.0087 in.			
TEST METHOD: SACMA SRM 1-88		MODULUS CALCULATION:		
NORMALIZED BY:	Specimen thickness and batch fiber volume to 57%			

Temperature (°F) Moisture Content (%) Equilibrium at T, RH Source Code		75 ambient 20		-65 ambient 20		180 ambient 20	
		Normalized	Measured	Normalized	Measured	Normalized	Measured
F_2^{cu} (ksi)	Mean	104	92.4	128	114	99.4	88.6
	Minimum	77.9	70.4	111	98.8	86.4	77.0
	Maximum	125	109	138	123	113	101
	C.V.(%)	13.1	12.6	8.0	8.1	12.0	12.0
	B-value	(1)	(1)	(1)	(1)	(1)	(1)
	Distribution	Weibull	Weibull	Normal	Normal	Normal	Normal
	C_1	110	97.4	128	114	99.4	88.6
	C_2	9.70	10.5	10.3	9.18	11.9	10.6
	No. Specimens	20		5		5	
	No. Batches	1		1		1	
	Approval Class	Screening		Screening		Screening	
E_2^c (Msi)	Mean	8.92	8.21	9.49	8.74	9.07	8.35
	Minimum	8.50	7.78	9.36	8.65	8.95	8.20
	Maximum	9.40	8.77	9.58	8.93	9.18	8.52
	C.V.(%)	2.5	3.4	0.9	1.3	1.3	1.7
	No. Specimens	20		5		5	
	No. Batches	1		1		1	
	Approval Class	Screening		Screening		Screening	
ν_{21}^c	Mean						
	No. Specimens						
	No. Batches						
	Approval Class						
ε_2^{cu} (με)	Mean		10900				
	Minimum		10400				
	Maximum		11400				
	C.V.(%)		2.4				
	B-value		(1)				
	Distribution		Weibull				
	C_1		11100				
	C_2		46.5				
	No. Specimens	20					
	No. Batches	1					
	Approval Class	Screening					

(1) B-values are presented only for fully approved data.

* DATA WERE SUBMITTED BEFORE THE ESTABLISHMENT OF DATA DOCUMENTATION REQUIREMENTS (JUNE 1989). ALL DOCUMENTATION PRESENTLY REQUIRED WAS NOT SUPPLIED FOR THIS MATERIAL.

MATERIAL:	HITEX 33 6k/E7K8 plain weave			**Table 4.2.10(f) C/Ep 195-PW HITEX 33/E7K8 Compression, 2-axis $[90_f]_6$ 75/W, 180/W Screening**
RESIN CONTENT:	39-41 wt%	COMP: DENSITY:	1.53 g/cm³	
FIBER VOLUME:	51-52 %	VOID CONTENT:	0.0%	
PLY THICKNESS:	0.0080-0.0083 in.			
TEST METHOD: SACMA SRM 1-88		MODULUS CALCULATION:		
NORMALIZED BY:	Specimen thickness and batch fiber volume to 57%			

		75		180			
Temperature (°F)		75		180			
Moisture Content (%)		wet		wet			
Equilibrium at T, RH		(1)		(1)			
Source Code		20		20			
		Normalized	Measured	Normalized	Measured	Normalized	Measured
F_2^{cu} (ksi)	Mean	99.2	88.5	84.0	74.9		
	Minimum	80.9	72.2	74.2	66.1		
	Maximum	112	100	88.8	79.2		
	C.V.(%)	12.1	12.1	7.0	6.9		
	B-value	(2)	(2)	(2)	(2)		
	Distribution	Normal	Normal	Normal	Normal		
	C_1	99.2	88.5	84.0	74.9		
	C_2	12.0	10.7	5.8	5.20		
	No. Specimens	5		5			
	No. Batches	1		1			
	Approval Class	Screening		Screening			
E_2^c (Msi)	Mean	9.30	8.56	8.96	8.25		
	Minimum	8.74	7.98	8.69	8.03		
	Maximum	9.56	8.78	9.31	8.43		
	C.V.(%)	3.5	3.9	2.9	2.0		
	No. Specimens	5		5			
	No. Batches	1		1			
	Approval Class	Screening		Screening			
ν_{21}^c	Mean						
	No. Specimens						
	No. Batches						
	Approval Class						
ε_2^{cu} (με)	Mean		10200				
	Minimum		9910				
	Maximum		10900				
	C.V.(%)		3.7				
	B-value		(2)				
	Distribution		Normal				
	C_1		10200				
	C_2		381				
	No. Specimens	5					
	No. Batches	1					
	Approval Class	Screening					

(1) Conditioned at 160°F, 85% relative humidity for 14 days.

(2) B-values are presented only for fully approved data.

* DATA WERE SUBMITTED BEFORE THE ESTABLISHMENT OF DATA DOCUMENTATION REQUIREMENTS (JUNE 1989). ALL DOCUMENTATION PRESENTLY REQUIRED WAS NOT SUPPLIED FOR THIS MATERIAL.

Table 4.2.10(g)
C/Ep 195-PW
HITEX 33/E7K8
Compression, 2-axis
$[90_f]_{12}$
75/A, -65/A, 180/A
Screening

MATERIAL:	HITEX 33 6k/E7K8 plain weave		
RESIN CONTENT:	45 wt%	COMP: DENSITY:	1.51 g/cm^3
FIBER VOLUME:	47 %	VOID CONTENT:	0.0%
PLY THICKNESS:	0.0080-0.0097 in.		
TEST METHOD: SACMA SRM 1-88		MODULUS CALCULATION:	
NORMALIZED BY:	Specimen thickness and batch fiber volume to 57%		

		75		-65		180	
Temperature (°F) / Moisture Content (%) / Equilibrium at T, RH / Source Code		75, ambient, 20		-65, ambient, 20		180, ambient, 20	
		Normalized	Measured	Normalized	Measured	Normalized	Measured
F_2^{cu} (ksi)	Mean	132	110	147	122	132	110
	Minimum	114	97.9	138	115	128	106
	Maximum	145	118	161	127	146	117
	C.V.(%)	5.7	5.3	6.0	4.1	5.9	4.7
	B-value	(1)	(1)	(1)	(1)	(1)	(1)
	Distribution	Weibull	Weibull	Normal	Normal	Normal	Normal
	C_1	136	113	147	122	132	110
	C_2	21.6	23.4	8.78	5.02	7.73	5.12
	No. Specimens	20		5		5	
	No. Batches	1		1		1	
	Approval Class	Screening		Screening		Screening	
E_2^c (Msi)	Mean	8.74	7.27	9.09	7.54	9.11	7.57
	Minimum	8.41	6.70	8.12	7.07	8.61	7.41
	Maximum	9.20	8.06	10.1	7.90	9.49	7.71
	C.V.(%)	2.6	4.7	9.1	5.6	3.8	1.5
	No. Specimens	20		5		5	
	No. Batches	1		1		1	
	Approval Class	Screening		Screening		Screening	
ν_{21}^c	Mean						
	No. Specimens						
	No. Batches						
	Approval Class						
ε_2^{cu} (με)	Mean		14100				
	Minimum		13400				
	Maximum		14700				
	C.V.(%)		2.6				
	B-value		(1)				
	Distribution		Weibull				
	C_1		14300				
	C_2		46.4				
	No. Specimens	20					
	No. Batches	1					
	Approval Class	Screening					

(1) B-values are presented only for fully approved data.

* DATA WERE SUBMITTED BEFORE THE ESTABLISHMENT OF DATA DOCUMENTATION REQUIREMENTS (JUNE 1989). ALL DOCUMENTATION PRESENTLY REQUIRED WAS NOT SUPPLIED FOR THIS MATERIAL.

Table 4.2.10(h)
C/Ep 195-PW
HITEX 33/E7K8
Compression, 2-axis
$[90_f]_{12}$
75/W, 180/W
Screening

MATERIAL: HITEX 33 6k/E7K8 plain weave

RESIN CONTENT: 45 wt% COMP: DENSITY: 1.51 g/cm³
FIBER VOLUME: 47 % VOID CONTENT: 0.0%
PLY THICKNESS: 0.0080-0.0097 in.

TEST METHOD: SACMA SRM 1-88 MODULUS CALCULATION:

NORMALIZED BY: Specimen thickness and batch fiber volume to 57%

Temperature (°F)		75		180			
Moisture Content (%)		wet		wet			
Equilibrium at T, RH		(1)		(1)			
Source Code		20		20			
		Normalized	Measured	Normalized	Measured	Normalized	Measured
F_2^{cu} (ksi)	Mean	117	97.4	61.1	50.8		
	Minimum	107	88.4	52.2	44.1		
	Maximum	132	105	66.4	57.2		
	C.V.(%)	9.1	6.9	9.9	9.9		
	B-value	(2)	(2)	(2)	(2)		
	Distribution	Normal	Normal	Normal	Normal		
	C_1	117	97.4	61.1	50.8		
	C_2	10.6	6.74	6.04	5.01		
	No. Specimens	5		5			
	No. Batches	1		1			
	Approval Class	Screening		Screening			
E_2^c (Msi)	Mean	8.99	7.48	9.26	7.71		
	Minimum	8.48	7.08	8.76	7.32		
	Maximum	9.54	7.8	9.69	8.39		
	C.V.(%)	4.5	4.0	4.0	6.2		
	No. Specimens	5		5			
	No. Batches	1		1			
	Approval Class	Screening		Screening			
ν_{21}^c	Mean						
	No. Specimens						
	No. Batches						
	Approval Class						
ε_2^{cu} (με)	Mean		13500				
	Minimum		12700				
	Maximum		14200				
	C.V.(%)		4.2				
	B-value		(2)				
	Distribution		Normal				
	C_1		13500				
	C_2		564				
	No. Specimens	5					
	No. Batches	1					
	Approval Class	Screening					

(1) Conditioned at 160°F, 85% relative humidity for 14 days.
(2) B-values are presented only for fully approved data.

* DATA WERE SUBMITTED BEFORE THE ESTABLISHMENT OF DATA DOCUMENTATION REQUIREMENTS (JUNE 1989). ALL DOCUMENTATION PRESENTLY REQUIRED WAS NOT SUPPLIED FOR THIS MATERIAL.

MATERIAL:	HITEX 33 6k/E7K8 Plain weave			**Table 4.2.10(i)** **C/Ep 195-PW** **HITEX 33/E7K8** **SBS, 31-plane** **$[90_f]_6$** **75/A, -65/A, 180/A** **Screening**
RESIN CONTENT:	44 wt%	COMP: DENSITY:	1.51 g/cm^3	
FIBER VOLUME:	48 %	VOID CONTENT:	0.18%	
PLY THICKNESS:	0.0077-0.0093 in.			
TEST METHOD:		MODULUS CALCULATION:		
ASTM D2344-76				
NORMALIZED BY:	Not normalized			

Temperature (°F) Moisture Content (%) Equilibrium at T, RH Source Code		75.0 ambient 20	-65.0 ambient 20	75.0 wet (1) 20	180.0 wet (1) 20	
	Mean	8.67	8.83	9.40	8.35	
	Minimum	7.77	8.14	9.20	7.83	
	Maximum	9.40	9.37	9.73	8.80	
	C.V.(%)	5.0	6.3	2.1	4.5	
	B-value	(2)	(2)	(2)	(2)	
	Distribution	Weibull	Normal	Normal	Normal	
F_{31}^{sbs}	C_1	8.86	8.83	9.40	8.35	
(ksi)	C_2	23.6	0.554	0.202	0.379	
	No. Specimens	20	5	5	5	
	No. Batches	1	1	1	1	
	Approval Class	Screening	Screening	Screening	Screening	

(1) Conditioned at 160°F, 85% relative humidity for 14 days.
(2) B-values are presented only for fully approved data.

4.2.11 AS4 3k/E7K8 plain weave fabric data set description

Material Description:

Material: AS4-3k/E7K8

Form: Plain weave fabric, areal weight of 195 g/m^2, typical cured resin content of 37-48%, typical cured ply thickness of 0.0087 inches.

Processing: Good drape. Autoclave cure; 290°F, 85 psi for 2 hours. Low exotherm profile for processing of thick parts.

General Supplier Information:

Fiber: AS4 fibers are continuous carbon filaments made from PAN precursor, surface treated to improve handling characteristics and structural properties. Filament count is 3000 filaments/tow. Typical tensile modulus is 34 x 10^6 psi. Typical tensile strength is 550,000 psi.

Matrix: E7K8 is a medium flow, low exotherm epoxy resin. Good tack; up to 20 days out-time at ambient temperature.

Maximum Short Term Service Temperature: >300°F (dry), >190°F (wet)

Typical applications: Primary and secondary structural applications on commercial and military aircraft, jet engine applications such as stationary airfoils and thrust reverser blocker doors.

4.2.11 AS4 3k/E7K8 plain weave*

MATERIAL:	AS4 3k/E7K8 plain weave		C/Ep AS4/E7K8 Summary
PREPREG:	U.S. Polymeric AS4/E7K8 Plain weave		
FIBER:	Hercules AS4 3k	MATRIX:	U.S. Polymeric E7K8
T_g(dry):	T_g(wet):	T_g METHOD:	
PROCESSING:	Autoclave: 290°F, 2 hours, 85 psig		

* DATA WERE SUBMITTED BEFORE THE ESTABLISHMENT OF DATA DOCUMENTATION REQUIREMENTS (JUNE 1989). ALL DOCUMENTATION PRESENTLY REQUIRED WERE NOT SUPPLIED FOR THIS MATERIAL.

Date of fiber manufacture		Date of testing	
Date of resin manufacture		Date of data submittal	1/88, 6/90
Date of prepreg manufacture	2/86 - 7/89	Date of analysis	1/93
Date of composite manufacture			

LAMINA PROPERTY SUMMARY

	75°F/A							
Tension, 1-axis								
Tension, 2-axis								
Tension, 3-axis								
Compression, 1-axis	II-I							
Compression, 2-axis								
Compression, 3-axis								
Shear, 12-plane								
Shear, 23-plane								
SB Strength, 31-plane	S---							

Classes of data: F - Fully approved, I - Interim, S - Screening in Strength/Modulus/Poisson's ratio/Strain-to-failure order.

* DATA WERE SUBMITTED BEFORE THE ESTABLISHMENT OF DATA DOCUMENTATION REQUIREMENTS (JUNE 1989). ALL DOCUMENTATION PRESENTLY REQUIRED WERE NOT SUPPLIED FOR THIS MATERIAL.

		Nominal	As Submitted	Test Method
Fiber Density	(g/cm^3)	1.77		
Resin Density	(g/cm^3)	1.28		
Composite Density	(g/cm^3)	1.56		
Fiber Areal Weight	(g/m^2)	195		
Ply Thickness	(in)	0.0087	0.0074 - 0.0088	

LAMINATE PROPERTY SUMMARY

Classes of data: F - Fully approved, I - Interim, S - Screening in Strength/Modulus/Poisson's ratio/Strain-to-failure order.

* DATA WERE SUBMITTED BEFORE THE ESTABLISHMENT OF DATA DOCUMENTATION REQUIREMENTS (JUNE 1989). ALL DOCUMENTATION PRESENTLY REQUIRED WAS NOT SUPPLIED FOR THIS MATERIAL.

MATERIAL:	AS4 3k/E7K8 Plain Weave			**Table 4.2.11(a)** **C/Ep 195-PW** **AS4/E7K8** **Compression, 1-axis** **$[0_f]_{12}$** **75/A** **Interim**
RESIN CONTENT:	37-48 wt%	COMP: DENSITY:	1.52-1.54 g/cm^3	
FIBER VOLUME:	48-55 %	VOID CONTENT:	0.0-1.9%	
PLY THICKNESS:	0.0074-0.0085 in.			
TEST METHOD: SACMA SRM 1-88		MODULUS CALCULATION:		
NORMALIZED BY:	Specimen thickness and batch fiber volume to 57%			

Temperature (°F) Moisture Content (%) Equilibrium at T, RH Source Code		75 ambient 20,27					
		Normalized	Measured	Normalized	Measured	Normalized	Measured
F_1^{cu} (ksi)	Mean	111	988				
	Minimum	64.4	58.0				
	Maximum	138	122				
	C.V.(%)	11.7	11.3				
	B-value	(1)	(1)				
	Distribution	ANOVA	ANOVA				
	C_1	13.3	11.3				
	C_2	1.81	1.80				
	No. Specimens	206					
	No. Batches	18					
	Approval Class	Interim					
E_1^c (Msi)	Mean	9.02	8.07				
	Minimum	7.87	7.07				
	Maximum	10.5	9.04				
	C.V.(%)	5.24	4.28				
	No. Specimens	210					
	No. Batches	18					
	Approval Class	Interim					
ν_{12}^c	Mean						
	No. Specimens						
	No. Batches						
	Approval Class						
ε_1^{cu} (με)	Mean		11600				
	Minimum		8820				
	Maximum		15000				
	C.V.(%)		14.5				
	B-value		(1)				
	Distribution		ANOVA				
	C_1		1730				
	C_2		1.97				
	No. Specimens	190					
	No. Batches	17					
	Approval Class	Interim					

(1) B-values are presented only for fully approved data.

* DATA WERE SUBMITTED BEFORE THE ESTABLISHMENT OF DATA DOCUMENTATION REQUIREMENTS (JUNE 1989). ALL DOCUMENTATION PRESENTLY REQUIRED WAS NOT SUPPLIED FOR THIS MATERIAL.

MATERIAL:	AS4/E7K8 Plain weave			**Table 4.2.11(b)** **C/Ep 195-PW** **AS4/E7K8** **SBS, 31-plane** **$[0_f]_{12}$** **75/A** **Screening**
RESIN CONTENT:	38-48 wt%	COMP: DENSITY:	1.52-1.54 g/cm^3	
FIBER VOLUME:	48-55 %	VOID CONTENT:	0.0-1.9%	
PLY THICKNESS:	0.0074-0.0085 in.			
TEST METHOD:		MODULUS CALCULATION:		
ASTM D2344-84				
NORMALIZED BY:	Not normalized			

Temperature (°F) Moisture Content (%) Equilibrium at T, RH Source Code		75 ambient 20,27					
	Mean	9.68					
	Minimum	7.53					
	Maximum	14.2					
	C.V.(%)	12.0					
	B-value	(1)					
F_{31}^{sbs}	Distribution	ANOVA					
(ksi)	C_1	1.20					
	C_2	1.95					
	No. Specimens	170					
	No. Batches	16					
	Approval Class	Screening					

(1) B-values are presented only for fully approved data.

4.2.12 AS4/3501-6 (bleed) unitape data set description

Material Description:

Material: AS4/3501-6

Form: Unidirectional tape, fiber areal weight of 145 g/m^2, typical cured resin content of 28%-34%, typical cured ply thickness of 0.0041-0.0062 inches.

Processing: Autoclave cure; 240°F, 85 psi for 1 hour, 350°F, 100 psi for 2 hours, bleed system.

General Supplier Information:

Fiber: AS4 fibers are continuous carbon filaments made from PAN precursor, surface treated to improve handling characteristics and structural properties. Typical tensile modulus is 34 x 10^6 psi. Typical tensile strength is 550,000 psi.

Matrix: 3501-6 is an amine-cured epoxy resin. It will retain light tack for a minimum of 10 days at room temperature.

Maximum Short Term Service Temperature: 300°F (dry), 180°F (wet)

Typical applications: General purpose structural applications.

4.2.12 AS4/3501-6 (bleed) unidirectional tape*

C/Ep
AS4/3501-6
Summary

MATERIAL:	AS4/3501-6 unidirectional tape		
PREPREG:	Hercules AS4/3501-6 unidirectional tape		
FIBER:	Hercules AS4	MATRIX:	Hercules 3501-6
T_g(dry):	390°F T_g(wet):	T_g METHOD:	TMA
PROCESSING:	Autoclave cure: 240 ± 10°F, 60 minutes, 85 psig; 350 ± 10°F, 120 ± 10 minutes, 100 ± 10 psig, bleed		

* ALL DOCUMENTATION PRESENTLY REQUIRED WERE NOT SUPPLIED FOR THIS MATERIAL.

Date of fiber manufacture		Date of testing	
Date of resin manufacture		Date of data submittal	6/90
Date of prepreg manufacture		Date of analysis	1/93
Date of composite manufacture			

LAMINA PROPERTY SUMMARY

	75°F/A		200°F/A		75°F/W	200°F/W		
Tension, 1-axis	II--							
Tension, 2-axis	SS--							
Tension, 3-axis								
Compression, 1-axis	IS--		II--		SS--	SS--		
Compression, 2-axis								
Compression, 3-axis								
Shear, 12-plane								
Shear, 23-plane								
SB Strength, 31-plane	S---		S---		S---	S---		

Classes of data: F - Fully approved, I - Interim, S - Screening in Strength/Modulus/Poisson's ratio/Strain-to-failure order.

** ALL DOCUMENTATION PRESENTLY REQUIRED WERE NOT SUPPLIED FOR THIS MATERIAL.

		Nominal	As Submitted	Test Method
Fiber Density	(g/cm^3)	1.8		
Resin Density	(g/cm^3)	1.27		
Composite Density	(g/cm^3)	1.59		
Fiber Areal Weight	(g/m^2)	145		
Ply Thickness	(in)		0.0041 - 0.0059	

LAMINATE PROPERTY SUMMARY

Classes of data: F - Fully approved, I - Interim, S - Screening in Strength/Modulus/Poisson's ratio/Strain-to-failure order.

* ALL DOCUMENTATION PRESENTLY REQUIRED WAS NOT SUPPLIED FOR THIS MATERIAL.

MATERIAL:	AS4/3501-6 (bleed) unidirectional tape		
RESIN CONTENT:	34-38 wt%	COMP: DENSITY:	1.56 g/cm^3
FIBER VOLUME:	58-65 %	VOID CONTENT:	
PLY THICKNESS:	0.0048-0.0057 in.		
TEST METHOD: ASTM D3039-76		MODULUS CALCULATION:	
NORMALIZED BY:	Specimen thickness and batch fiber volume to 60%		

Table 4.2.12(a)
C/Ep 145-UT
AS4/3501-6
Tension, 1-axis
$[0]_8$
75/A
Interim

Temperature (°F)		75					
Moisture Content (%)		ambient					
Equilibrium at T, RH							
Source Code		26					
		Normalized	Measured	Normalized	Measured	Normalized	Measured
F_1^{tu} (ksi)	Mean	291	295				
	Minimum	263	271				
	Maximum	326	326				
	C.V.(%)	6.09	5.05				
	B-value	(1)	(1)				
	Distribution	Weibull	Weibull				
	C_1	300	302				
	C_2	18.4	20.3				
	No. Specimens	21					
	No. Batches	7					
	Approval Class	Interim					
E_1^t (Msi)	Mean	19.6	19.9				
	Minimum	18.0	18.3				
	Maximum	21.1	22.6				
	C.V.(%)	3.73	6.48				
	No. Specimens	21					
	No. Batches	7					
	Approval Class	Interim					
ν_{12}^t	Mean						
	No. Specimens						
	No. Batches						
	Approval Class						
ε_1^{tu} (με)	Mean						
	Minimum						
	Maximum						
	C.V.(%)						
	B-value						
	Distribution						
	C_1						
	C_2						
	No. Specimens						
	No. Batches						
	Approval Class						

(1) B-values are presented only for fully approved data.

** ALL DOCUMENTATION PRESENTLY REQUIRED WERE NOT SUPPLIED FOR THIS MATERIAL.

MATERIAL:	AS4/3501-6 (bleed) unidirectional tape			**Table 4.2.12(b)** **C/Ep 145-UT** **AS4/3501-6** **Tension, 2-axis** **$[90]_8$** **75/A** **Screening**
RESIN CONTENT:	28-29 wt%	COMP: DENSITY:	1.60-1.61 g/cm³	
FIBER VOLUME:	63-64 %	VOID CONTENT:		
PLY THICKNESS:	0.0048-0.0057 in.			
TEST METHOD: ASTM D3039-76		MODULUS CALCULATION:		
NORMALIZED BY:	Not normalized			

Temperature (°F) Moisture Content (%) Equilibrium at T, RH Source Code		75 ambient 26					
F_2^{tu} (ksi)	Mean	7.78					
	Minimum	7.00					
	Maximum	9.50					
	C.V.(%)	12.1					
	B-value	(1)					
	Distribution	Normal					
	C_1	7.78					
	C_2	0.941					
	No. Specimens	6					
	No. Batches	2					
	Approval Class	Screening					
E_2^t (Msi)	Mean	1.48					
	Minimum	1.40					
	Maximum	1.50					
	C.V.(%)	2.75					
	No. Specimens	6					
	No. Batches	2					
	Approval Class	Screening					
ν_{12}^t	Mean						
	No. Specimens						
	No. Batches						
	Approval Class						
ε_2^{tu} (με)	Mean						
	Minimum						
	Maximum						
	C.V.(%)						
	B-value						
	Distribution						
	C_1						
	C_2						
	No. Specimens						
	No. Batches						
	Approval Class						

(1) B-values are presented only for fully approved data.

** ALL DOCUMENTATION PRESENTLY REQUIRED WAS NOT SUPPLIED FOR THIS MATERIAL.

MATERIAL:	AS4/3501-6 (bleed) unidirectional tape			**Table 4.2.12(c) C/Ep 145-UT AS4/3501-6 Compression, 1-axis $[0]_s$ 75/A, 200/A, 75/W Interim, Screening**
RESIN CONTENT:	28-34 wt%	COMP: DENSITY:	1.58-1.61 g/cm^3	
FIBER VOLUME:	58-65 %	VOID CONTENT:		
PLY THICKNESS:	0.0041-0.0055 in.			
TEST METHOD: SACMA SRM 1-88		MODULUS CALCULATION:		
NORMALIZED BY:	Specimen thickness and batch fiber volume to 60%			

Temperature (°F) Moisture Content (%) Equilibrium at T, RH Source Code		75 ambient 26		200 ambient 26		75 wet (1) 26	
		Normalized	Measured	Normalized	Measured	Normalized	Measured
F_1^{cu} (ksi)	Mean	210	214	196	201	202	213
	Minimum	144	161	148	165	165	179
	Maximum	269	260	242	237	274	266
	C.V.(%)	16.0	13.5	13.6	10.7	18.0	14.1
	B-value	(2)	(2)	(2)	(2)	(2)	(2)
	Distribution	ANOVA	ANOVA	ANOVA	ANOVA	Weibull	Weibull
	C_1	34.7	27.7	27.7	22.3	217	226
	C_2	2.39	2.52	2.52	2.35	5.89	7.82
	No. Specimens	26		27		10	
	No. Batches	7		7		2	
	Approval Class	Interim		Interim		Screening	
E_1^c (Msi)	Mean	17.8	18.8	16.3	17.4	17.4	18.5
	Minimum	15.1	16.4	13.0	14.3	15.6	17.1
	Maximum	20.3	20.0	18.7	19.6	20.3	20.6
	C.V.(%)	7.50	7.18	10.7	10.1	9.14	5.84
	No. Specimens	14		15		10	
	No. Batches	3		3		2	
	Approval Class	Screening		Interim		Screening	
ν_{12}^c	Mean						
	No. Specimens						
	No. Batches						
	Approval Class						
ε_1^{cu} ($\mu\varepsilon$)	Mean						
	Minimum						
	Maximum						
	C.V.(%)						
	B-value						
	Distribution						
	C_1						
	C_2						
	No. Specimens						
	No. Batches						
	Approval Class						

(1) Conditioned at 140°F, 95% relative humidity for 30 days.
(2) B-values are presented only for fully approved data.

** ALL DOCUMENTATION PRESENTLY REQUIRED WAS NOT SUPPLIED FOR THIS MATERIAL.

MATERIAL:	AS4/3501-6 (bleed) unidirectional tape			Table 4.2.12(d) C/Ep 145-UT AS4/3501-6 Compression, 1-axis $[0]_s$ 200/W Screening
RESIN CONTENT:	28-34 wt%	COMP: DENSITY:	1.58-1.61 g/cm^3	
FIBER VOLUME:	58-65 %	VOID CONTENT:		
PLY THICKNESS:	0.0041-0.0055 in.			
TEST METHOD: SACMA SRM 1-88		MODULUS CALCULATION:		
NORMALIZED BY:	Specimen thickness and batch fiber volume to 60%			

Temperature (°F) Moisture Content (%) Equilibrium at T, RH Source Code		200 wet 26					
		Normalized	Measured	Normalized	Measured	Normalized	Measured
	Mean	169	179				
	Minimum	100	107				
	Maximum	212	226				
	C.V.(%)	22.2	22.9				
	B-value	(1)	(1)				
F_1^{cu}	Distribution	ANOVA	ANOVA				
(ksi)	C_1	41.7	46.6				
	C_2	5.28	5.72				
	No. Specimens	10					
	No. Batches	3					
	Approval Class	Screening					
	Mean	17.7	18.7				
	Minimum	12.1	13.4				
	Maximum	27.2	25.5				
E_1^c	C.V.(%)	21.6	15.8				
(Msi)	No. Specimens	10					
	No. Batches	3					
	Approval Class	Screening					
	Mean						
	No. Specimens						
ν_{12}^c	No. Batches						
	Approval Class						
	Mean						
	Minimum						
	Maximum						
	C.V.(%)						
	B-value						
ε_1^{cu}	Distribution						
($\mu\varepsilon$)	C_1						
	C_2						
	No. Specimens						
	No. Batches						
	Approval Class						

(1) B-values are presented only for fully approved data.

** ALL DOCUMENTATION PRESENTLY REQUIRED WERE NOT SUPPLIED FOR THIS MATERIAL.

MATERIAL: AS4/3501-6 (bleed) unidirectional tape RESIN CONTENT: 30-34 wt% COMP: DENSITY: 1.58-1.60 g/cm^3 FIBER VOLUME: 58-62 % VOID CONTENT: PLY THICKNESS: 0.0047-0.0055 in. TEST METHOD: ASTM D2344 MODULUS CALCULATION: NORMALIZED BY: Not normalized	**Table 4.2.12(e)** **C/Ep 145-UT** **AS4/3501-6** **SBS, 31-plane** **$[0]_8$** **75/A, 200/A, 75/W, 200/W** **Screening**

		75	200	75	200	
Temperature (°F)		75	200	75	200	
Moisture Content (%)		ambient	ambient	wet	wet	
Equilibrium at T, RH				(1)	(1)	
Source Code		26	26	26	26	
	Mean	17.3	13.0	13.9	9.0	
	Minimum	14.1	11.1	13.1	8.3	
	Maximum	19.4	14.9	15.5	10.1	
	C.V.(%)	7.63	11.6	6.13	6.4	
	B-value	(2)	(2)	(2)	(2)	
	Distribution	ANOVA	ANOVA	Normal	Normal	
F_{31}^{sbs}	C_1	1.38	1.59	13.9	9.0	
(ksi)	C_2	2.62	2.77	0.852	0.58	
	No. Specimens	21	21	6	9	
	No. Batches	7	7	2	3	
	Approval Class	Screening	Screening	Screening	Screening	

(1) Conditioned at 140°F, 95% relative humidity for 30 days.
(2) B-values are presented only for fully approved data.

MATERIAL:	AS4/3501-6 (bleed) unidirectional tape			**Table 4.2.12(f)** **C/Ep 145-UT** **AS4/3501-6** **Tension, x-axis** **[0/45/90/-45]$_s$** **75/A** **Screening**
RESIN CONTENT:	29-32 wt%	COMP: DENSITY:	1.59-1.60 g/cm^3	
FIBER VOLUME:	60-63 %	VOID CONTENT:		
PLY THICKNESS:	0.0055-0.0062 in.			
TEST METHOD: ASTM D3039-76		MODULUS CALCULATION: Linear portion of curve		
NORMALIZED BY:	NA			

Temperature (°F)		75					
Moisture Content (%)		ambient					
Equilibrium at T, RH							
Source Code		26					
	Mean	95.8					
	Minimum	90.6					
	Maximum	106					
	C.V.(%)	5.95					
	B-value	(1)					
F_x^{tu}	Distribution	ANOVA					
(ksi)	C_1	29.9					
	C_2	14.5					
	No. Specimens	6					
	No. Batches	2					
	Approval Class	Screening					
	Mean	7.22					
	Minimum	6.60					
	Maximum	8.40					
E_x^t	C.V.(%)	9.74					
(Msi)	No. Specimens	6					
	No. Batches	2					
	Approval Class	Screening					
	Mean						
	No. Specimens						
ν_{xy}^t	No. Batches						
	Approval Class						
	Mean						
	Minimum						
	Maximum						
	C.V.(%)						
	B-value						
ε_x^{tu}	Distribution						
($\mu\varepsilon$)	C_1						
	C_2						
	No. Specimens						
	No. Batches						
	Approval Class						

(1) B-values are presented only for fully approved data.

MATERIAL:	AS4/3501-6 (bleed) unidirectional tape			Table 4.2.12(g) C/Ep 145-UT AS4/3501-6 Open Hole Tension, x-axis $[0/45/90/-45]_s$ 75/A Screening
RESIN CONTENT:	29-32 wt%	COMP: DENSITY:	1.59-1.60 g/cm^3	
FIBER VOLUME:	60-63 %	VOID CONTENT:		
PLY THICKNESS:	0.0055-0.0057 in.			
TEST METHOD: SACMA SRM 5-88 (1)		MODULUS CALCULATION:		
NORMALIZED BY:	NA			

Temperature (°F)		75				
Moisture Content (%)		ambient				
Equilibrium at T, RH						
Source Code		26				
	Mean	62.0				
	Minimum	59.2				
	Maximum	65.1				
	C.V.(%)	3.13				
	B-value	(2)				
	Distribution	Normal				
F_x^{oht}	C_1	62.0				
(ksi)	C_2	1.94				
	No. Specimens	6				
	No. Batches	2				
	Approval Class	Screening				
	Mean					
	Minimum					
	Maximum					
E_x^{oht}	C.V.(%)					
(Msi)	No. Specimens					
	No. Batches					
	Approval Class					
	Mean					
	Minimum					
	Maximum					
	C.V.(%)					
	B-value					
ε_x^{oht}	Distribution					
($\mu\varepsilon$)	C_1					
	C_2					
	No. Specimens					
	No. Batches					
	Approval Class					

(1) Note SACMA SRM 5-88 uses a $[+45/0/-45/90]_{2S}$ layup.
(2) B-values are presented only for fully approved data.

4.2.13 AS4/3501-6 (no bleed) unitape data set description

Material Description:

Material: AS4/3501-6

Form: Unidirectional tape, fiber areal weight of 145 g/m^2, typical cured resin content of 36%-39%, typical cured ply thickness of 0.0055-0.0063 inches.

Processing: Autoclave cure; 240°F, 85 psi for 1 hour; 350°F, 100 psi for 2 hours, no bleed.

General Supplier Information:

Fiber: AS4 fibers are continuous carbon filaments made from PAN precursor, surface treated to improve handling characteristics and structural properties. Typical tensile modulus is 34 x 10^6 psi. Typical tensile strength is 550,000 psi.

Matrix: 3501-6 is an amine-cured epoxy resin. It will retain light tack for a minimum of 10 days at room temperature.

Maximum Short Term Service Temperature: 300°F (dry), 180°F (wet)

Typical applications: General purpose structural applications.

4.2.13 AS4/3501-6 (no bleed) unidirectional tape*

MATERIAL:	AS4/3501-6 unidirectional tape			**C/Ep AS4/3501-6 Summary**
PREPREG:	Hercules AS4/3501-6 unidirectional tape			
FIBER:	Hercules AS4, unsized	MATRIX:	Hercules 3501-6	
T_g(dry):	390°F T_g(wet):	T_g METHOD:	TMA	
PROCESSING:	Autoclave cure: 240 ± 10°F, 60 minutes, 85 psig; 350 ± 10°F, 120 ± 10 minutes, 100 ± 10 psig, no bleed			

** ALL DOCUMENTATION PRESENTLY REQUIRED WERE NOT SUPPLIED FOR THIS MATERIAL.

Date of fiber manufacture	~12/82-8/89	Date of testing	~6/83 - ~4/91
Date of resin manufacture		Date of data submittal	6/90
Date of prepreg manufacture	1/83 - 11/89	Date of analysis	1/93
Date of composite manufacture			

LAMINA PROPERTY SUMMARY

	75°F/A		-65°F/A	200°F/A		200°F/W		
Tension, 1-axis	II--		SS--	SS--				
Tension, 2-axis	SS--							
Tension, 3-axis								
Compression, 1-axis	II--			I---		II--		
Compression, 2-axis								
Compression, 3-axis								
Shear, 12-plane								
Shear, 23-plane								
SB Strength, 31-plane	S---			S---				

Classes of data: F - Fully approved, I - Interim, S - Screening in Strength/Modulus/Poisson's ratio/Strain-to-failure order.

** ALL DOCUMENTATION PRESENTLY REQUIRED WERE NOT SUPPLIED FOR THIS MATERIAL.

		Nominal	As Submitted	Test Method
Fiber Density	(g/cm^3)	1.8		
Resin Density	(g/cm^3)	1.27		
Composite Density	(g/cm^3)	1.59		
Fiber Areal Weight	(g/m^2)	145	142 - 149	
Ply Thickness	(in)		0.0055 - 0.0063	

LAMINATE PROPERTY SUMMARY

	75°F/A							
[0/45/90/-45] family								
Tension, x-axis	S---							
OHT, x-axis	S---							

Classes of data: F - Fully approved, I - Interim, S - Screening in Strength/Modulus/Poisson's ratio/Strain-to-failure order.

** ALL DOCUMENTATION PRESENTLY REQUIRED WERE NOT SUPPLIED FOR THIS MATERIAL.

MATERIAL:	AS4/3501-6 (no bleed)			**Table 4.2.13(a) C/Ep 145-UT AS4/3501-6 Tension, 1-axis $[0]_8$ 75/A, -65/A, 200/A Interim, Screening**
RESIN CONTENT:	36-39 wt%	COMP: DENSITY:	1.55-1.57 g/cm^3	
FIBER VOLUME:	52-56 %	VOID CONTENT:		
PLY THICKNESS:	0.0055-0.0060 in.			
TEST METHOD:		MODULUS CALCULATION:		
ASTM D3039-76		Initial tangent		
NORMALIZED BY:	Specimen thickness and batch fiber volume to 60%			

		75		-65		200	
Temperature (°F) Moisture Content (%) Equilibrium at T, RH Source Code		75 ambient 26		-65 ambient 26		200 ambient 26	
		Normalized	Measured	Normalized	Measured	Normalized	Measured
F_1^{tu} (ksi)	Mean	290	262	261	237	315	286
	Minimum	262	235	207	187	278	247
	Maximum	322	286	300	274	330	297
	C.V.(%)	5.62	5.38	12.4	12.8	4.89	5.59
	B-value	(1)	(1)	(1)	(1)	(1)	(1)
	Distribution	ANOVA	ANOVA	ANOVA	ANOVA	Nonpara.	Nonpara.
	C_1	16.5	14.3	34.9	33.1	6	6
	C_2	2.05	2.01	4.69	5.05	2.25	2.25
	No. Specimens	30		9		9	
	No. Batches	10		3		3	
	Approval Class	Interim		Screening		Screening	
E_1^t (Msi)	Mean	18.9	17.1	21.1	19.2	20.8	18.9
	Minimum	17.0	15.5	19.7	17.7	19.4	17.4
	Maximum	20.3	17.9	22.3	21.4	22.0	20.2
	C.V.(%)	4.0	3.20	4.60	5.78	4.72	4.70
	No. Specimens	30		9		9	
	No. Batches	10		3		3	
	Approval Class	Interim		Screening		Screening	
ν_{12}^t	Mean						
	No. Specimens						
	No. Batches						
	Approval Class						
ε_1^{tu} ($\mu\varepsilon$)	Mean						
	Minimum						
	Maximum						
	C.V.(%)						
	B-value						
	Distribution						
	C_1						
	C_2						
	No. Specimens						
	No. Batches						
	Approval Class						

(1) B-values are presented only for fully approved data.

** ALL DOCUMENTATION PRESENTLY REQUIRED WERE NOT SUPPLIED FOR THIS MATERIAL.

MATERIAL:	AS4/3501-6 (no bleed)		
RESIN CONTENT:	37 wt%	COMP: DENSITY:	1.56 g/cm³
FIBER VOLUME:	54-55 %	VOID CONTENT:	
PLY THICKNESS:	0.0060-0.0062 in.		
TEST METHOD: ASTM D3039-76		MODULUS CALCULATION: Initial tangent	
NORMALIZED BY:	Not normalized		

Table 4.2.13(b)
C/Ep 145-UT
AS4/3501-6
Tension, 2-axis
$[90]_8$
75/A
Screening

Temperature (°F)		75					
Moisture Content (%)		ambient					
Equilibrium at T, RH							
Source Code		26					
F_2^{tu} (ksi)	Mean	8.0					
	Minimum	6.8					
	Maximum	9.3					
	C.V.(%)	10					
	B-value	(1)					
	Distribution	Normal					
	C_1	8.0					
	C_2	0.81					
	No. Specimens	9					
	No. Batches	3					
	Approval Class	Screening					
E_2^t (Msi)	Mean	1.2					
	Minimum	1.1					
	Maximum	1.4					
	C.V.(%)	8.9					
	No. Specimens	9					
	No. Batches	3					
	Approval Class	Screening					
ν_{21}^t	Mean						
	No. Specimens						
	No. Batches						
	Approval Class						
ε_2^{tu} (με)	Mean						
	Minimum						
	Maximum						
	C.V.(%)						
	B-value						
	Distribution						
	C_1						
	C_2						
	No. Specimens						
	No. Batches						
	Approval Class						

(1) B-values are presented only for fully approved data.

** ALL DOCUMENTATION PRESENTLY REQUIRED WERE NOT SUPPLIED FOR THIS MATERIAL.

MATERIAL:	AS4/3501-6 (no bleed)			**Table 4.2.13(c) C/Ep 145-UT AS4/3501-6 Compression, 1-axis $[0]_8$ 75/A, 200/A, 20/W Interim**
RESIN CONTENT:	36-39 wt%	COMP: DENSITY:	1.55-1.57 g/cm^3	
FIBER VOLUME:	52-56 %	VOID CONTENT:		
PLY THICKNESS:	0.0056-0.0060 in.			
TEST METHOD: SACMA SRM 1-88		MODULUS CALCULATION: Initial tangent		
NORMALIZED BY:	Specimen thickness and batch fiber volume to 60%			

Temperature (°F) Moisture Content (%) Equilibrium at T, RH Source Code		75 ambient 26		200 ambient 26		200 wet (1) 26	
		Normalized	Measured	Normalized	Measured	Normalized	Measured
F_1^{cu} (ksi)	Mean	233	211	213	193	191	173
	Minimum	200	186	174	157	142	128
	Maximum	260	234	267	243	220	201
	C.V.(%)	6.39	6.16	9.74	10.0	11.0	11.4
	B-value	(2)	(2)	(2)	(2)	(2)	(2)
	Distribution	ANOVA	ANOVA	ANOVA	ANOVA	ANOVA	ANOVA
	C_1	15.2	13.4	21.0	19.6	22.4	21.1
	C_2	2.21	2.23	2.00	2.03	4.17	4.25
	No. Specimens	30		30		15	
	No. Batches	8		10		3	
	Approval Class	Interim		Interim		Interim	
E_1^c (Msi)	Mean	18.8	17.0			18.3	16.6
	Minimum	17.9	16.2			17.5	15.7
	Maximum	19.7	17.8			19.1	17.3
	C.V.(%)	3.21	3.53			2.62	3.16
	No. Specimens	15				15	
	No. Batches	3				3	
	Approval Class	Interim				Interim	
ν_{12}^c	Mean						
	No. Specimens						
	No. Batches						
	Approval Class						
ε_1^{cu} (με)	Mean						
	Minimum						
	Maximum						
	C.V.(%)						
	B-value						
	Distribution						
	C_1						
	C_2						
	No. Specimens						
	No. Batches						
	Approval Class						

(1) Conditioned at 140°F, 95% relative humidity for 30 days.
(2) B-values are presented only for fully approved data.

** ALL DOCUMENTATION PRESENTLY REQUIRED WERE NOT SUPPLIED FOR THIS MATERIAL.

MATERIAL:	AS4/3501-6 (no bleed)			Table 4.2.13(d) C/Ep 145-UT AS4/3501-6 SBS, 31-plane $[0]_8$ 75/A, 200/A Screening
RESIN CONTENT:	36-39 wt%	COMP: DENSITY:	1.55-1.57 g/cm³	
FIBER VOLUME:	52-56 %	VOID CONTENT:		
PLY THICKNESS:	0.0057-0.0063 in.			
TEST METHOD: ASTM D2344-76		MODULUS CALCULATION: Initial tangent		
NORMALIZED BY:	Not normalized			

Temperature (°F) Moisture Content (%) Equilibrium at T, RH Source Code		75 ambient 26	200 ambient 26			
	Mean	17.9	14.0			
	Minimum	16.5	12.9			
	Maximum	19.0	15.4			
	C.V.(%)	4.46	4.73			
	B-value	(1)	(1)			
	Distribution	ANOVA	ANOVA			
F_{31}^{sbs}	C_1	0.824	0.683			
(ksi)	C_2	2.36	2.34			
	No. Specimens	30	30			
	No. Batches	8	10			
	Approval Class	Screening	Screening			

(1) B-values are presented only for fully approved data.

MATERIAL:	AS4/3501-6 (no bleed)			**Table 4.2.13(e)** **C/Ep 145-UT** **AS4/3501-6** **Tension, x-axis** **$[0/45/90/-45]_s$** **75/A** **Screening**
RESIN CONTENT:	36-37 wt%	COMP: DENSITY:	1.56-1.57 g/cm^3	
FIBER VOLUME:	54-56 %	VOID CONTENT:		
PLY THICKNESS:	0.0057-0.0062 in.			
TEST METHOD: ASTM D3039-76		MODULUS CALCULATION:		
NORMALIZED BY:	NA			

Temperature (°F) Moisture Content (%) Equilibrium at T, RH Source Code		75 ambient 26					
F_x^{tu} (ksi)	Mean	87.4					
	Minimum	83.2					
	Maximum	92.8					
	C.V.(%)	3.43					
	B-value	(1)					
	Distribution	Normal					
	C_1	87.4					
	C_2	3.00					
	No. Specimens	9					
	No. Batches	3					
	Approval Class	Screening					
E_x^t (Msi)	Mean						
	Minimum						
	Maximum						
	C.V.(%)						
	No. Specimens						
	No. Batches						
	Approval Class						
ν_{xy}^t	Mean						
	No. Specimens						
	No. Batches						
	Approval Class						
ε_x^{tu} (με)	Mean						
	Minimum						
	Maximum						
	C.V.(%)						
	B-value						
	Distribution						
	C_1						
	C_2						
	No. Specimens						
	No. Batches						
	Approval Class						

(1) B-values are presented only for fully approved data.

MATERIAL:	AS4/3501-6 (no bleed)			**Table 4.2.13(f)** **C/Ep 145-UT** **AS4/3501-6** **Open Hole Tension, x-axis** **[0/45/90/-45]$_s$** **75/A** **Screening**
RESIN CONTENT:	36-37 wt%	COMP: DENSITY:	1.56-1.57 g/cm^3	
FIBER VOLUME:	54-56 %	VOID CONTENT:		
PLY THICKNESS:	0.0060-0.0064 in			
TEST METHOD: SACMA SRM 5-88 (1)		MODULUS CALCULATION:		
NORMALIZED BY:	NA			

Temperature (°F)		75					
Moisture Content (%)		ambient					
Equilibrium at T, RH							
Source Code		26					
F_x^{oht} (ksi)	Mean	56.8					
	Minimum	54.4					
	Maximum	60.8					
	C.V.(%)	3.75					
	B-value	(2)					
	Distribution	Normal					
	C_1	56.8					
	C_2	2.13					
	No. Specimens	9					
	No. Batches	3					
	Approval Class	Screening					
E_x^{oht} (Msi)	Mean						
	Minimum						
	Maximum						
	C.V.(%)						
	No. Specimens						
	No. Batches						
	Approval Class						
ν_{xy}^t	Mean						
	No. Specimens						
	No. Batches						
	Approval Class						
ε_x^{oht} (με)	Mean						
	Minimum						
	Maximum						
	C.V.(%)						
	B-value						
	Distribution						
	C_1						
	C_2						
	No. Specimens						
	No. Batches						
	Approval Class						

(1) Note SACMA SRM 5-88 uses a $[45/0/-45/90]_{2S}$ layup.
(2) B-values are presented only for fully approved data.

4.2.14 AS4 3k/3501-6 plain weave data set description

Material Description:

Material: AS4-3k/3501-6

Form: Plain weave fabric, areal weight of 193 g/m^2, typical cured resin content of 37-41%, typical cured ply thickness of 0.0074-0.0086 inches.

Processing: Autoclave cure; 240°F, 85 psi for 1 hour; 350°F, 100 psi for 2 hours, no bleed.

General Supplier Information:

Fiber: AS4 fibers are continuous carbon filaments made from PAN precursor, surface treated to improve handling characteristics and structural properties. Filament count is 3000 filaments/tow. Typical tensile modulus is 34 x 10^6 psi. Typical tensile strength is 550,000 psi.

Matrix: 3501-6 is an amine-cured epoxy resin. It will retain light tack for a minimum of 10 days at room temperature.

Maximum Short Term Service Temperature: 300°F (dry), 180°F (wet)

Typical applications: General purpose structural applications.

4.2.14 AS4 3k/3501-6 plain weave*

				C/Ep AS4/3501-6 Summary
MATERIAL:	AS4 3k/3501-6 plain weave			
PREPREG:	Hercules AW193P			
FIBER:	Hercules AS4 3k W	MATRIX:	Hercules 3501-6	
T_g(dry):	T_g(wet):	T_g METHOD:		
PROCESSING:	Autoclave cure: 240 ± 10°F, 60 minutes, 85 psig; 350 ± 10°F, 120 ± 10 minutes, 100 ± 10 psig, no bleed			

** ALL DOCUMENTATION PRESENTLY REQUIRED WERE NOT SUPPLIED FOR THIS MATERIAL.

Date of fiber manufacture	Date of testing	
Date of resin manufacture	Date of data submittal	6/88
Date of prepreg manufacture	Date of analysis	1/93
Date of composite manufacture		

LAMINA PROPERTY SUMMARY

	75°F/A		-65°A/F	200°F/A		75°F/W	200°F/W	
Tension, 1-axis	SS--		SS--	SS--				
Tension, 2-axis								
Tension, 3-axis								
Compression, 1-axis	II--			II--		II--	II--	
Compression, 2-axis								
Compression, 3-axis								
Shear, 12-plane								
Shear, 23-plane								
SB Strength, 31-plane	S---			S---		S---	S---	

Classes of data: F - Fully approved, I - Interim, S - Screening in Strength/Modulus/Poisson's ratio/Strain-to-failure order.

** ALL DOCUMENTATION PRESENTLY REQUIRED WERE NOT SUPPLIED FOR THIS MATERIAL.

		Nominal	As Submitted	Test Method
Fiber Density	(g/cm^3)	1.80		
Resin Density	(g/cm^3)	1.28		
Composite Density	(g/cm^3)	1.58	1.54 - 1.56	
Fiber Areal Weight	(g/m^2)	193	193	
Ply Thickness	(in)	0.0070	0.0074 - 0.0086	

LAMINATE PROPERTY SUMMARY

	75°F/A							
$[0_f/90_f/\pm45_f]$ Family								
Tension, x-axis	SS--							
$[\pm45_f/0_f/90_f]$ Family								
OHT, x-axis	S---							

Classes of data: F - Fully approved, I - Interim, S - Screening in Strength/Modulus/Poisson's ratio/Strain-to-failure order.

** ALL DOCUMENTATION PRESENTLY REQUIRED WERE NOT SUPPLIED FOR THIS MATERIAL.

Table 4.2.14(a)
C/Ep 193-PW
AS4/3501-6
Tension, 1-axis
$[0_f]_8$
75/A, -65/A, 200/A
Screening

MATERIAL:	AS4 3k/3501-6 plain weave		
RESIN CONTENT:	38 wt%	COMP: DENSITY:	1.56 g/cm^3
FIBER VOLUME:	53-54 %	VOID CONTENT:	
PLY THICKNESS:	0.0074-0.0080 in.		
TEST METHOD: ASTM D3039-76		MODULUS CALCULATION:	
NORMALIZED BY:	Specimen thickness and batch fiber volume to 57%		

		75		-65		200	
Temperature (°F) Moisture Content (%) Equilibrium at T, RH Source Code		75 ambient 26		-65 ambient 26		200 ambient 26	
		Normalized	Measured	Normalized	Measured	Normalized	Measured
F_1^{tu} (ksi)	Mean	124	117	112	105	126	119
	Minimum	117	111	103	98.1	116	108
	Maximum	133	124	120	112	133	126
	C.V.(%)	4.18	3.56	4.63	4.00	4.79	5.88
	B-value	(2)	(2)	(2)	(2)	(2)	(2)
	Distribution	Normal	Normal	Normal	Normal	Normal	Normal
	C_1	124	117	112	105	126	119
	C_2	5.17	4.15	5.17	4.21	6.05	7.00
	No. Specimens	9		9		9	
	No. Batches	3		3		3	
	Approval Class	Screening		Screening		Screening	
E_1^t (Msi)	Mean	9.8	9.2	10.5	9.9	10.1	9.5
	Minimum	9.4	8.8	9.7	9.1	7.1	6.7
	Maximum	10.2	9.5	11.1	10.4	10.7	10.1
	C.V.(%)	3,0	2.5	4.6	4.2	11	11
	No. Specimens	9		9		9	
	No. Batches	3		3		3	
	Approval Class	Screening		Screening		Screening	
ν_{12}^t	Mean						
	No. Specimens						
	No. Batches						
	Approval Class						
ε_1^{tu} ($\mu\varepsilon$)	Mean						
	Minimum						
	Maximum						
	C.V.(%)						
	B-value						
	Distribution						
	C_1						
	C_2						
	No. Specimens						
	No. Batches						
	Approval Class						

(1) B-values are presented only for fully approved data.

** ALL DOCUMENTATION PRESENTLY REQUIRED WERE NOT SUPPLIED FOR THIS MATERIAL.

MATERIAL: AS4 3k/3501-6 plain weave		**Table 4.2.14(b)** **C/Ep 193-PW** **AS4/3501-6** **Compression, 1-axis** **$[0_f]_{14}$** **75/A, 200/A, 75/W** **Interim**
RESIN CONTENT: 39-41 wt% FIBER VOLUME: 51-52 % PLY THICKNESS: 0.0081-0.0086 in.	COMP: DENSITY: 1.54-1.55 g/cm³ VOID CONTENT:	
TEST METHOD: SACMA SRM 1-88	MODULUS CALCULATION:	
NORMALIZED BY: Specimen thickness and batch fiber volume to 57%		

Temperature (°F) Moisture Content (%) Equilibrium at T, RH Source Code		75 ambient 26		200 ambient 26		75 (1) wet 26	
		Normalized	Measured	Normalized	Measured	Normalized	Measured
F_1^{cu} (ksi)	Mean	130	117	108	97.3	112	101
	Minimum	115	104	92.8	83.0	99.6	88.0
	Maximum	140	127	121	109	122	109
	C.V.(%)	6.45	6.49	7.44	7.71	5.56	5.65
	B-value	(2)	(2)	(2)	(2)	(2)	(2)
	Distribution	Nonpara.	Nonpara.	Weibull	Normal	ANOVA	ANOVA
	C_1	8	8	112	97.3	6.83	6.32
	C_2	1.54	1.54	15.1	7.51	4.85	5.09
	No. Specimens	15		15		15	
	No. Batches	3		3		3	
	Approval Class	Interim		Interim		Interim	
E_1^c (Msi)	Mean	9.2	8.3	9.8	8.8	9.4	8.4
	Minimum	8.5	7.7	9.2	8.4	8.8	8.1
	Maximum	9.8	8.8	10.2	9.1	9.9	8.8
	C.V.(%)	3.4	4.3	3.5	2.5	3.0	2.4
	No. Specimens	15		15		15	
	No. Batches	3		3		3	
	Approval Class	Interim		Interim		Interim	
ν_{12}^c	Mean						
	No. Specimens						
	No. Batches						
	Approval Class						
ε_1^{cu} (με)	Mean						
	Minimum						
	Maximum						
	C.V.(%)						
	B-value						
	Distribution						
	C_1						
	C_2						
	No. Specimens						
	No. Batches						
	Approval Class						

(1) Conditioned at 140°F, 95% relative humidity for 30 days.
(2) B-values are presented only for fully approved data.

** ALL DOCUMENTATION PRESENTLY REQUIRED WERE NOT SUPPLIED FOR THIS MATERIAL.

MATERIAL:	AS4 3k/3501-6 plain weave			**Table 4.2.14(c)** **C/Ep 193-PW** **AS4/3501-6** **Compression, 1-axis** **$[0_f]_{14}$** **200/W** **Interim**
RESIN CONTENT:	39-41 wt%	COMP: DENSITY:	1.54-1.55 g/cm^3	
FIBER VOLUME:	51-52 %	VOID CONTENT:		
PLY THICKNESS:	0.0081-0.0086 in.			
TEST METHOD: SACMA SRM 1-88		MODULUS CALCULATION:		
NORMALIZED BY:	Specimen thickness and batch fiber volume to 57%			

Temperature (°F) Moisture Content (%) Equilibrium at T, RH Source Code		200 (1) wet 26					
		Normalized	Measured	Normalized	Measured	Normalized	Measured
F_1^{cu} (ksi)	Mean	58.7	52.7				
	Minimum	51.7	46.2				
	Maximum	65.4	59.7				
	C.V.(%)	7.27	7.58				
	B-value	(2)	(2)				
	Distribution	Weibull	Weibull				
	C_1	60.6	54.5				
	C_2	15.6	15.2				
	No. Specimens	15					
	No. Batches	3					
	Approval Class	Interim					
E_1^c (Msi)	Mean	9.1	8.1				
	Minimum	8.7	7.8				
	Maximum	9.4	8.5				
	C.V.(%)	2.4	2.9				
	No. Specimens	15					
	No. Batches	3					
	Approval Class	Interim					
ν_{12}^c	Mean						
	No. Specimens						
	No. Batches						
	Approval Class						
ε_1^{cu} (με)	Mean						
	Minimum						
	Maximum						
	C.V.(%)						
	B-value						
	Distribution						
	C_1						
	C_2						
	No. Specimens						
	No. Batches						
	Approval Class						

(1) Conditioned at 140°F, 95% relative humidity for 30 days.
(2) B-values are presented only for fully approved data.

** ALL DOCUMENTATION PRESENTLY REQUIRED WERE NOT SUPPLIED FOR THIS MATERIAL.

MATERIAL:	AS4 3k/3501-6 plain weave			**Table 4.2.14(d)** **C/Ep 193-PW** **AS4/3501-6** **SBS, 31-plane** **$[0_f]_{14}$** **75/A, 200/A, 75/W, 200/W** **Screening**
RESIN CONTENT:	39-41 wt%	COMP: DENSITY:	1.54-1.55 g/cm³	
FIBER VOLUME:	51-52 %	VOID CONTENT:		
PLY THICKNESS:	0.0077-0.0082 in.			
TEST METHOD: ASTM D2344		MODULUS CALCULATION:		
NORMALIZED BY:	Not normalized			

Temperature (°F) Moisture Content (%) Equilibrium at T, RH Source Code	75 ambient 26	200 ambient 26	75 wet (1) 26	200 wet (1) 26	
Mean	10.9	8.4	10.9	5.3	
Minimum	9.7	8.1	10.0	5.2	
Maximum	11.9	8.8	11.4	5.5	
C.V.(%)	6.09	2.5	3.47	2.3	
B-value	(2)	(2)	(2)	(2)	
Distribution	Weibull	Normal	Weibull	Nonpara.	
F_{31}^{sbs} (ksi) C_1	11.2	8.4	11.0	7	
C_2	20.1	0.21	35.4	1.81	
No. Specimens	15	9	15	12	
No. Batches	3	3	3	3	
Approval Class	Screening	Screening	Screening	Screening	

(1) Conditioned at 140°F, 95% relative humidity for 30 days.
(2) B-values are presented only for fully approved data.

MATERIAL:	AS4 3k/3501-6 plain weave			Table 4.2.14(e) C/Ep 193-PW AS4/3501-6 Tension, x-axis $[0_f/90_f/\pm45_f]_{2s}$ 75/A Screening
RESIN CONTENT:	37-38 wt%	COMP: DENSITY:	0.056 lb/in^3	
FIBER VOLUME:	53-54 %	VOID CONTENT:		
PLY THICKNESS:	0.0080-0.0085 in.			
TEST METHOD: ASTM D3039-76		MODULUS CALCULATION:		
NORMALIZED BY:	NA			

Temperature (°F)		75					
Moisture Content (%)		ambient					
Equilibrium at T, RH							
Source Code		26					
F_x^{tu} (ksi)	Mean	68.5					
	Minimum	62.0					
	Maximum	75.1					
	C.V.(%)	7.60					
	B-value	(1)					
	Distribution	Normal					
	C_1	68.5					
	C_2	5.21					
	No. Specimens	9					
	No. Batches	3					
	Approval Class	Screening					
E_x^t (Msi)	Mean	6.0					
	Minimum	5.6					
	Maximum	6.3					
	C.V.(%)	3.6					
	No. Specimens	9					
	No. Batches	3					
	Approval Class	Screening					
ν_{xy}^t	Mean						
	No. Specimens						
	No. Batches						
	Approval Class						
ε_x^{tu} (με)	Mean						
	Minimum						
	Maximum						
	C.V.(%)						
	B-value						
	Distribution						
	C_1						
	C_2						
	No. Specimens						
	No. Batches						
	Approval Class						

(1) B-values are presented only for fully approved data.

MATERIAL: AS4 3k/3501-6 plain weave					Table 4.2.14(f) C/Ep 193-PW AS4/3501-6 Open Hole Tension, x-axis $[\pm45_f/0_f/90_f]_{2s}$ 75/A Screening
RESIN CONTENT: 37-38 wt% FIBER VOLUME: 53-54 %		COMP: DENSITY: 0.056 lb/in^3 VOID CONTENT:			
PLY THICKNESS: 0.0080-0.0085 in.					
TEST METHOD: SACMA SRM 5-88 (1)		MODULUS CALCULATION:			
NORMALIZED BY: NA					
Temperature (°F) Moisture Content (%) Equilibrium at T, RH Source Code	75 ambient 26				
F_x^{oht} (ksi) Mean	51.4				
Minimum	48.6				
Maximum	53.8				
C.V.(%)	3.40				
B-value	(2)				
Distribution	ANOVA				
C_1	2.46				
C_2	1.20				
No. Specimens	9				
No. Batches	3				
Approval Class	Screening				
E_x^t (Msi) Mean					
Minimum					
Maximum					
C.V.(%)					
No. Specimens					
No. Batches					
Approval Class					
ε_x^{tu} ($\mu\varepsilon$) Mean					
Minimum					
Maximum					
C.V.(%)					
B-value					
Distribution					
C_1					
C_2					
No. Specimens					
No. Batches					
Approval Class					

(1) Note SACMA SRM 5-88 uses a $[45/0/-45/90]_s$ layup.
(2) B-values are presented only for fully approved data.

4.2.15 AS4 3k/3501-6S 5-harness satin weave fabric data set description

Material Description:

Material: AS4-3k/3501-6S

Form: 5 harness satin weave fabric, areal weight of 280 g/m^2, typical cured resin content of 33-35%, typical cured ply thickness of 0.0106 -0.0107 inches.

Processing: Autoclave cure; 240°F, 85 psi for 1 hour, 350°F, 100 psi for 2 hours, no bleed.

General Supplier Information:

Fiber: AS4 fibers are continuous carbon filaments made from PAN precursor, surface treated to improve handling characteristics and structural properties. Filament count is 3000 filaments/tow. Typical tensile modulus is 34 x 10^6 psi. Typical tensile strength is 550,000 psi.

Matrix: 3501-6S is an amine-cured epoxy resin. This resin is a solvated material. It results in a more drapeable prepreg for use on highly complex parts. This resin is also amenable to occurring. The hot/wet strengths are slightly lower than the non-solvated resin. It will retain light tack for a minimum of 10 days at room temperature.

Maximum Short Term Service Temperature: 300°F (dry), 180°F (wet)

Typical Applications: General purpose structural applications.

4.2.15 AS4 3k/3501-6S 5-harness satin weave*

MATERIAL:	AS4 3k/3501-6S 5-harness satin weave		**C/Ep AS4/3501-6S Summary**
PREPREG:	Hercules AW280-5HS		
FIBER:	Hercules AS4 3k W	MATRIX:	Hercules 3501-6S
T_g(dry):	T_g(wet):	T_g METHOD:	
PROCESSING:	Autoclave cure: 240 ± 10°F, 60 minutes, 85 psig; 350 ± 10°F, 120 ± 10 minutes, 100 ± 10 psig, no bleed		

** ALL DOCUMENTATION PRESENTLY REQUIRED WERE NOT SUPPLIED FOR THIS MATERIAL.

Date of fiber manufacture		Date of testing	
Date of resin manufacture		Date of data submittal	6/88
Date of prepreg manufacture		Date of analysis	1/93
Date of composite manufacture			

LAMINA PROPERTY SUMMARY

	75°F/A		200°F/A					
Tension, 1-axis	II--							
Tension, 2-axis								
Tension, 3-axis								
Compression, 1-axis	I---		I---					
Compression, 2-axis								
Compression, 3-axis								
Shear, 12-plane								
Shear, 23-plane								
SB Strength, 31-plane	S---		S---					

Classes of data: F - Fully approved, I - Interim, S - Screening in Strength/Modulus/Poisson's ratio/Strain-to-failure order.

** ALL DOCUMENTATION PRESENTLY REQUIRED WERE NOT SUPPLIED FOR THIS MATERIAL.

		Nominal	As Submitted	Test Method
Fiber Density	(g/cm^3)	1.80		
Resin Density	(g/cm^3)	1.28		
Composite Density	(g/cm^3)	1.58	1.58 - 1.59	
Fiber Areal Weight	(g/m^2)	280	279 - 284	
Ply Thickness	(in)		0.0106 - 0.0107	

LAMINATE PROPERTY SUMMARY

Classes of data: F - Fully approved, I - Interim, S - Screening in Strength/Modulus/Poisson's ratio/Strain-to-failure order.

** ALL DOCUMENTATION PRESENTLY REQUIRED WERE NOT SUPPLIED FOR THIS MATERIAL.

MATERIAL: AS4 3k/3501-6S 5-harness satin weave	**Table 4.2.15(a)** **C/Ep 280-5HS** **AS4/3501-6S** **Tension, 1-axis** **$[0_f]_6$** **75/A** **Interim**
RESIN CONTENT: 33-35 wt% COMP: DENSITY: 1.58-1.59 g/cm^3 FIBER VOLUME: 57-60 % VOID CONTENT: PLY THICKNESS: 0.0106-0.0107 in.	
TEST METHOD: ASTM D3039-76 MODULUS CALCULATION:	
NORMALIZED BY: Specimen thickness and batch fiber volume to 57%	

Temperature (°F)		75					
Moisture Content (%)		ambient					
Equilibrium at T, RH							
Source Code		26					
		Normalized	Measured	Normalized	Measured	Normalized	Measured
	Mean	112	115				
	Minimum	97.6	100				
	Maximum	123	126				
	C.V.(%)	5.78	5.55				
	B-value	(1)	(1)				
F_1^{tu}	Distribution	ANOVA	ANOVA				
(ksi)	C_1	6.63	6.55				
	C_2	2.26	2.25				
	No. Specimens	30					
	No. Batches	10					
	Approval Class	Interim					
	Mean	9.73	10.0				
	Minimum	8.93	9.20				
	Maximum	10.1	10.3				
E_1^t	C.V.(%)	2.48	2.31				
(Msi)	No. Specimens	30					
	No. Batches	10					
	Approval Class	Interim					
	Mean						
	No. Specimens						
ν_{12}^t	No. Batches						
	Approval Class						
	Mean						
	Minimum						
	Maximum						
	C.V.(%)						
	B-value						
ε_1^{tu}	Distribution						
(με)	C_1						
	C_2						
	No. Specimens						
	No. Batches						
	Approval Class						

(1) B-values are presented only for fully approved data.

** ALL DOCUMENTATION PRESENTLY REQUIRED WERE NOT SUPPLIED FOR THIS MATERIAL.

MATERIAL:	AS4 3k/3501-6S 5-harness satin weave		
RESIN CONTENT:	33-35 wt%	COMP: DENSITY:	1.58-1.59 g/cm^3
FIBER VOLUME:	57-60 %	VOID CONTENT:	
PLY THICKNESS:	0.0106-0.0107 in.		
TEST METHOD: SACMA SRM 1-88		MODULUS CALCULATION:	
NORMALIZED BY:	Specimen thickness and batch fiber volume to 57%		

Table 4.2.15(b)
C/Ep 280-5HS
AS4/3501-6S
Compression, 1-axis
$[0_f]_8$
75/A, 200/A
Interim

Temperature (°F) / Moisture Content (%) / Equilibrium at T, RH / Source Code		75 / ambient / 26		200 / ambient / 26			
		Normalized	Measured	Normalized	Measured	Normalized	Measured
F_1^{cu} (ksi)	Mean	124	128	110	113		
	Minimum	108	111	96.1	99.0		
	Maximum	144	148	122	125		
	C.V.(%)	6.73	6.74	6.31	6.24		
	B-value	(1)	(1)	(1)	(1)		
	Distribution	Weibull	Weibull	ANOVA	ANOVA		
	C_1	128	132	7.04	7.15		
	C_2	15.4	15.3	2.10	2.09		
	No. Specimens	30		30			
	No. Batches	10		10			
	Approval Class	Interim		Interim			
E_1^c (Msi)	Mean						
	Minimum						
	Maximum						
	C.V.(%)						
	No. Specimens						
	No. Batches						
	Approval Class						
ν_{12}^c	Mean						
	No. Specimens						
	No. Batches						
	Approval Class						
ε_1^{cu} ($\mu\varepsilon$)	Mean						
	Minimum						
	Maximum						
	C.V.(%)						
	B-value						
	Distribution						
	C_1						
	C_2						
	No. Specimens						
	No. Batches						
	Approval Class						

(1) B-values are presented only for fully approved data.

** ALL DOCUMENTATION PRESENTLY REQUIRED WERE NOT SUPPLIED FOR THIS MATERIAL.

MATERIAL: AS4 3k/3501-6S 5-harness satin weave RESIN CONTENT: 33-35 wt% COMP: DENSITY: 1.58-1.59 g/cm^3 FIBER VOLUME: 57-60 % VOID CONTENT: PLY THICKNESS: 0.0106-0.0107 in. TEST METHOD: MODULUS CALCULATION: ASTM D2344 NORMALIZED BY: Not normalized		**Table 4.2.15(c)** **C/Ep 280-5HS** **AS4/3501-6S** **SBS, 31-plane** **$[0_f]_6$** **75/A, 200/A** **Screening**

Temperature (°F) Moisture Content (%) Equilibrium at T, RH Source Code		75 ambient 26	200 ambient 26			
	Mean	11.0	9.53			
	Minimum	9.00	8.40			
	Maximum	13.2	10.8			
	C.V.(%)	10.8	6.70			
	B-value	(1)	(1)			
	Distribution	ANOVA	ANOVA			
F_{31}^{sbs}	C_1	1.22	0.66			
(ksi)	C_2	2.18	2.32			
	No. Specimens	30	30			
	No. Batches	10	10			
	Approval Class	Screening	Screening			

(1) B-values are presented only for fully approved data.

4.2.16 AS4 6k/3502-6S 5HS fabric data set description

Material Description:

Material: AS4-6k/3502-6S

Form: 5 harness satin weave fabric, fiber areal weight of 365 g/m^2, typical cured resin content of 56-57%, typical cured ply thickness of 0.0142-0.0157 inches.

Processing: Autoclave cure; 275°F, 85 psi for 45 minutes; 350°F, 85 psi, hold for two hours. Post cure at 400°F to develop optimum 350°F properties.

General Supplier Information:

Fiber: AS4 fibers are continuous high strength, high strain, standard modulus carbon filaments made from PAN precursor. The fibers are surface treated to improve handling characteristics and structural properties. Filament count is 6,000 filaments/tow. Typical tensile modulus is 34 x 10^6 psi. Typical tensile strength is 550,000 psi.

Matrix: 3502 is an epoxy resin. This is a solvated resin formulated to improve drapeability over complex shapes. The hot/wet strengths will be slightly lower than the non-solvated resin. Good tack up to 10 days out-time at ambient temperature.

Maximum Short Term Service Temperature: 350°F (dry), 180°F (wet)

Typical applications: Primary and secondary structural applications on commercial and military aircraft.

4.2.16 AS4 6k/3502-6S 5-harness satin weave*

MATERIAL:	AS4 6k/3502 5-harness satin weave				**C/Ep AS4/3502 Summary**
PREPREG:	Hercules A370-5H/3502, 5-harness satin weave, 11 x 11 tow/in.				
FIBER:	Hercules AS4 6k, surface-treated "W"**, 0 twist		MATRIX:	Hercules 3502	
T_g(dry):	404°F	T_g(wet): 313°F	T_g METHOD:	TMA	
PROCESSING:	Autoclave cure: 280 ± 5°F, 90 minutes, 85+15-0 psi; 350°F, 120 minutes.				

** now "G"

Date of fiber manufacture	10/82-3/83	Date of testing	9/83-1/84
Date of resin manufacture	5/83	Date of data submittal	12/93, 5/94
Date of prepreg manufacture	5/83	Date of analysis	8/94
Date of composite manufacture	8/83-9/83		

LAMINA PROPERTY SUMMARY

	75°F/A		-65°F/A		180°F/W	250°F/W		
Tension, 1-axis	FF--		FF--		FF--	FF--		
Tension, 2-axis								
Tension, 3-axis								
Compression, 1-axis	FF--		IS--		FF--	FF--		
Compression, 2-axis								
Compression, 3-axis								
Shear, 12-plane	FF--		FF--		FS--	FS--		
Shear, 23-plane								
Shear, 31-plane								

Classes of data: F - Fully approved, I - Interim, S - Screening in Strength/Modulus/Poisson's ratio/Strain-to-failure order.

		Nominal	As Submitted	Test Method
Fiber Density	(g/cm^3)	1.79		
Resin Density	(g/cm^3)	1.26		
Composite Density	(g/cm^3)	1.57	1.55 - 1.60	
Fiber Areal Weight	(g/m^2)	365	361 - 372	
Ply Thickness	(in)	0.0145	0.0142 - 0.0158	

LAMINATE PROPERTY SUMMARY

Classes of data: F - Fully approved, I - Interim, S - Screening in Strength/Modulus/Poisson's ratio/Strain-to-failure order.

MATERIAL:	AS4 6k/3502 5-harness satin weave			**Table 4.2.16(a)** **C/Ep - 365 - 5HS** **AS4/3502** **Tension, 1-axis** **[0/90/0/90/90/0]$_s$** **75/A, -65/A, 180/W** **Fully Approved**
RESIN CONTENT:	36-37 wt%	COMP: DENSITY:	1.55-1.56 g/cm^3	
FIBER VOLUME:	56-57 %	VOID CONTENT:	0.0-0.2%	
PLY THICKNESS:	0.0146-0.0157 in.			
TEST METHOD: BMS 8-168D		MODULUS CALCULATION: Linear portion of curve		
NORMALIZED BY:	Ply thickness to 0.0145 in. (57%)			

		75		-65		180	
Temperature (°F) Moisture Content (%) Equilibrium at T, RH Source Code		75 ambient 49		-65 ambient 49		180 1.1 - 1.3 (1) 49	
		Normalized	Measured	Normalized	Measured	Normalized	Measured
F_1^{tu} (ksi)	Mean	114	109	105	101	117	111
	Minimum	97.1	91.9	87.9	84.0	102	96.1
	Maximum	126	121	116	112	128	123
	C.V.(%)	6.87	7.01	5.33	5.53	5.29	5.36
	B-value	91.9		95.0		102	
	Distribution	ANOVA		Normal		ANOVA	
	C_1	8.15		104.9		6.31	
	C_2	2.70		5.59		2.33	
	No. Specimens	30		30		30	
	No. Batches	5		5		5	
	Approval Class	Fully Approved		Fully Approved		Fully Approved	
E_1^t (Msi)	Mean	9.61		9.67		10.5	
	Minimum	9.29		9.09		9.74	
	Maximum	10.4		10.1		10.9	
	C.V.(%)	3.08		2.35		2.75	
	No. Specimens	30		30		30	
	No. Batches	5		5		5	
	Approval Class	Fully Approved		Fully Approved		Fully Approved	
ν_{12}^t	Mean						
	No. Specimens						
	No. Batches						
	Approval Class						
ε_1^{tu} ($\mu\varepsilon$)	Mean						
	Minimum						
	Maximum						
	C.V.(%)						
	B-value						
	Distribution						
	C_1						
	C_2						
	No. Specimens						
	No. Batches						
	Approval Class						

(1) Conditioned at 160°F, 95-100% relative humidity until the moisture content was between 1.1 and 1.3%.

MATERIAL:	AS4 6k/3502 5-harness satin weave			**Table 4.2.16(b)** **C/Ep - 365 - 5HS** **AS4/3502** **Tension, 1-axis** **[0/90/0/90/90/0]$_f$** **250/W** **Fully Approved**
RESIN CONTENT:	36-37 wt%	COMP: DENSITY:	1.55-1.56 g/cm^3	
FIBER VOLUME:	56-57 %	VOID CONTENT:	0.0-0.2%	
PLY THICKNESS:	0.0150-0.0157 in.			
TEST METHOD: BMS 8-168D		MODULUS CALCULATION: Linear portion of curve		
NORMALIZED BY:	Ply thickness to 0.0145 in. (57%)			

Temperature (°F)		250					
Moisture Content (%)		1.1 - 1.3					
Equilibrium at T, RH							
Source Code		49					
		Normalized	Measured	Normalized	Measured	Normalized	Measured
F_1^{tu} (ksi)	Mean	108	102				
	Minimum	96.8	91.8				
	Maximum	119	113				
	C.V.(%)	4.62	4.86				
	B-value	96.6					
	Distribution	Weibull					
	C_1	111					
	C_2	23.1					
	No. Specimens	30					
	No. Batches	5					
	Approval Class	Fully Approved					
E_1^t (Msi)	Mean	10.1					
	Minimum	9.29					
	Maximum	10.7					
	C.V.(%)	3.65					
	No. Specimens	30					
	No. Batches	5					
	Approval Class	Fully Approved					
ν_{12}^t	Mean						
	No. Specimens						
	No. Batches						
	Approval Class						
ε_1^{tu} (με)	Mean						
	Minimum						
	Maximum						
	C.V.(%)						
	B-value						
	Distribution						
	C_1						
	C_2						
	No. Specimens						
	No. Batches						
	Approval Class						

(1) Conditioned at 160°F, 95-100% relative humidity until the moisture content was between 1.1 and 1.3%.

MATERIAL:	AS4 6k/3502 5-harness satin weave			**Table 4.2.16(c)** **C/Ep - 365 - 5HS** **AS4/3502** **Compression, 1-axis** **[0/90/0/90/90/0],** **75/A, -65/A, 180/W** **Fully Approved, Interim**
RESIN CONTENT:	36-37 wt%	COMP: DENSITY:	1.55-1.56 g/cm^3	
FIBER VOLUME:	56-57 %	VOID CONTENT:	0.0-0.2%	
PLY THICKNESS:	0.0142-0.0157 in.			
TEST METHOD: ASTM D695M (1) (4)		MODULUS CALCULATION: Linear portion of curve		
NORMALIZED BY:	Ply thickness to 0.0145 in. (57%)			

		75		-65		180	
Temperature (°F)		75		-65		180	
Moisture Content (%)		ambient		ambient		1.1 - 1.3	
Equilibrium at T, RH		(2)		(2)		(2)	
Source Code		1		49		49	
		Normalized	Measured	Normalized	Measured	Normalized	Measured
	Mean	104	99.7	108	103	65.9	63.3
	Minimum	79.7	75.5	85.0	79.0	52.1	49.7
	Maximum	122	116	118	114	76.7	74.0
	C.V.(%)	10.1	10.7	8.62	9.56	9.81	10.0
	B-value	83.7		(3)		52.4	
F_1^{cu}	Distribution	Weibull		Weibull		Weibull	
(ksi)	C_1	109		111		68.7	
	C_2	12.1		16.4		11.7	
	No. Specimens	30		15		30	
	No. Batches	5		5		5	
	Approval Class	Fully Approved		Interim		Fully Approved	
	Mean	8.49		8.90		9.21	
	Minimum	8.15		7.70		6.25	
	Maximum	8.86		11.0		12.5	
E_1^c	C.V.(%)	2.13		10.3		18.2	
(Msi)	No. Specimens	30		14		30	
	No. Batches	5		5		5	
	Approval Class	Fully Approved		Interim		Fully Approved	
	Mean						
	No. Specimens						
ν_{12}^c	No. Batches						
	Approval Class						
	Mean						
	Minimum						
	Maximum						
	C.V.(%)						
	B-value						
ε_1^{cu}	Distribution						
($\mu\varepsilon$)	C_1						
	C_2						
	No. Specimens						
	No. Batches						
	Approval Class						

(1) Tabbed specimen, length 3.12 inch, width 0.050 inch, gage length 0.50 inch.
(2) Conditioned at 160°F, 95-100% relative humidity until the moisture content was between 1.1 and 1.3%.
(3) B-values are presented only for fully approved data.
(4) The test method, ASTM D695M-96, was withdrawn on July 10, 1996.

MATERIAL:	AS4 6k/3502 5-harness satin weave			**Table 4.2.16(d)** **C/Ep - 365 - 5HS** **AS4/3502** **Compression, 1-axis** **[0/90/0/90/90/0]$_s$** **250/W** **Fully Approved**
RESIN CONTENT:	36-37 wt%	COMP: DENSITY:	1.55-1.56 g/cm^3	
FIBER VOLUME:	56-57 %	VOID CONTENT:	0.0-0.2%	
PLY THICKNESS:	0.0142-0.0157 in.			
TEST METHOD: ASTM D695M (1) (3)		MODULUS CALCULATION: Linear portion of curve		
NORMALIZED BY:	Ply thickness to 0.0145 in. (57%)			

Temperature (°F)		250					
Moisture Content (%)		1.1 - 1.3					
Equilibrium at T, RH							
Source Code		49					
		Normalized	Measured	Normalized	Measured	Normalized	Measured
	Mean	56.3	53.2				
	Minimum	45.5	42.6				
	Maximum	75.2	70.8				
	C.V.(%)	16.0	16.2				
	B-value	30.5					
F_1^{cu}	Distribution	ANOVA					
(ksi)	C_1	9.41					
	C_2	2.75					
	No. Specimens	30					
	No. Batches	5					
	Approval Class	Fully Approved					
	Mean	10.3					
	Minimum	8.88					
	Maximum	12.4					
E_1^c	C.V.(%)	6.60					
(Msi)	No. Specimens	30					
	No. Batches	5					
	Approval Class	Fully Approved					
	Mean						
	No. Specimens						
ν_{12}^c	No. Batches						
	Approval Class						
	Mean						
	Minimum						
	Maximum						
	C.V.(%)						
	B-value						
ε_1^{cu}	Distribution						
(με)	C_1						
	C_2						
	No. Specimens						
	No. Batches						
	Approval Class						

(1) Tabbed specimen, length 3.12 inch, width 0.050 inch, gage length 0.50 inch.
(2) Conditioned at 160°F, 95-100% relative humidity until the moisture content was between 1.1 and 1.3%.
(3) The test method, ASTM D695M-96, was withdrawn on July 10, 1996.

MATERIAL:	AS4 6k/3502 5-harness satin weave			**Table 4.2.16(e)** **C/Ep - 365 - 5HS** **AS4/3502** **Shear, 12-plane** **[±45/±45/∓45]$_s$** **75/A, -65/A, 180/W, 250/W** **Fully Approved, Screening**
RESIN CONTENT:	36-37 wt%	COMP: DENSITY:	1.55-1.56 g/cm^3	
FIBER VOLUME:	56-57 %	VOID CONTENT:	0.0-0.2%	
PLY THICKNESS:	0.0145-0.0158 in.			
TEST METHOD:		MODULUS CALCULATION:		
ASTM D3518-76		Linear portion of curve		
NORMALIZED BY:	Not normalized			

Temperature (°F)		75	-65	180	250	
Moisture Content (%)		ambient	ambient	1.1 - 1.3	1.1 - 1.3	
Equilibrium at T, RH				(1)	(1)	
Source Code		49	49	49	49	
F_{12}^{su} (ksi)	Mean	12.6	14.0	11.7	9.30	
	Minimum	11.4	12.1	10.7	8.27	
	Maximum	13.7	15.4	12.9	10.5	
	C.V.(%)	5.61	7.47	5.24	6.76	
	B-value	10.1	10.1	9.53	6.95	
	Distribution	ANOVA	ANOVA	ANOVA	ANOVA	
	C_1	0.775	1.16	0.669	0.698	
	C_2	3.21	3.36	3.20	3.37	
	No. Specimens	36	36	36	36	
	No. Batches	5	5	5	5	
	Approval Class	Fully Approved	Fully Approved	Fully Approved	Fully Approved	
G_{12}^{s} (Msi)	Mean	0.514	0.682	0.204	0.174	
	Minimum	0.485	0.638	0.196	0.147	
	Maximum	0.553	0.731	0.212	0.203	
	C.V.(%)	3.68	3.40	2.82	11.8	
	No. Specimens	36	36	6	5	
	No. Batches	5	5	1	1	
	Approval Class	Fully Approved	Fully Approved	Screening	Screening	
γ_{12}^{su} (με)	Mean					
	Minimum					
	Maximum					
	C.V.(%)					
	B-value					
	Distribution					
	C_1					
	C_2					
	No. Specimens					
	No. Batches					
	Approval Class					

(1) Conditioned at 160°F, 95-100% relative humidity until the moisture content was between 1.1 and 1.3%.

4.2.17 T-300 15k/976 unitape data set description

Material Description:

Material: T-300 15k/976

Form: Unidirectional tape, fiber areal weight of 152 g/m^2, typical cured resin content of 25-35%, typical cured ply thickness of 0.0051 inches.

Processing: Autoclave cure; 250°F, 100 psi for 45 mins.; 350°F, 2 hours.

General Supplier Information:

Fiber: T-300 fibers are continuous carbon filaments made from PAN precursor, surface treated to improve handling characteristics and structural properties. Filament count is 15,000 filaments/tow. Typical tensile modulus is 33 x 10^6 psi. Typical tensile strength is 530,000 psi.

Matrix: 976 is a high flow, modified epoxy resin that meets the NASA outgassing requirements. 10 days out-time at 72°F.

Maximum Short Term Service Temperature: 350°F (dry), 250°F (wet)

Typical applications: General purpose commercial and military structural applications, good hot/wet properties.

4.2.17 T-300 15k/976 unidirectional tape*

MATERIAL:	T300 15k/976 unidirectional tape			**C/Ep T300 15k/976 Summary**
PREPREG:	Fiberite T300/976 unidirectional tape			
FIBER:	Union Carbide T300 15k	MATRIX:	Fiberite 976	
T_g(dry):	518°F T_g(wet): 493°F	T_g METHOD:	DMA	
PROCESSING:	Autoclave cure: 250°F, 100 psi, 45 minutes; 350°F, 2 hours			

** DATA WERE SUBMITTED BEFORE THE ESTABLISHMENT OF DATA DOCUMENTATION REQUIREMENTS (JUNE 1989). ALL DOCUMENTATION PRESENTLY REQUIRED WERE NOT SUPPLIED FOR THIS MATERIAL.

Date of fiber manufacture		Date of testing	
Date of resin manufacture		Date of data submittal	2/82
Date of prepreg manufacture	7/80	Date of analysis	9/94
Date of composite manufacture			

LAMINA PROPERTY SUMMARY

	72°F/A		-67°F/A	260°F/A	350°F/A			
Tension, 1-axis	SSSS		SSSS	SSSS	SSSS			
Tension, 2-axis	SS-S		SS-S	SS-S	SS-S			
Tension, 3-axis								
Compression, 1-axis	SS-S		SS-S	SS-S	SS-S			
Compression, 2-axis	SS-S		SS-S	SS-S	SS-S			
Compression, 3-axis								
Shear, 12-plane	SS--		SS--	SS--	SS--			
Shear, 23-plane								
SB Strength, 31-plane	S---		S---	S---	S---			

Classes of data: F - Fully approved, I - Interim, S - Screening in Strength/Modulus/Poisson's ratio/Strain-to-failure order.

** DATA WERE SUBMITTED BEFORE THE ESTABLISHMENT OF DATA DOCUMENTATION REQUIREMENTS (JUNE 1989). ALL DOCUMENTATION PRESENTLY REQUIRED WERE NOT SUPPLIED FOR THIS MATERIAL.

		Nominal	As Submitted	Test Method
Fiber Density	(g/cm^3)	1.78		
Resin Density	(g/cm^3)	1.28		
Composite Density	(g/cm^3)	1.62	1.58 - 1.65	
Fiber Areal Weight	(g/m^2)	152		
Ply Thickness	(in)		0.0049 - 0.0053	

LAMINATE PROPERTY SUMMARY

Classes of data: F - Fully approved, I - Interim, S - Screening in Strength/Modulus/Poisson's ratio/Strain-to-failure order.

** DATA WERE SUBMITTED BEFORE THE ESTABLISHMENT OF DATA DOCUMENTATION REQUIREMENTS (JUNE 1989). ALL DOCUMENTATION PRESENTLY REQUIRED WAS NOT SUPPLIED FOR THIS MATERIAL.

MATERIAL:	T300 15k/976 unidirectional tape			**Table 4.2.17(a)** **C/Ep - UT** **T300 15k/976** **Tension, 1-axis** **$[0]_6$** **72/A, -67/A, 260/A** **Screening**
RESIN CONTENT:	35 wt%	COMP: DENSITY:	1.60 g/cm^3	
FIBER VOLUME:	59 %	VOID CONTENT:	approx. 0.0%	
PLY THICKNESS:	0.0053 in.			
TEST METHOD: ASTM D3039-76		MODULUS CALCULATION: Linear portion of curve		
NORMALIZED BY:	Fiber volume to 60%			

		72		-67		260	
Temperature (°F) / Moisture Content (%) / Equilibrium at T, RH / Source Code		72 / ambient / / 48		-67 / ambient / / 48		260 / ambient / / 48	
		Normalized	Measured	Normalized	Measured	Normalized	Measured
F_1^{tu} (ksi)	Mean	211	207	199	197	236	232
	Minimum	185	191	187	173	205	212
	Maximum	235	219	220	214	256	255
	C.V.(%)	11.2	6.47	6.83	7.67	9.88	6.84
	B-value	(1)		(1)		(1)	
	Distribution	Normal		Normal		Normal	
	C_1	211		199		236	
	C_2	23.6		13.6		23.3	
	No. Specimens	5		5		5	
	No. Batches	1		1		1	
	Approval Class	Screening		Screening		Screening	
E_1^t (Msi)	Mean	19.6	19.3	20.8	20.4	22.6	22.4
	Minimum	17.8	18.2	19.5	19.6	20.5	21.2
	Maximum	21.2	20.4	22.6	21.0	24.9	22.9
	C.V.(%)	6.09	5.18	5.88	2.74	8.97	2.19
	No. Specimens	5		5		5	
	No. Batches	1		1		1	
	Approval Class	Screening		Screening		Screening	
ν_{12}^t	Mean		0.318		0.318		0.312
	No. Specimens	5		5		5	
	No. Batches	1		1		1	
	Approval Class	Screening		Screening		Screening	
ε_1^{tu} (με)	Mean		10400		8600		9900
	Minimum		10000		8000		9500
	Maximum		10800		9000		10500
	C.V.(%)		3.42		5.29		4.46
	B-value		(1)		(1)		(1)
	Distribution		Normal		Normal		Normal
	C_1		10400		8600		9900
	C_2		356		454		442
	No. Specimens	5		4		5	
	No. Batches	1		1		1	
	Approval Class	Screening		Screening		Screening	

(1) B-values are presented only for fully approved data.

** DATA WERE SUBMITTED BEFORE THE ESTABLISHMENT OF DATA DOCUMENTATION REQUIREMENTS (JUNE 1989). ALL DOCUMENTATION PRESENTLY REQUIRED WAS NOT SUPPLIED FOR THIS MATERIAL.

Table 4.2.17(b)
C/Ep - UT
T300 15k/976
Tension, 1-axis
$[0]_6$
350/A
Screening

MATERIAL:	T300 15k/976 unidirectional tape		
RESIN CONTENT:	35 wt%	COMP: DENSITY:	1.60 g/cm^3
FIBER VOLUME:	59 %	VOID CONTENT:	approx. 0.0%
PLY THICKNESS:	0.0053 in.		
TEST METHOD: ASTM D3039-76		MODULUS CALCULATION: Linear portion of curve	
NORMALIZED BY:	Fiber volume to 60%		

		Normalized	Measured	Normalized	Measured	Normalized	Measured
Temperature (°F)		350					
Moisture Content (%)		ambient					
Equilibrium at T, RH							
Source Code		48					
F_1^{tu} (ksi)	Mean	232	228				
	Minimum	212	219				
	Maximum	248	242				
	C.V.(%)	7.11	3.77				
	B-value	(1)					
	Distribution	Normal					
	C_1	232					
	C_2	16.5					
	No. Specimens	5					
	No. Batches	1					
	Approval Class	Screening					
E_1^t (Msi)	Mean	22.4	22.1				
	Minimum	21.0	20.2				
	Maximum	24.2	23.9				
	C.V.(%)	5.59	6.19				
	No. Specimens	5					
	No. Batches	1					
	Approval Class	Screening					
ν_{12}^t	Mean		0.348				
	No. Specimens	5					
	No. Batches	1					
	Approval Class	Screening					
ε_1^{tu} ($\mu\varepsilon$)	Mean		9930				
	Minimum		9600				
	Maximum		10700				
	C.V.(%)		5.29				
	B-value		(2)				
	Distribution		Normal				
	C_1		9930				
	C_2		525				
	No. Specimens	4					
	No. Batches	1					
	Approval Class	Screening					

(1) B-values are presented only for fully approved data.

** DATA WERE SUBMITTED BEFORE THE ESTABLISHMENT OF DATA DOCUMENTATION REQUIREMENTS (JUNE 1989). ALL DOCUMENTATION PRESENTLY REQUIRED WAS NOT SUPPLIED FOR THIS MATERIAL.

Table 4.2.17(c)
C/Ep - UT
T300 15k/976
Tension, 2-axis
$[90]_{15}$
72/A, -67/A, 260/A, 350/A
Screening

MATERIAL:	T300 15k/976 unidirectional tape
RESIN CONTENT:	25 wt%
COMP: DENSITY:	1.64 g/cm^3
FIBER VOLUME:	69 %
VOID CONTENT:	approx. 0.0%
PLY THICKNESS:	0.0049 in.
TEST METHOD:	ASTM D3039-76
MODULUS CALCULATION:	Linear portion of curve
NORMALIZED BY:	Not normalized

		72	-67	260	350		
Temperature (°F)		72	-67	260	350		
Moisture Content (%)		ambient	ambient	ambient	ambient		
Equilibrium at T, RH							
Source Code		48	48	48	48		
	Mean	5.66	4.73	3.81	3.47		
	Minimum	4.53	3.23	2.87	2.67		
	Maximum	6.52	6.29	4.68	3.83		
	C.V.(%)	15.4	25.1	17.4	13.2		
	B-value	(1)	(1)	(1)	(1)		
F_2^{tu}	Distribution	Normal	Normal	Normal	Normal		
(ksi)	C_1	5.66	4.73	3.812	3.47		
	C_2	0.870	1.19	0.664	0.458		
	No. Specimens	5	5	5	5		
	No. Batches	1	1	1	1		
	Approval Class	Screening	Screening	Screening	Screening		
	Mean	1.34	1.69	1.37	1.30		
	Minimum	1.28	1.49	1.16	1.25		
	Maximum	1.39	1.88	1.55	1.43		
E_2^t	C.V.(%)	3.13	9.01	10.1	5.83		
(Msi)	No. Specimens	5	5	5	5		
	No. Batches	1	1	1	1		
	Approval Class	Screening	Screening	Screening	Screening		
	Mean						
	No. Specimens						
ν_{21}^t	No. Batches						
	Approval Class						
	Mean	3900	2760	2640	2620		
	Minimum	3200	1900	2100	2200		
	Maximum	4600	3300	3400	3000		
	C.V.(%)	14.6	20.4	19.1	13.3		
	B-value	(1)	(1)	(1)	(1)		
ε_2^{tu}	Distribution	Normal	Normal	Normal	Normal		
(με)	C_1	3900	2760	2640	2620		
	C_2	570	564	503	349		
	No. Specimens	5	5	5	5		
	No. Batches	1	1	1	1		
	Approval Class	Screening	Screening	Screening	Screening		

(1) B-values are presented only for fully approved data.

** DATA WERE SUBMITTED BEFORE THE ESTABLISHMENT OF DATA DOCUMENTATION REQUIREMENTS (JUNE 1989). ALL DOCUMENTATION PRESENTLY REQUIRED WAS NOT SUPPLIED FOR THIS MATERIAL.

MATERIAL:	T300 15k/976 unidirectional tape			**Table 4.2.17(d)** **C/Ep - UT** **T300 15k/976** **Compression, 1-axis** **$[0]_{20}$** **72/A, -67/A, 260/A** **Screening**
RESIN CONTENT:	24 wt%	COMP: DENSITY:	1.63 g/cm^3	
FIBER VOLUME:	70 %	VOID CONTENT:	approx. 0.0%	
PLY THICKNESS:	0.0050 in.			
TEST METHOD:		MODULUS CALCULATION:		
ASTM D3410A-75		Linear portion of curve		
NORMALIZED BY:	Fiber volume to 60%			

Temperature (°F) Moisture Content (%) Equilibrium at T, RH Source Code		72 ambient 48		-67 ambient 48		260 ambient 48	
		Normalized	Measured	Normalized	Measured	Normalized	Measured
F_1^{cu} (ksi)	Mean	188	218	192	223	147	171
	Minimum	139	162	169	196	95.6	111
	Maximum	214	248	218	254	177	205
	C.V.(%)	15.9	15.9	9.76	9.76	21.7	21.7
	B-value	(1)		(1)		(1)	
	Distribution	Normal		Normal		Normal	
	C_1	188		192		147	
	C_2	29.9		18.8		31.9	
	No. Specimens	5		5		5	
	No. Batches	1		1		1	
	Approval Class	Screening		Screening		Screening	
E_1^c (Msi)	Mean	18.7	21.8	18.8	21.9	18.4	21.4
	Minimum	14.9	17.3	16.2	18.8	10.8	12.6
	Maximum	21.9	25.5	25.5	29.6	22.6	26.2
	C.V.(%)	13.4	13.4	20.1	20.1	26.5	26.5
	No. Specimens	5		5		5	
	No. Batches	1		1		1	
	Approval Class	Screening		Screening		Screening	
ν_{12}^c	Mean						
	No. Specimens						
	No. Batches						
	Approval Class						
ε_1^{cu} (με)	Mean		12500		14500		8860
	Minimum		9500		9900		6300
	Maximum		19600		20000		12600
	C.V.(%)		32.2		31.5		30.2
	B-value		(1)		(1)		(1)
	Distribution		Normal		Normal		Normal
	C_1		12500		14500		8860
	C_2		404		4560		2670
	No. Specimens	5		5		5	
	No. Batches	1		1		1	
	Approval Class	Screening		Screening		Screening	

(1) B-values are presented only for fully approved data.

** DATA WERE SUBMITTED BEFORE THE ESTABLISHMENT OF DATA DOCUMENTATION REQUIREMENTS (JUNE 1989). ALL DOCUMENTATION PRESENTLY REQUIRED WAS NOT SUPPLIED FOR THIS MATERIAL.

Table 4.2.17(e)
C/Ep - UT
T300 15k/976
Compression, 1-axis
$[0]_{20}$
350/A
Screening

MATERIAL: T300 15k/976 unidirectional tape

RESIN CONTENT: 24 wt%
FIBER VOLUME: 70 %
PLY THICKNESS: 0.0050 in.

COMP: DENSITY: 1.63 g/cm^3
VOID CONTENT: approx. 1.0%

TEST METHOD: ASTM D3410A-75

MODULUS CALCULATION: Linear portion of curve

NORMALIZED BY: Fiber volume to 60%

		Normalized	Measured	Normalized	Measured	Normalized	Measured
Temperature (°F)		350					
Moisture Content (%)		ambient					
Equilibrium at T, RH							
Source Code		48					
F_1^{cu} (ksi)	Mean	136	158				
	Minimum	107	124				
	Maximum	160	186				
	C.V.(%)	18.5	18.5				
	B-value	(1)					
	Distribution	Normal					
	C_1	136					
	C_2	25.2					
	No. Specimens	5					
	No. Batches	1					
	Approval Class	Screening					
E_1^c (Msi)	Mean	19.7	22.9				
	Minimum	16.5	19.1				
	Maximum	23.0	26.7				
	C.V.(%)	13.2	13.2				
	No. Specimens	5					
	No. Batches	1					
	Approval Class	Screening					
ν_{12}^c	Mean						
	No. Specimens						
	No. Batches						
	Approval Class						
ε_1^{cu} (με)	Mean		9400				
	Minimum		5000				
	Maximum		14000				
	C.V.(%)		39.7				
	B-value		(2)				
	Distribution		Normal				
	C_1		9400				
	C_2		3730				
	No. Specimens	5					
	No. Batches	1					
	Approval Class	Screening					

(1) B-values are presented only for fully approved data.

** DATA WERE SUBMITTED BEFORE THE ESTABLISHMENT OF DATA DOCUMENTATION REQUIREMENTS (JUNE 1989). ALL DOCUMENTATION PRESENTLY REQUIRED WAS NOT SUPPLIED FOR THIS MATERIAL.

Table 4.2.17(f)
C/Ep - UT
T300 15k/976
Compression, 2-axis
$[90]_{20}$
72/A, -67/A, 260/A, 350/A
Screening

MATERIAL: T300 15k/976 unidirectional tape

RESIN CONTENT: 24 wt% COMP: DENSITY: 1.63 g/cm^3
FIBER VOLUME: 70 % VOID CONTENT: approx 0.0%
PLY THICKNESS: 0.0050 in.

TEST METHOD: ASTM D3410A-75

MODULUS CALCULATION: Linear portion of curve

NORMALIZED BY: Not normalized

Temperature (°F) Moisture Content (%) Equilibrium at T, RH Source Code		72 ambient 48	-67 ambient 48	260 ambient 48	350 ambient 48		
F_2^{cu} (ksi)	Mean	30.0	35.1	22.6	19.1		
	Minimum	26.7	26.7	19.4	17.3		
	Maximum	31.9	44.9	25.7	22.8		
	C.V.(%)	7.10	18.9	10.7	11.7		
	B-value	(1)	(1)	(1)	(1)		
	Distribution	Normal	Normal	Normal	Normal		
	C_1	30.0	35.1	22.6	19.1		
	C_2	2.13	6.62	2.42	2.24		
	No. Specimens	5	5	5	5		
	No. Batches	1	1	1	1		
	Approval Class	Screening	Screening	Screening	Screening		
E_2^c (Msi)	Mean	1.46	1.84	1.84	1.64		
	Minimum	1.32	1.46	1.37	1.25		
	Maximum	1.73	2.18	3.03	2.02		
	C.V.(%)	11.1	17.0	36.7	19.6		
	No. Specimens	5	5	5	5		
	No. Batches	1	1	1	1		
	Approval Class	Screening	Screening	Screening	Screening		
ν_{21}^c	Mean						
	No. Specimens						
	No. Batches						
	Approval Class						
ε_2^{cu} ($\mu\varepsilon$)	Mean	32300	22100	14900	14200		
	Minimum	7900	13000	9600	6900		
	Maximum	46300	27700	21400	21300		
	C.V.(%)	44.7	31.1	40.1	47.2		
	B-value	(1)	(1)	(2)	(1)		
	Distribution	Normal	Normal		Normal		
	C_1	32300	22100		14200		
	C_2	14400	6880		6720		
	No. Specimens	5	5	3	5		
	No. Batches	1	1	1	1		
	Approval Class	Screening	Screening	Screening	Screening		

(1) B-values are presented only for fully approved data.
(2) The statistical analysis is not completed for less than four specimens.

** DATA WERE SUBMITTED BEFORE THE ESTABLISHMENT OF DATA DOCUMENTATION REQUIREMENTS (JUNE 1989). ALL DOCUMENTATION PRESENTLY REQUIRED WAS NOT SUPPLIED FOR THIS MATERIAL.

MATERIAL: T300 15k/976 unidirectional tape	**Table 4.2.17(g)** **C/Ep - UT** **T300 15k/976** **Shear, 12-plane** **$[\pm 45]_{2S}$** **72/A, -67/A, 260/A, 350/A** **Screening**
RESIN CONTENT: 25 wt% — COMP: DENSITY: 1.63 g/cm^3	
FIBER VOLUME: 69 % — VOID CONTENT: approx. 0.1%	
PLY THICKNESS: 0.0052 in.	
TEST METHOD: ASTM D3518-76 — MODULUS CALCULATION: Linear portion of curve	
NORMALIZED BY: Not normalized	

		72	-67	260	350	
Temperature (°F) Moisture Content (%) Equilibrium at T, RH Source Code		72 ambient 48	-67 ambient 48	260 ambient 48	350 ambient 48	
F_{12}^{su} (ksi)	Mean	11.1	13.7	8.25	8.30	
	Minimum	11.0	13.2	7.78	7.67	
	Maximum	11.4	15.5	8.72	9.36	
	C.V.(%)	1.23	6.99	4.78	7.80	
	B-value	(1)	(1)	(1)	(1)	
	Distribution	Normal	Nonpara.	Normal	Normal	
	C_1	11.1	4	8.25	8.30	
	C_2	0.137	4.10	0.394	0.647	
	No. Specimens	5	5	5	5	
	No. Batches	1	1	1	1	
	Approval Class	Screening	Screening	Screening	Screening	
G_{12}^{s} (Msi)	Mean	0.91	1.0	0.89	0.77	
	Minimum	0.84	0.89	0.82	0.70	
	Maximum	0.96	1.08	0.94	0.82	
	C.V.(%)	5.1	7.1	5.3	7.4	
	No. Specimens	5	5	5	5	
	No. Batches	1	1	1	1	
	Approval Class	Screening	Screening	Screening	Screening	
γ_{12}^{su} (με)	Mean					
	Minimum					
	Maximum					
	C.V.(%)					
	B-value					
	Distribution					
	C_1					
	C_2					
	No. Specimens					
	No. Batches					
	Approval Class					

(1) B-values are presented only for fully approved data.

** DATA WERE SUBMITTED BEFORE THE ESTABLISHMENT OF DATA DOCUMENTATION REQUIREMENTS (JUNE 1989). ALL DOCUMENTATION PRESENTLY REQUIRED WAS NOT SUPPLIED FOR THIS MATERIAL.

MATERIAL:	T300 15k/976 unidirectional tape			**Table 4.2.17(h) C/Ep - UT T300 15k/976 SBS, 31-plane $[0]_{15}$ 72/A, -67/A, 260/A, 350/A Screening**
RESIN CONTENT:	25 wt%	COMP: DENSITY:	1.63 g/cm^3	
FIBER VOLUME:	69 %	VOID CONTENT:	approx. 0.1%	
PLY THICKNESS:	0.0052 in.			
TEST METHOD: ASTM D2344-76		MODULUS CALCULATION: Linear portion of curve		
NORMALIZED BY:	Not normalized			

Temperature (°F) Moisture Content (%) Equilibrium at T, RH Source Code		72 ambient 48	-67 ambient 48	260 ambient 48	350 ambient 48	
	Mean	12.9	16.6	9.36	8.60	
	Minimum	9.42	14.2	8.59	7.71	
	Maximum	17.1	19.6	10.8	9.56	
	C.V.(%)	18.4	12.8	10.1	8.06	
	B-value	(1)	(1)	(1)	(1)	
	Distribution	Weibull	Normal	Normal	Normal	
F_{31}^{sbs}	C_1	13.8	16.6	9.36	8.60	
(ksi)	C_2	6.17	2.12	0.949	0.693	
	No. Specimens	10	5	5	5	
	No. Batches	1	1	1	1	
	Approval Class	Screening	Screening	Screening	Screening	

(1) B-values are presented only for fully approved data.

4.2.18 IM7 12k/8551-7A unitape

These data are presented in the MIL-HDBK-17-2E Annex A.

4.2.19 AS4 3k/3501-6 5 HS fabric data set description

Material Description:

Material: AS4-3k/3501-6

Form: 5 harness satin weave fabric, areal weight of 280 g/m^2, typical cured resin content of 28-30%, typical cured ply thickness of 0.0099 -0.0109 inches.

Processing: Autoclave cure; 240°F, 85 psi for 1 hour; 350°F, 100 psi for 2 hours, bleed.

General Supplier Information:

Fiber: AS4 fibers are continuous carbon filaments made from PAN precursor, surface treated to improve handling characteristics and structural properties. Filament count is 3000 filaments/tow, no twist. Typical tensile modulus is 34 x 10^6 psi. Typical tensile strength is 550,000 psi.

Matrix: 3501-6 is an amine-cured epoxy resin. It will retain light tack for a minimum of 10 days at room temperature.

Maximum Short Term Service Temperature: 300°F (dry), 180°F (wet)

Typical applications: General purpose structural applications.

4.2.19 AS4 3k/3501-6 5-harness satin weave (bleed)*

MATERIAL:	AS4 3k/3501-6 5-harness satin weave (Bleed)		**C/Ep 280-5HS AS4/3501-6 (Bleed) Summary**
PREPREG:	Hercules AW280-5H/3501-6		
FIBER:	Hercules AS4 3k, no twist	MATRIX:	Hercules 3501-6
T_g(dry):	T_g(wet):	T_g METHOD:	
PROCESSING:	Autoclave cure, 240 ± 10°F at 85 psig for 60 minutes; 350 ± 10°F for 120 ± 10 minutes at 100 ± 5 psig		

** ALL DOCUMENTATION PRESENTLY REQUIRED WERE NOT SUPPLIED FOR THIS MATERIAL.

Date of fiber manufacture		Date of testing	
Date of resin manufacture		Date of data submittal	6/90
Date of prepreg manufacture		Date of analysis	2/95
Date of composite manufacture			

LAMINA PROPERTY SUMMARY

	75°F/A		200°F/A		75°F/W	200°F/W		
Tension, 1-axis	SS--							
Tension, 2-axis								
Tension, 3-axis								
Compression, 1-axis	SS--		SS--		SS--	II--		
Compression, 2-axis								
Compression, 3-axis								
Shear, 12-plane								
Shear, 23-plane								
SB Strength, 31-plane	S---		S---		S---			

Classes of data: F - Fully approved, I - Interim, S - Screening in Strength/Modulus/Poisson's ratio/Strain-to-failure order.

** ALL DOCUMENTATION PRESENTLY REQUIRED WERE NOT SUPPLIED FOR THIS MATERIAL.

		Nominal	As Submitted	Test Method
Fiber Density	(g/cm^3)	1.80		
Resin Density	(g/cm^3)	1.26		
Composite Density	(g/cm^3)		1.59 - 1.60	
Fiber Areal Weight	(g/m^2)	280		
Ply Thickness	(in)		0.0099 - 0.0171	

LAMINATE PROPERTY SUMMARY

	75°F/A							
0/±45/90 Family								
Tension, x-axis	SS--							
OHT, x-axis	S---							

Classes of data: F - Fully approved, I - Interim, S - Screening in Strength/Modulus/Poisson's ratio/Strain-to-failure order.

** ALL DOCUMENTATION PRESENTLY REQUIRED WERE NOT SUPPLIED FOR THIS MATERIAL.

MATERIAL:	AS4 3k/3501-6 (Bleed) 5-harness satin weave			**Table 4.2.19(a)** **C/Ep 280-5HS** **AS4/3501-6 (Bleed)** **Tension, 1-axis** **$[0_f]_8$** **75/A** **Screening**
RESIN CONTENT:	29 wt%	COMP: DENSITY:	1.61 g/cm^3	
FIBER VOLUME:	61 vol %	VOID CONTENT:		
PLY THICKNESS:	0.0100-0.0106 in.			
TEST METHOD: ASTM D3039-76		MODULUS CALCULATION:		
NORMALIZED BY:	Specimen thickness and batch fiber volume to 57% (0.019 in. CPT)			

		Normalized	Measured	Normalized	Measured	Normalized	Measured
Temperature (°F)		75					
Moisture Content (%)		ambient					
Equilibrium at T, RH							
Source Code		43					
F_1^{tu} (ksi)	Mean	108	115				
	Minimum	93.3	98.8				
	Maximum	128	137				
	C.V.(%)	12.2	12.2				
	B-value	(1)	(1)				
	Distribution	ANOVA	ANOVA				
	C_1	14.9	15.8				
	C_2	5.74	5.72				
	No. Specimens	9					
	No. Batches	3					
	Approval Class	Screening					
E_1^t (Msi)	Mean	9.83	10.4				
	Minimum	8.25	8.80				
	Maximum	12.0	13.1				
	C.V.(%)	9.88	10.8				
	No. Specimens	9					
	No. Batches	3					
	Approval Class	Screening					
ν_{12}^t	Mean						
	No. Specimens						
	No. Batches						
	Approval Class						
ε_1^{tu} (με)	Mean						
	Minimum						
	Maximum						
	C.V.(%)						
	B-value						
	Distribution						
	C_1						
	C_2						
	No. Specimens						
	No. Batches						
	Approval Class						

(1) B-values are presented only for fully approved data.

** ALL DOCUMENTATION PRESENTLY REQUIRED WERE NOT SUPPLIED FOR THIS MATERIAL.

Table 4.2.19(b)
C/Ep 280-5HS
AS4/3501-6 (Bleed)
Compression, 1-axis
$[0_f]_8$
75/A, 200/A, 75/W
Screening

MATERIAL: AS4 3k/3501-6 (Bleed) 5-harness satin weave

RESIN CONTENT: 29 wt% | COMP: DENSITY: 1.61 g/cm³
FIBER VOLUME: 61 vol % | VOID CONTENT:
PLY THICKNESS: 0.0099-0.0104 in.

TEST METHOD: SACMA SRM 1-88 | MODULUS CALCULATION:

NORMALIZED BY: Specimen thickness and batch fiber volume to 57% (0.019 in. CPT)

		75		200		75	
Temperature (°F)		75		200		75	
Moisture Content (%)		ambient		ambient		wet	
Equilibrium at T, RH						(1)	
Source Code		43		43		43	
		Normalized	Measured	Normalized	Measured	Normalized	Measured
F_1^{cu} (ksi)	Mean	106	113	80.8	86.1	95.8	102
	Minimum	91.0	97.7	67.6	73.7	79.3	84.7
	Maximum	115	123	93.1	99.9	106	113
	C.V.(%)	6.52	6.65	8.84	8.69	9.43	9.42
	B-value	(2)	(2)	(2)	(2)	(2)	(2)
	Distribution	ANOVA	Weibull	Weibull	Weibull	Normal	Normal
	C_1	7.21	116	83.9	89.4	95.8	102
	C_2	3.73	18.4	13.6	13.4	9.03	9.64
	No. Specimens	13		13		9	
	No. Batches	3		3		2	
	Approval Class	Screening		Screening		Screening	
E_1^c (Msi)	Mean	8.7	9.3	8.48	9.04	9.23	9.87
	Minimum	7.6	8.2	6.42	7.00	9.07	9.70
	Maximum	9.4	9.9	9.43	10.0	9.44	10.2
	C.V.(%)	8.2	8.4	10.6	10.4	1.55	1.68
	No. Specimens	13		13		9	
	No. Batches	3		3		2	
	Approval Class	Screening		Screening		Screening	
ν_{12}^c	Mean						
	No. Specimens						
	No. Batches						
	Approval Class						
ε_1^{cu} (με)	Mean						
	Minimum						
	Maximum						
	C.V.(%)						
	B-value						
	Distribution						
	C_1						
	C_2						
	No. Specimens						
	No. Batches						
	Approval Class						

(1) Conditioned at 140°F, 95% relative humidity for 30 days.
(2) B-values are presented only for fully approved data.

** ALL DOCUMENTATION PRESENTLY REQUIRED WERE NOT SUPPLIED FOR THIS MATERIAL.

MATERIAL: AS4 3k/3501-6 (Bleed) 5-harness satin weave RESIN CONTENT: 29 wt% COMP: DENSITY: 1.59 g/cm³ FIBER VOLUME: 61 vol % VOID CONTENT: PLY THICKNESS: 0.0111-0.0171 in. TEST METHOD: MODULUS CALCULATION: SACMA SRM 1-88 NORMALIZED BY: Specimen thickness and batch fiber volume to 57% (0.019 in. CPT)	**Table 4.2.19(c)** **C/Ep 280-5HS** **AS4/3501-6 (Bleed)** **Compression, 1-axis** **$[0]_8$** **200/W** **Interim**

Temperature (°F) Moisture Content (%) Equilibrium at T, RH Source Code		200 wet (1) 43					
		Normalized	Measured	Normalized	Measured	Normalized	Measured
F_1^{cu} (ksi)	Mean	57.0	60.8				
	Minimum	49.8	53.8				
	Maximum	67.8	72.2				
	C.V.(%)	8.85	8.82				
	B-value	(2)	(2)				
	Distribution	ANOVA	ANOVA				
	C_1	5.46	5.761				
	C_2	4.57	4.38				
	No. Specimens	15					
	No. Batches	3					
	Approval Class	Interim					
E_1^c (Msi)	Mean	8.1	8.6				
	Minimum	6.5	7.0				
	Maximum	9.0	9.4				
	C.V.(%)	10	10				
	No. Specimens	15					
	No. Batches	3					
	Approval Class	Interim					
ν_{12}^c	Mean						
	No. Specimens						
	No. Batches						
	Approval Class						
ε_1^{cu} (με)	Mean						
	Minimum						
	Maximum						
	C.V.(%)						
	B-value						
	Distribution						
	C_1						
	C_2						
	No. Specimens						
	No. Batches						
	Approval Class						

(1) Conditioned at 140°F, 95% relative humidity for 30 days.
(2) B-values are presented only for fully approved data.

** ALL DOCUMENTATION PRESENTLY REQUIRED WERE NOT SUPPLIED FOR THIS MATERIAL.

MATERIAL:	AS4 3k/3501-6 (Bleed) 5-harness satin weave			Table 4.2.19(d) C/Ep 280-5HS AS4/3501-6 (Bleed) SBS, 31-plane $[0]_8$ 75/A, 200/A, 75/W Screening
RESIN CONTENT:	28-30 wt%	COMP: DENSITY:	1.59-1.60 g/cm^3	
FIBER VOLUME:	60-62 vol %	VOID CONTENT:		
PLY THICKNESS:	0.0099-0.0104 in.			
TEST METHOD:		MODULUS CALCULATION:		
ASTM D2344-84		N/A		
NORMALIZED BY:	Not normalized			

Temperature (°F)		75	200	75		
Moisture Content (%)		ambient	ambient	wet		
Equilibrium at T, RH				(1)		
Source Code		43	43	43		
	Mean	9.93	7.94	9.35		
	Minimum	8.50	7.60	9.00		
	Maximum	10.7	8.40	9.60		
	C.V.(%)	7.38	3.89	2.22		
	B-value	(2)	(2)	(2)		
	Distribution	Normal	ANOVA	Normal		
F_{31}^{sbs}	C_1	9.93	0.353	9.35		
(ksi)	C_2	0.733	6.02	0.207		
	No. Specimens	9	9	6		
	No. Batches	3	3	2		
	Approval Class	Screening	Screening	Screening		

(1) Conditioned at 140°F, 95% relative humidity for 30 days.
(2) B-values are presented only for fully approved data.

** ALL DOCUMENTATION PRESENTLY REQUIRED WERE NOT SUPPLIED FOR THIS MATERIAL.

MATERIAL:	AS4 3k/3501-6 (Bleed) 5-harness satin weave			**Table 4.2.19(e)** **C/Ep 280-5HS** **AS4/3501-6 (Bleed)** **Tension, x-axis** **[(0/±45/90)f]$_s$** **75/A** **Screening**
RESIN CONTENT:	29 wt%	COMP: DENSITY:	1.59 g/cm^3	
FIBER VOLUME:	61 vol %	VOID CONTENT:		
PLY THICKNESS:	0.0105-0.0106 in.			
TEST METHOD: ASTM D3039-76		MODULUS CALCULATION:		
NORMALIZED BY:	Specimen thickness and batch fiber volume to 57% (0.019 in. CPT)			

Temperature (°F) Moisture Content (%) Equilibrium at T, RH Source Code		75 ambient 43					
		Normalized	Measured	Normalized	Measured	Normalized	Measured
F_x^{tu} (ksi)	Mean	83.4	88.6				
	Minimum	75.7	81.3				
	Maximum	88.2	94.2				
	C.V.(%)	5.28	4.86				
	B-value	(1)	(1)				
	Distribution	Normal	Normal				
	C_1	83.4	88.6				
	C_2	4.41	4.30				
	No. Specimens	6					
	No. Batches	2					
	Approval Class	Screening					
E_x^t (Msi)	Mean	6.9	7.3				
	Minimum	6.6	7.0				
	Maximum	7.0	7.5				
	C.V.(%)	2.8	2.9				
	No. Specimens	6					
	No. Batches	2					
	Approval Class	Screening					
ν_{xy}^t	Mean						
	No. Specimens						
	No. Batches						
	Approval Class						
ε_x^{tu} (με)	Mean						
	Minimum						
	Maximum						
	C.V.(%)						
	B-value						
	Distribution						
	C_1						
	C_2						
	No. Specimens						
	No. Batches						
	Approval Class						

(1) B-values are presented only for fully approved data.

** ALL DOCUMENTATION PRESENTLY REQUIRED WERE NOT SUPPLIED FOR THIS MATERIAL.

MATERIAL:	AS4 3k/3501-6 (Bleed) 5-harness satin weave			**Table 4.2.19(f)** **C/Ep 280-5HS** **AS4/3501-6 (Bleed)** **OHT, x-axis** **[(0/±45/90)f]$_s$** **75/A** **Screening**
RESIN CONTENT:	29-30 wt%	COMP: DENSITY:	1.59-1.60 g/cm^3	
FIBER VOLUME:	61-62 vol %	VOID CONTENT:		
PLY THICKNESS:	0.0105-0.0109 in.			
TEST METHOD: SACMA SRM 5-88		MODULUS CALCULATION:		
NORMALIZED BY:	Specimen thickness and batch fiber volume to 57% (0.019 in. CPT)			

Temperature (°F) Moisture Content (%) Equilibrium at T, RH Source Code		75 ambient 43					
		Normalized	Measured	Normalized	Measured	Normalized	Measured
F_x^{oht} (ksi)	Mean	58.4	63.0				
	Minimum	57.0	60.9				
	Maximum	61.0	64.5				
	C.V.(%)	2.57	2.43				
	B-value	(1)	(1)				
	Distribution	Normal	Normal				
	C_1	58.4	63.0				
	C_2	1.50	1.53				
	No. Specimens	6					
	No. Batches	2					
	Approval Class	Screening					
E_x^{oht} (Msi)	Mean						
	Minimum						
	Maximum						
	C.V.(%)						
	No. Specimens						
	No. Batches						
	Approval Class						
ε_x^{oht} ($\mu\varepsilon$)	Mean						
	Minimum						
	Maximum						
	C.V.(%)						
	B-value						
	Distribution						
	C_1						
	C_2						
	No. Specimens						
	No. Batches						
	Approval Class						

(1) B-values are presented only for fully approved data.

4.2.20 AS4 3k/3501-6 5 HS fabric data set description

Material Description:

Material: AS4-3k/3501-6

Form: 5 harness satin weave fabric, areal weight of 280 g/m^2, typical cured resin content of 36-39%, typical cured ply thickness of 0.0110 -0.0121 inches.

Processing: Autoclave cure; 240°F, 85 psi for 1 hour, 350°F, 100 psi for 2 hours, no bleed.

General Supplier Information:

Fiber: AS4 fibers are continuous carbon filaments made from PAN precursor, surface treated to improve handling characteristics and structural properties. Filament count is 3000 filaments per tow, no twist. Typical tensile modulus is 34 x 10^6 psi. Typical tensile strength is 550,000 psi.

Matrix: 3501-6 is an amine-cured epoxy resin. It will retain light tack for a minimum of 10 days at room temperature.

Maximum Short Term Service Temperature: 300°F (dry), 180°F (wet)

Typical applications: General purpose structural applications.

4.2.20 AS4 3k/3501-6 (no bleed) 5-harness satin*

				C/Ep AS4/3501-6 (No Bleed) Summary
MATERIAL:	AS4 3k/3501-6 (No Bleed) 5-harness satin			
PREPREG:	Hercules AW280-5H/3501-6			
FIBER:	Hercules AS4 3k, no twist	MATRIX:	Hercules 3501-6	
T_g(dry):	T_g(wet):	T_g METHOD:		
PROCESSING:	Autoclave cure, 240 ± 10°F at 85 psig for 60 minutes; 350 ± 10°F at 100 ± 5 psig for 120 ± 10 minutes.			

** ALL DOCUMENTATION PRESENTLY REQUIRED WERE NOT SUPPLIED FOR THIS MATERIAL.

Date of fiber manufacture		Date of testing	
Date of resin manufacture		Date of data submittal	6/90
Date of prepreg manufacture		Date of analysis	2/95-3/95
Date of composite manufacture			

LAMINA PROPERTY SUMMARY

	75°F/A		-65°F/A	200°F/A				
Tension, 1-axis	SS--		SS--	SS--				
Tension, 2-axis								
Tension, 3-axis								
Compression, 1-axis	SS--							
Compression, 2-axis								
Compression, 3-axis								
Shear, 12-plane								
Shear, 23-plane								
SB Strength, 31-plane	S---							

Classes of data: F - Fully approved, I - Interim, S - Screening in Strength/Modulus/Poisson's ratio/Strain-to-failure order.

** ALL DOCUMENTATION PRESENTLY REQUIRED WERE NOT SUPPLIED FOR THIS MATERIAL.

		Nominal	As Submitted	Test Method
Fiber Density	(g/cm^3)	1.80		
Resin Density	(g/cm^3)	1.27		
Composite Density	(g/cm^3)	1.55	1.55 - 1.56	
Fiber Areal Weight	(g/m^2)	280		
Ply Thickness	(in)	0.011	0.011 - 0.017	

LAMINATE PROPERTY SUMMARY

	75°F/A							
0/±45/90 Family								
Tension, x-axis	SS--							
OHT, x-axis	S---							

Classes of data: F - Fully approved, I - Interim, S - Screening in Strength/Modulus/Poisson's ratio/Strain-to-failure order.

** ALL DOCUMENTATION PRESENTLY REQUIRED WERE NOT SUPPLIED FOR THIS MATERIAL.

MATERIAL:	AS4 3k/3501-6 (No Bleed) 5-harness satin			**Table 4.2.20(a)** **C/EP 280-5HS** **AS4/3501-6 (No Bleed)** **Tension, 1-axis** **$[0_f]_8$** **75/A, -65/A, 200/A** **Screening**
RESIN CONTENT:	36-39 wt%	COMP: DENSITY:	1.55-1.56 g/cm^3	
FIBER VOLUME:	52-55 vol %	VOID CONTENT:		
PLY THICKNESS:	0.0111-0.0171 in.			
TEST METHOD: ASTM D3039-76		MODULUS CALCULATION:		
NORMALIZED BY:	Specimen thickness and batch fiber volume to 57% (0.011 in. CPT)			

Temperature (°F) Moisture Content (%) Equilibrium at T, RH Source Code		75 ambient 43		-65 ambient 43		200 ambient 43	
		Normalized	Measured	Normalized	Measured	Normalized	Measured
	Mean	134	125	125	117	130	121
	Minimum	129	117	120	109	124	116
	Maximum	146	136	136	127	141	136
	C.V.(%)	3.79	4.85	3.85	4.89	4.49	5.11
	B-value	(1)	(1)	(1)	(1)	(1)	(1)
F_1^{tu}	Distribution	Normal	ANOVA	Normal	ANOVA	Lognormal	Nonpara.
(ksi)	C_1	134	6.56	125	6.07	4.86	6
	C_2	5.07	4.77	4.81	4.40	0.0440	2.25
	No. Specimens	9		9		9	
	No. Batches	3		3		3	
	Approval Class	Screening		Screening		Screening	
	Mean	9.67	9.06	10.2	9.57	10.8	10.1
	Minimum	9.39	8.60	9.63	8.80	9.88	9.00
	Maximum	9.88	9.50	11.0	10.3	11.8	11.3
E_1^t	C.V.(%)	1.65	3.63	4.26	5.68	6.74	8.23
(Msi)	No. Specimens	9		9		9	
	No. Batches	3		3		3	
	Approval Class	Screening		Screening		Screening	
	Mean						
	No. Specimens						
ν_{12}^t	No. Batches						
	Approval Class						
	Mean						
	Minimum						
	Maximum						
	C.V.(%)						
	B-value						
ε_1^{tu}	Distribution						
($\mu\varepsilon$)	C_1						
	C_2						
	No. Specimens						
	No. Batches						
	Approval Class						

(1) B-values are presented only for fully approved data.

** ALL DOCUMENTATION PRESENTLY REQUIRED WERE NOT SUPPLIED FOR THIS MATERIAL.

MATERIAL: AS4 3k/3501-6 (No Bleed) 5-harness satin RESIN CONTENT: 36-39 wt% COMP: DENSITY: 1.55-1.56 g/cm^3 FIBER VOLUME: 52-55 vol % VOID CONTENT: PLY THICKNESS: 0.0114-0.0121 in. TEST METHOD: MODULUS CALCULATION: SACMA SRM 1-88 NORMALIZED BY: Specimen thickness and batch fiber volume to 57% (0.011 in. CPT)	**Table 4.2.20(b)** **C/EP 280-5HS** **AS4/3501-6 (No Bleed)** **Compression, 1-axis** **$[0_f]_8$** **75/A** **Interim**

Temperature (°F) Moisture Content (%) Equilibrium at T, RH Source Code		75 ambient 43					
		Normalized	Measured	Normalized	Measured	Normalized	Measured
	Mean	129	121				
	Minimum	121	111				
	Maximum	145	137				
	C.V.(%)	5.02	6.03				
	B-value	(1)	(1)				
F_1^{cu}	Distribution	Weibull	ANOVA				
(ksi)	C_1	133	7.84				
	C_2	18.9	4.39				
	No. Specimens	15					
	No. Batches	3					
	Approval Class	Interim					
	Mean	9.42	8.81				
	Minimum	8.71	8.30				
	Maximum	10.0	9.50				
E_1^c	C.V.(%)	4.25	5.35				
(Msi)	No. Specimens	15					
	No. Batches	3					
	Approval Class	Interim					
	Mean						
	No. Specimens						
ν_{12}^c	No. Batches						
	Approval Class						
	Mean						
	Minimum						
	Maximum						
	C.V.(%)						
	B-value						
ε_1^{cu}	Distribution						
(με)	C_1						
	C_2						
	No. Specimens						
	No. Batches						
	Approval Class						

(1) B-values are presented only for fully approved data.

** ALL DOCUMENTATION PRESENTLY REQUIRED WERE NOT SUPPLIED FOR THIS MATERIAL.

Table 4.2.20(c)
C/Ep 280-5HS
AS4/3501-6 (No Bleed)
SBS, 31-plane
$[0_f]_8$
75/A
Screening

MATERIAL: AS4 3k/3501-6 (No Bleed) 5-harness satin

RESIN CONTENT: 36-39 wt% COMP: DENSITY: 1.55-1.56 g/cm³
FIBER VOLUME: 52-55 vol % VOID CONTENT:
PLY THICKNESS: 0.0110-0.0114 in.

TEST METHOD: ASTM D2344-84

MODULUS CALCULATION: N/A

NORMALIZED BY: Not normalized

Temperature (°F) Moisture Content (%) Equilibrium at T, RH Source Code		75 ambient 43				
	Mean	11.3				
	Minimum	10.1				
	Maximum	12.1				
	C.V.(%)	5.05				
	B-value	(1)				
	Distribution	ANOVA				
F_{31}^{sbs}	C_1	0.611				
(ksi)	C_2	4.35				
	No. Specimens	15				
	No. Batches	3				
	Approval Class	Screening				

(1) B-values are presented only for fully approved data.

** ALL DOCUMENTATION PRESENTLY REQUIRED WERE NOT SUPPLIED FOR THIS MATERIAL.

MATERIAL: AS4 3k/3501-6 (No Bleed) 5-harness satin

RESIN CONTENT: 36-39 wt% COMP: DENSITY: 1.55-1.56 g/cm^3
FIBER VOLUME: 52-55 vol % VOID CONTENT:
PLY THICKNESS: 0.0113-0.0116 in.

TEST METHOD: MODULUS CALCULATION:
ASTM D3039-76

NORMALIZED BY: Specimen thickness and batch fiber volume to 57% (0.011 in. CPT)

Table 4.2.20(d)
C/EP 280-5HS
AS4/3501-6 (No Bleed)
Tension, x-axis
[(0/45/90/-45)$_s$]$_s$
75/A
Screening

Temperature (°F) Moisture Content (%) Equilibrium at T, RH Source Code	75 ambient 43					
	Normalized	Measured	Normalized	Measured	Normalized	Measured
F_x^{tu} (ksi) Mean	80.4	75.3				
Minimum	77.1	68.8				
Maximum	86.4	82.0				
C.V.(%)	3.85	5.41				
B-value	(1)	(1)				
Distribution	Normal	ANOVA				
C_1	80.4	4.45				
C_2	3.09	5.07				
No. Specimens	9					
No. Batches	3					
Approval Class	Screening					
E_x^t (Msi) Mean	6.94	6.50				
Minimum	6.73	6.30				
Maximum	7.13	6.60				
C.V.(%)	1.87	2.04				
No. Specimens	9					
No. Batches	3					
Approval Class	Screening					
ν_{xy}^t Mean						
No. Specimens						
No. Batches						
Approval Class						
ε_x^{tu} (με) Mean						
Minimum						
Maximum						
C.V.(%)						
B-value						
Distribution						
C_1						
C_2						
No. Specimens						
No. Batches						
Approval Class						

(1) B-values are presented only for fully approved data.

** ALL DOCUMENTATION PRESENTLY REQUIRED WERE NOT SUPPLIED FOR THIS MATERIAL.

Table 4.2.20(e)
C/EP 280-5HS
AS4/3501-6 (No Bleed)
OHT, x-axis
$[(0/\pm45/90)_f]_s$
75/A
Screening

MATERIAL: AS4 3k/3501-6 (No Bleed) 5-harness satin

RESIN CONTENT: 36-39 wt% COMP: DENSITY: 1.55-1.56 g/cm³
FIBER VOLUME: 52-55 vol % VOID CONTENT:
PLY THICKNESS: 0.0113-0.0116 in.

TEST METHOD: SACMA SRM 5-88 MODULUS CALCULATION:

NORMALIZED BY: Specimen thickness and batch fiber volume to 57% (0.011 in. CPT)

Temperature (°F) Moisture Content (%) Equilibrium at T, RH Source Code		75 ambient 43					
		Normalized	Measured	Normalized	Measured	Normalized	Measured
F_x^{oht} (ksi)	Mean	54.4	55.5				
	Minimum	51.4	52.9				
	Maximum	57.7	58.7				
	C.V.(%)	4.58	3.72				
	B-value	(1)	(1)				
	Distribution	ANOVA	Normal				
	C_1	2.80	55.5				
	C_2	5.64	2.06				
	No. Specimens	9					
	No. Batches	3					
	Approval Class	Screening					
E_x^{oht} (Msi)	Mean						
	Minimum						
	Maximum						
	C.V.(%)						
	No. Specimens						
	No. Batches						
	Approval Class						
ε_x^{oht} ($\mu\varepsilon$)	Mean						
	Minimum						
	Maximum						
	C.V.(%)						
	B-value						
	Distribution						
	C_1						
	C_2						
	No. Specimens						
	No. Batches						
	Approval Class						

(1) B-values are presented only for fully approved data.

4.2.21 IM6 3501-6 unitape

These data are presented in the MIL-HDBK-17-2E Annex A.

4.2.22 IM7 12k/8552 unitape

These data are presented in the MIL-HDBK-17-2E Annex A.

4.2.23 T300 3k/977-2 plain weave fabric

These data are presented in the MIL-HDBK-17-2E Annex A.

4.2.24 T-300 3k/977-2 8HS

These data are presented in the MIL-HDBK-17-2E Annex A.

4.2.25 IM7 12k/977-2 unitape

These data are presented in the MIL-HDBK-17-2E Annex A.

4.2.26 AS4 6k/PR500 5HS data set description

Material Description:

Material: AS4 6k/PR500

Form: 5 harness satin weave fabric, with 4% PT500 tackifier resin, fiber areal weight of 370 g/m^2, injected with PR500 resin by Resin Transfer Molding (RTM); typical cured resin content of 28-34%, typical cured ply thickness of 0.013 - 0.0145 inches.

Processing: RTM injection at > 320°F, cure for 2 hours at 350°F

General Supplier Information:

Fiber: Hercules/Hexcel AS4 fibers are continuous carbon filaments made from a PAN precursor woven into 5HS fabric. Typical tensile modulus is 34 x 10^6 psi. Typical tensile strength is 550,000 psi.

Matrix: 3M PR 500 is a one part, 350°F curing epoxy resin system especially suited to RTM processing. Characteristics include: excellent toughness with 300°F wet mechanical performance, several weeks of room temperature stability and low viscosity at recommended injection temperature.

Maximum Short Term Service Temperature: 350°F (dry), 300°F (wet)

Typical applications: Primary and secondary aircraft structure (commercial and military) and other applications requiring unusual hot/wet properties and impact resistance where RTM advantages such as precise dimensional tolerances, part consolidation, complex lay-ups and replicated surface finishes are desired.

4.2.26 AS4 6k/PR500 5-harness satin weave*

MATERIAL:	AS4 6k/PR 500 5HS		**C/Ep 370-5HS AS4/PR 500 Summary**
PREPREG:	Fiberite 5HS 12 tows/in., 4% PT-500		
FIBER:	Hercules AS4 6K, GP sizing, 0 twist	MATRIX:	3M PR 500 RTM
T_g(dry):	378°F T_g(wet): 340°F	T_g METHOD:	SRM 18-94, RDA G′ knee
PROCESSING:	Resin transfer molding: 360±10°F, 120 minutes, press pressure 175 psi, internal cure pressure 80 psi, mold temperature during injection 320°F, pump plate temperature 140-5, pump hose temperate 160-5		

Date of fiber manufacture	12/93-5/94	Date of testing	5/95-11/95
Date of resin manufacture	8/94-9/94	Date of data submittal	6/96
Date of prepreg manufacture	11/94-12/94	Date of analysis	8/96
Date of composite manufacture	1/95-10/95		

LAMINA PROPERTY SUMMARY

	72°F/A	-75°F/A	180°F/A	300°F/A	350°F/A	180°F/ W	240°F/W	300°F/W
Tension, 1-axis	II-I		II-I	SS-S	IS-S	II-S	II-S	II-I
Tension, 2-axis								
Tension, 3-axis								
Compression, 1-axis	II--	-I--	II--	I---	S---	I---	S---	S---
Compression, 2-axis								
Compression, 3-axis								
Shear, 12-plane	II--	II--	SS--	II--	SS--	II--	SS--	SS--
Shear, 23-plane								
Shear, 31-plane	I---		I---	I---		I---		I---
SB Strength, 31-plane	S---		S---	S---		S---		S---

Classes of data: F - Fully approved, I - Interim, S - Screening in Strength/Modulus/Poisson's ratio/Strain-to-failure order.

Data are also included for 240°F/W and 12-plane shear for four fluids in addition to water.

		Nominal	As Submitted	Test Method
Fiber Density	(g/cm^3)	1.787		ASTM C693
Resin Density	(g/cm^3)	1.25		ASTM D792
Composite Density	(g/cm^3)		1.55-1.60*	
Fiber Areal Weight	(g/m^2)	370	375	SRM 23-94
Fiber Volume	(% vol)		55.5-64.8	
Ply Thickness	(in)	0.014	0.0128-0.0149	

* Throughout this section, resin content and composite density have been calculated assuming zero void content.

LAMINATE PROPERTY SUMMARY

	72°F/A	-75°F/A	180°F/A	300°F/A	350°F/A	180°F/W	240°F/W	300°F/W
[0/45/90/-45]								
OHT, x-axis	IS-S	IS-S	IS-S	IS-S	IS-S	IS-S	IS-S	S---
OHC, x-axis	IS-S		IS-S	II-I		IS-S	II-I	II-I
CAI, x-axis	I---							
G_{Ic}	S---							
G_{IIc}	I---							

Classes of data: F - Fully approved, I - Interim, S - Screening in Strength/Modulus/Poisson's ratio/Strain-to-failure order.

Data are also included for 240/W and five impact energy levels for CAI.

MATERIAL:	AS4 6k/PR 500 RTM 5HS			Table 4.2.26(a) C/Ep 370-5HS AS4/PR 500 Tension, 1-axis $[0_f]_{3s}$ 72/A, 180/A, 240/A Screening, Interim
RESIN CONTENT:	30 - 34 wt%	COMP: DENSITY:	1.56 - 1.58 g/cm³	
FIBER VOLUME:	57.6 - 62.0 vol %	VOID CONTENT:	NA	
PLY THICKNESS:	0.0133 - 0.0142 in.			
TEST METHOD: SRM 4R-94		MODULUS CALCULATION: Chord between 1000 and 3000 $\mu\varepsilon$		
NORMALIZED BY:	Specimen thickness and batch FAW to 57% fiber volume (0.0145 in. CPT)			

Temperature (°F) Moisture Content (%) Equilibrium at T, RH Source Code		72 ambient 61		180 ambient 61		240 ambient 61	
		Normalized	Measured	Normalized	Measured	Normalized	Measured
	Mean	115	120	115	118	117	122
	Minimum	105	111	102	105	103	106
	Maximum	124	129	126	128	125	133
	C.V.(%)	4.50	4.74	5.48	4.94	4.79	5.15
	B-value	(1)	(1)	(1)	(1)	(1)	(1)
F_1^{tu}	Distribution	ANOVA	ANOVA	ANOVA	Weibull	ANOVA	ANOVA
(ksi)	C_1	5.71	6.44	7.01	121	6.03	6.67
	C_2	4.43	4.83	4.65	23.5	4.42	4.06
	No. Specimens	17		16		15	
	No. Batches	3		3		3	
	Approval Class	Interim		Interim		Interim	
	Mean	9.54	9.97	9.44	9.73	9.53	9.94
	Minimum	9.15	9.46	9.01	9.09	9.26	9.46
	Maximum	9.86	10.5	9.80	10.2	9.88	10.2
E_1^t	C.V.(%)	1.78	3.64	2.62	3.35	2.13	2.43
(Msi)	No. Specimens	15		16		15	
	No. Batches	3		3		3	
	Approval Class	Interim		Interim		Interim	
	Mean						
	No. Specimens						
ν_{12}^t	No. Batches						
	Approval Class						
	Mean		11900		11800		11600
	Minimum		10800		10200		10000
	Maximum		13700		16400		13100
	C.V.(%)		6.17		12.4		7.68
	B-value		(1)		(1)		(1)
ε_1^{tu}	Distribution		Nonpara		ANOVA		Weibull
($\mu\varepsilon$)	C_1		8		1510		12000
	C_2		1.54		3.294		16.2
	No. Specimens	15		15		13	
	No. Batches	3		3		3	
	Approval Class	Interim		Interim		Screening	

(1) B-values are presented only for fully approved data.

MATERIAL:	AS4 6k/PR 500 RTM 5HS			**Table 4.2.26(b)** **C/Ep 370-5HS** **AS4/PR 500** **Tension, 1-axis** **$[0]_8$** **300/A, 350/A, 180/W** **Screening, Interim**
RESIN CONTENT:	30 - 34 wt%	COMP: DENSITY:	1.56 - 1.58 g/cm^3	
FIBER VOLUME:	57.6 - 62.0 vol %	VOID CONTENT:	NA	
PLY THICKNESS:	0.0133 - 0.0142 in.			
TEST METHOD: SRM 4R-94		MODULUS CALCULATION: Chord between 1000 and 3000 $\mu\varepsilon$		
NORMALIZED BY:	Specimen thickness and batch FAW to 57% fiber volume (0.0145 in. CPT)			

Temperature (°F) Moisture Content (%) Equilibrium at T, RH Source Code		300 ambient 61		350 ambient 61		180 (2) 160°F water 61	
		Normalized	Measured	Normalized	Measured	Normalized	Measured
F_1^{tu} (ksi)	Mean	111	117	105	114	112	114
	Minimum	104	111	94.6	103	103	109
	Maximum	118	122	112	123	119	119
	C.V.(%)	3.97	2.82	4.39	4.75	4.66	2.57
	B-value	(1)	(1)	(1)	(1)	(1)	(1)
	Distribution	ANOVA	Weibull	ANOVA	Weibull	ANOVA	ANOVA
	C_1	4.91	119	5.19	117	5.89	3.25
	C_2	5.14	49.5	5.34	25.9	5.48	5.03
	No. Specimens	14		15		15	
	No. Batches	3		3		3	
	Approval Class	Screening		Interim		Interim	
E_1^t (Msi)	Mean	9.51	10.0	9.07	9.88	9.70	9.92
	Minimum	9.14	9.79	8.46	9.28	9.40	9.47
	Maximum	9.79	10.5	9.76	10.5	10.2	10.4
	C.V.(%)	2.16	2.21	4.50	3.76	2.25	2.78
	No. Specimens	14		12		15	
	No. Batches	3		3		3	
	Approval Class	Screening		Screening		Interim	
ν_{12}^t	Mean						
	No. Specimens						
	No. Batches						
	Approval Class						
ε_1^{tu} ($\mu\varepsilon$)	Mean		11500		11800		11000
	Minimum		10900		10900		9700
	Maximum		12800		12400		11900
	C.V.(%)		4.78		3.88		5.88
	B-value		(1)		(1)		(1)
	Distribution		Normal		Weibull		ANOVA
	C_1		11500		12000		691.
	C_2		550.		34.4		4.32
	No. Specimens	13		12		14	
	No. Batches	3		3		3	
	Approval Class	Screening		Screening		Screening	

(1) B-values are presented only for fully approved data.
(2) Held in 160°F water bath until full saturation or 95% of equilibrium once full saturation was established.

MATERIAL:	AS4 6k/PR 500 RTM 5HS			Table 4.2.26(c) C/Ep 370-5HS AS4/PR 500 Tension, 1-axis $[0]_8$ 240/W, 300/W Screening, Interim
RESIN CONTENT:	30 - 34 wt%	COMP: DENSITY:	1.56 - 1.58 g/cm³	
FIBER VOLUME:	57.6 - 62.0 vol %	VOID CONTENT:	NA	
PLY THICKNESS:	0.0133 - 0.0142 in.			
TEST METHOD: SRM 4R-94		MODULUS CALCULATION: Chord between 1000 and 3000 $\mu\varepsilon$		
NORMALIZED BY:	Specimen thickness and batch FAW to 57% fiber volume (0.0145 in. CPT)			

		240		300			
Temperature (°F) / Moisture Content (%) / Equilibrium at T, RH / Source Code		240 / (2) / 160°F water / 61		300 / (2) / 160°F water / 61			
		Normalized	Measured	Normalized	Measured	Normalized	Measured
F_1^{tu} (ksi)	Mean	109	114	102	110		
	Minimum	98.0	104	98.1	102		
	Maximum	118	120	110	116		
	C.V.(%)	5.65	4.13	2.81	3.46		
	B-value	(1)	(1)	(1)	(1)		
	Distribution	ANOVA	ANOVA	Nonpara.	Weibull		
	C_1	6.82	5.05	8	112		
	C_2	4.98	4.32	1.43	35.4		
	No. Specimens	15		17			
	No. Batches	3		3			
	Approval Class	Interim		Interim			
E_1^t (Msi)	Mean	9.42	9.84	9.24	9.96		
	Minimum	9.04	9.45	8.69	9.20		
	Maximum	9.82	10.5	9.60	10.5		
	C.V.(%)	2.47	3.11	2.60	3.62		
	No. Specimens	15		15			
	No. Batches	3		3			
	Approval Class	Interim		Interim			
ν_{12}^t	Mean						
	No. Specimens						
	No. Batches						
	Approval Class						
ε_1^{tu} ($\mu\varepsilon$)	Mean		11200		11000		
	Minimum		10400		10100		
	Maximum		13500		12000		
	C.V.(%)		7.43		4.38		
	B-value		(1)		(1)		
	Distribution		Nonpara.		Weibull		
	C_1		7		11300		
	C_2		1.81		23.7		
	No. Specimens	12		15			
	No. Batches	3		3			
	Approval Class	Screening		Interim			

(1) B-values are presented only for fully approved data.
(2) Held in 160°F water bath until full saturation or 95% of equilibrium once full saturation was established.

MATERIAL:	AS4 6k/PR 500 RTM 5HS			**Table 4.2.26(d)** **C/Ep 370-5HS** **AS4/PR 500** **Compression, 1-axis** **$[0_f]_{3s}$** **72/A, -75/A, 180/A** **Interim**
RESIN CONTENT:	30 - 35 wt%	COMP: DENSITY:	1.55 - 1.58 g/cm^3	
FIBER VOLUME:	56.5 - 61.8 vol %	VOID CONTENT:	NA	
PLY THICKNESS:	0.0134 - 0.0146 in.			
TEST METHOD:		MODULUS CALCULATION:		
SRM 1R-94		Chord between 1000 and 3000 $\mu\varepsilon$		
NORMALIZED BY:	Specimen thickness and batch FAW to 57% fiber volume (0.0145 in. CPT)			

Temperature (°F) Moisture Content (%) Equilibrium at T, RH Source Code		72 ambient 61		-75 ambient 61		180 ambient 61	
		Normalized	Measured	Normalized	Measured	Normalized	Measured
F_1^{cu} (ksi)	Mean	118	127			105	110
	Minimum	103	110			92.1	94.4
	Maximum	136	141			116	126
	C.V.(%)	7.91	7.41			5.86	7.02
	B-value	(1)	(1)			(1)	(1)
	Distribution	ANOVA	Weibull			Weibull	Weibull
	C_1	9.99	131			108	114
	C_2	3.81	16.1			19.8	15.8
	No. Specimens	17				15	
	No. Batches	3				3	
	Approval Class	Interim				Interim	
E_1^c (Msi)	Mean	8.88	8.95	8.85	8.90	8.99	9.00
	Minimum	8.30	8.28	8.19	8.10	8.69	7.99
	Maximum	9.41	9.86	9.30	9.72	9.30	9.48
	C.V.(%)	3.16	5.41	3.09	4.71	2.16	5.08
	No. Specimens	17		15		15	
	No. Batches	3		3		3	
	Approval Class	Interim		Interim		Interim	
ν_{12}^c	Mean						
	No. Specimens						
	No. Batches						
	Approval Class						
ε_1^{cu} ($\mu\varepsilon$)	Mean						
	Minimum						
	Maximum						
	C.V.(%)						
	B-value						
	Distribution						
	C_1						
	C_2						
	No. Specimens						
	No. Batches						
	Approval Class						

(1) B-values are presented only for fully approved data.

MATERIAL:	AS4 6k/PR 500 RTM 5HS			**Table 4.2.26(e)** **C/Ep 370-5HS** **AS4/PR 500** **Compression, 1-axis** **$[0_f]_{3s}$** **240/A, 300/A, 350/A** **Screening, Interim**
RESIN CONTENT:	30 - 35 wt%	COMP: DENSITY:	1.55 - 1.58 g/cm³	
FIBER VOLUME:	56.5 - 61.8 vol %	VOID CONTENT:	NA	
PLY THICKNESS:	0.0134 - 0.0146 in.			
TEST METHOD: SRM 1R-94		MODULUS CALCULATION: Chord between 1000 and 3000 με		
NORMALIZED BY:	Specimen thickness and batch FAW to 57% fiber volume (0.0145 in. CPT)			

		240		300		350	
Temperature (°F)		240		300		350	
Moisture Content (%)		ambient		ambient		ambient	
Equilibrium at T, RH							
Source Code		61		61		61	
		Normalized	Measured	Normalized	Measured	Normalized	Measured
F_1^{cu} (ksi)	Mean	103	106	80.1	84.2	51.0	53.5
	Minimum	98.2	99.5	69.5	71.2	42.2	44.4
	Maximum	110	114	87.5	93.0	61.6	64.8
	C.V.(%)	3.36	4.37	6.69	7.31	9.72	10.6
	B-value	(1)	(1)	(1)	(1)	(1)	(1)
	Distribution	Weibull	ANOVA	Weibull	ANOVA	Weibull	ANOVA
	C_1	104	4.94	82.5	6.68	53.3	6.10
	C_2	29.3	4.14	18.0	4.18	10.7	4.30
	No. Specimens	15		16		12	
	No. Batches	3		3		3	
	Approval Class	Interim		Interim		Screening	
E_1^c (Msi)	Mean						
	Minimum						
	Maximum						
	C.V.(%)						
	No. Specimens						
	No. Batches						
	Approval Class						
ν_{12}^c	Mean						
	No. Specimens						
	No. Batches						
	Approval Class						
ε_1^{cu} (με)	Mean						
	Minimum						
	Maximum						
	C.V.(%)						
	B-value						
	Distribution						
	C_1						
	C_2						
	No. Specimens						
	No. Batches						
	Approval Class						

(1) B-values are presented only for fully approved data.

MATERIAL:	AS4 6k/PR 500 RTM 5HS			**Table 4.2.26(f)** **C/Ep 370-5HS** **AS4/PR 500** **Compression, 1-axis** **$[0_f]_{3s}$** **180/W, 240/W, 300/W** **Screening, Interim**
RESIN CONTENT:	30 - 35 wt%	COMP: DENSITY:	1.55 - 1.58 g/cm³	
FIBER VOLUME:	56.5 - 61.8 vol %	VOID CONTENT:	NA	
PLY THICKNESS:	0.0134 - 0.0146 in.			
TEST METHOD: SRM 1R-94		MODULUS CALCULATION: Chord between 1000 and 3000 $\mu\varepsilon$		
NORMALIZED BY:	Specimen thickness and batch FAW to 57% fiber volume (0.0145 in. CPT)			

Temperature (°F) Moisture Content (%) Equilibrium at T, RH Source Code		180 (2) 160°F water 61		240 (2) 160°F water 61		300 (2) 160°F water 61	
		Normalized	Measured	Normalized	Measured	Normalized	Measured
F_1^{cu} (ksi)	Mean	100	106	77.5	79.3	67.0	71.7
	Minimum	87.9	87.7	67.4	66.1	62.2	65.5
	Maximum	114	126	87.1	93.4	71.6	78.2
	C.V.(%)	7.08	10.2	8.97	12.3	4.43	6.05
	B-value	(1)	(1)	(1)	(1)	(1)	(1)
	Distribution	ANOVA	ANOVA	Normal	ANOVA	ANOVA	ANOVA
	C_1	7.53	12.3	77.5	11.9	3.33	5.33
	C_2	3.67	4.89	6.95	16.8	11.7	16.2
	No. Specimens	17		9		11	
	No. Batches	3		2		2	
	Approval Class	Interim		Screening		Screening	
E_1^c (Msi)	Mean						
	Minimum						
	Maximum						
	C.V.(%)						
	No. Specimens						
	No. Batches						
	Approval Class						
ν_{12}^c	Mean						
	No. Specimens						
	No. Batches						
	Approval Class						
ε_1^{cu} ($\mu\varepsilon$)	Mean						
	Minimum						
	Maximum						
	C.V.(%)						
	B-value						
	Distribution						
	C_1						
	C_2						
	No. Specimens						
	No. Batches						
	Approval Class						

(1) B-values are presented only for fully approved data.
(2) Held in 160°F water bath until full saturation or 95% of equilibrium once full saturation was established.

MATERIAL:	AS4 6k/PR 500 RTM 5HS			**Table 4.2.26(g) C/Ep 370-5HS AS4/PR 500 Shear, 12-plane $[45_f]_{2s}$ 72/A, -75/A, 180/A, 240/A, 300/A Screening, Interim**
RESIN CONTENT:	29 - 35 wt%	COMP: DENSITY:	1.55 - 1.59 g/cm³	
FIBER VOLUME:	56.0 - 63.6 vol %	VOID CONTENT:	NA	
PLY THICKNESS:	0.0130 - 0.0148 in.			
TEST METHOD: SRM 7R-94		MODULUS CALCULATION: Chord axial modulus between 1000 and 4000 με		
NORMALIZED BY:	Not normalized			

Temperature (°F) Moisture Content (%) Equilibrium at T, RH Source Code		72 ambient 61	-75 ambient 61	180 ambient 61	240 ambient 61	300 ambient 61
	Mean	14.8	15.4	13.5	11.5	9.25
	Minimum	13.0	14.5	12.6	10.7	7.97
	Maximum	18.2	18.0	14.4	13.1	10.3
	C.V.(%)	8.63	5.50	4.15	5.37	7.28
	B-value	(1)	(1)	(1)	(1)	(1)
	Distribution	Normal	Nonpara	ANOVA	Normal	Weibull
F_{12}^{s}	C_1	14.8	8	0.632	11.5	9.55
(ksi)	C_2	1.28	1.54	5.37	0.618	15.6
	No. Specimens	16	15	14	15	16
	No. Batches	3	3	3	3	3
	Approval Class	Interim	Interim	Screening	Interim	Interim
	Mean	0.639	0.838	0.513	0.432	0.361
	Minimum	0.585	0.795	0.451	0.388	0.331
	Maximum	0.703	0.893	0.593	0.505	0.381
G_{12}^{s}	C.V.(%)	6.56	4.28	7.17	7.56	3.92
(Msi)	No. Specimens	16	15	14	15	16
	No. Batches	3	3	3	3	3
	Approval Class	Interim	Interim	Screening	Interim	Interim

(1) B-values are presented only for fully approved data.

MATERIAL:	AS4 6k/PR 500 RTM 5HS			**Table 4.2.26(h)** **C/Ep 370-5HS** **AS4/PR 500** **Shear, 12-plane** **$[45_f]_{2s}$** **350/A, 180/W, 240/W, 300/W** **Interim, Screening**
RESIN CONTENT:	29 - 35 wt%	COMP: DENSITY:	1.55 - 1.59 g/cm^3	
FIBER VOLUME:	56.0 - 63.6 vol %	VOID CONTENT:	NA	
PLY THICKNESS:	0.0130 - 0.0148 in.			
TEST METHOD:		MODULUS CALCULATION:		
SRM 7R-94		Chord axial modulus between 1000 and 4000 $\mu\varepsilon$		
NORMALIZED BY:	Not normalized			

Temperature (°F) Moisture Content (%) Equilibrium at T, RH Source Code		350 ambient 61		180 (2) 160°F water 61	240 (2) 160°F water 61	300 (2) 160°F water 61
	Mean	7.75		12.2	10.2	7.82
	Minimum	7.37		11.3	9.61	7.03
	Maximum	8.15		13.0	11.4	8.45
	C.V.(%)	4.36		4.76	4.78	6.35
	B-value	(1)		(1)	(1)	(1)
	Distribution	Normal		ANOVA	ANOVA	Weibull
F_{12}^{s}	C_1	7.75		0.656	0.529	8.04
(ksi)	C_2	0.338		5.36	4.62	19.6
	No. Specimens	8		15	14	11
	No. Batches	2		3	3	3
	Approval Class	Screening		Interim	Screening	Screening
	Mean	0.252		0.506	0.400	0.235
	Minimum	0.216		0.450	0.352	0.190
	Maximum	0.264		0.577	0.450	0.274
G_{12}^{s}	C.V.(%)	6.02		5.80	6.95	12.0
(Msi)	No. Specimens	8		15	14	11
	No. Batches	2		3	3	3
	Approval Class	Screening		Interim	Screening	Screening

(1) B-values are presented only for fully approved data.
(2) Held in 160°F water bath until full saturation or 95% of equilibrium once full saturation was established.

MATERIAL:	AS4 6k/PR 500 RTM 5HS			**Table 4.2.26(i) C/Ep 370-5HS AS4/PR 500 Shear, 12-plane $[45_f]_{2s}$ 72/Fluids Screening**
RESIN CONTENT: FIBER VOLUME: PLY THICKNESS:	29 - 35 wt% 56.0 - 63.6 vol % 0.0130 - 0.0148 in.	COMP: DENSITY: VOID CONTENT:	1.55 - 1.59 g/cm³ NA	
TEST METHOD: SRM 7R-94		MODULUS CALCULATION: Chord axial modulus between 1000 and 3000 με		
NORMALIZED BY:	Not normalized			

Temperature (°F) Moisture Content (%) Equilibrium at T, RH Source Code		72 (2) 61	72 (3) 61	72 (4) 61	72 (5) 61	
	Mean	13.5	14.6	15.0	14.8	
	Minimum	12.4	13.4	13.5	13.7	
	Maximum	14.9	16.7	16.7	15.8	
	C.V.(%)	6.46	8.44	8.41	6.88	
	B-value	(1)	(1)	(1)	(1)	
	Distribution	Normal	Normal	Normal	Normal	
F_{12}^{s}	C_1	13.5	14.6	15.0	14.8	
(ksi)	C_2	0.872	1.23	1.26	1.02	
	No. Specimens	7	7	6	6	
	No. Batches	1	1	1	1	
	Approval Class	Screening	Screening	Screening	Screening	
	Mean	0.601	0.678	0.651	0.666	
	Minimum	0.560	0.639	0.633	0.650	
	Maximum	0.638	0.716	0.677	0.701	
G_{12}^{s}	C.V.(%)	5.65	4.45	2.64	2.77	
(Msi)	No. Specimens	7	7	6	6	
	No. Batches	1	1	1	1	
	Approval Class	Screening	Screening	Screening	Screening	

(1) B-values are presented only for fully approved data.
(2) Held for 6 days at room temperature in MEK cleaning solvent.
(3) Held for 6 days at 160°F in Skydrol hydraulic fluid.
(4) Held for 6 days at room temperature in JP-4 jet fuel.
(5) Held for 6 days at room temperature in deicing fluid.

MATERIAL:	AS4 6k/PR 500 RTM 5HS			**Table 4.2.26(j)** **C/Ep 370-5HS** **AS4/PR 500** **SBS, 31-plane** **$[0_f]_{3s}$** **72/A, 180/A, 300/A,** **180/W, 300/W** **Screening**
RESIN CONTENT:	30 - 34 wt%	COMP: DENSITY:	1.56 - 1.58 g/cm^3	
FIBER VOLUME:	57.6 - 62.0 vol %	VOID CONTENT:	NA	
PLY THICKNESS:	0.0133 - 0.0142 in.			
TEST METHOD:		MODULUS CALCULATION:		
SRM 8R-94		Chord axial modulus between 1000 and 3000 $\mu\varepsilon$		
NORMALIZED BY:	Not normalized			

Temperature (°F) Moisture Content (%) Equilibrium at T, RH Source Code		72 ambient 61	180 ambient 61	300 ambient 61	180 (2) 160°F water 61	300 (2) 160°F water 61
	Mean	11.6	9.6	6.8	8.0	5.47
	Minimum	10.4	9.0	6.5	7.2	5.2
	Maximum	12.7	10.2	7.3	8.4	5.7
	C.V.(%)	5.36	3.4	3.2	4.6	3.3
	B-value	(1)	(1)	(1)	(1)	(1)
	Distribution	Weibull	ANOVA	Normal	Weibull	Normal
F_{31}^{sbs}	C_1	11.9	0.35	6.8	8.1	5.5
(ksi)	C_2	22.2	3.5	0.22	30.	0.18
	No. Specimens	19	19	19	12	7
	No. Batches	3	3	3	2	1
	Approval Class	Screening	Screening	Screening	Screening	Screening

(1) B-values are presented only for fully approved data.
(2) Held in 160°F water bath until full saturation or 95% of equilibrium once full saturation was established.

MATERIAL:	AS4 6k/PR 500 RTM 5HS			**Table 4.2.26(k)** **C/Ep 370-5HS** **AS4/PR 500** **OHT, x-axis** **[0/45/90/-45]$_s$** **72/A, -75/A, 180/A** **Screening, Interim**
RESIN CONTENT:	28 - 36 wt%	COMP: DENSITY:	1.55 - 1.60 g/cm^3	
FIBER VOLUME:	55.5 - 64.8 vol %	VOID CONTENT:	NA	
PLY THICKNESS:	0.0128 - 0.0149 in.			
TEST METHOD:		MODULUS CALCULATION:		
SRM 5R-94		Chord between 1000 and 3000 $\mu\varepsilon$		
NORMALIZED BY:	Specimen thickness and batch FAW to 57% fiber volume (0.0145 in. CPT)			

Temperature (°F) Moisture Content (%) Equilibrium at T, RH Source Code		72 ambient 61		-75 ambient 61		180 ambient 61	
		Normalized	Measured	Normalized	Measured	Normalized	Measured
F_x^{ohtu} (ksi)	Mean	47.5	49.4	47.7	49.9	46.9	48.3
	Minimum	42.5	41.7	41.7	40.6	43.8	44.9
	Maximum	51.5	54.0	51.6	54.8	48.8	51.5
	C.V.(%)	5.49	7.03	5.73	7.82	3.46	4.66
	B-value	(1)	(1)	(1)	(1)	(1)	(1)
	Distribution	Weibull	Weibull	Weibull	Weibull	ANOVA	ANOVA
	C_1	48.7	51.0	48.8	51.5	1.69	2.20
	C_2	21.8	17.6	22.6	17.6	3.61	3.81
	No. Specimens	15		15		15	
	No. Batches	3		3		3	
	Approval Class	Interim		Interim		Interim	
E_x^{oht} (Msi)	Mean	6.86	7.24	7.25	7.77	6.75	7.04
	Minimum	6.72	7.09	7.08	7.63	6.55	6.71
	Maximum	7.07	7.41	7.34	7.94	7.14	7.45
	C.V.(%)	1.94	1.59	1.42	1.90	3.26	3.48
	No. Specimens	5		5		6	
	No. Batches	1		1		1	
	Approval Class	Screening		Screening		Screening	
ε_x^{ohtu} ($\mu\varepsilon$)	Mean		7100		6700		7100
	Minimum		6500		6600		6800
	Maximum		7500		7000		7400
	C.V.(%)		5.7		2.5		3.8
	B-value		(1)		(1)		(1)
	Distribution		Normal		Normal		Normal
	C_1		7100		6700		7100
	C_2		400		170		270
	No. Specimens	5		5		5	
	No. Batches	1		1		1	
	Approval Class	Screening		Screening		Screening	

(1) B-values are presented only for fully approved data.

MATERIAL:	AS4 6k/PR 500 RTM 5HS			Table 4.2.26(l) C/Ep 370-5HS AS4/PR 500 OHT, x-axis $[0/45/90/-45]_s$ 240/A, 300/A, 350/A Screening, Interim
RESIN CONTENT:	28 - 36 wt%	COMP: DENSITY:	1.55 - 1.60 g/cm³	
FIBER VOLUME:	55.5 - 64.8 vol %	VOID CONTENT:	NA	
PLY THICKNESS:	0.0128 - 0.0149 in.			
TEST METHOD: SRM 5R-94		MODULUS CALCULATION: Chord between 1000 and 3000 με		
NORMALIZED BY:	Specimen thickness and batch FAW to 57% fiber volume (0.0145 in. CPT)			

	Temperature (°F) Moisture Content (%) Equilibrium at T, RH Source Code	240 ambient 61		300 ambient 61		350 ambient 61	
		Normalized	Measured	Normalized	Measured	Normalized	Measured
F_x^{ohtu} (ksi)	Mean	48.6	51.2	47.5	49.7	44.1	45.4
	Minimum	45.4	47.8	45.9	46.6	41.6	41.4
	Maximum	52.8	56.1	51.2	53.3	46.7	48.4
	C.V.(%)	3.89	4.96	3.20	4.11	3.61	3.86
	B-value	(1)	(1)	(1)	(1)	(1)	(1)
	Distribution	Weibull	Normal	Nonpara.	Weibull	ANOVA	Weibull
	C_1	49.5	51.2	8	50.7	1.70	46.3
	C_2	25.6	2.54	1.49	26.1	3.84	29.3
	No. Specimens	16		16		16	
	No. Batches	3		3		3	
	Approval Class	Interim		Interim		Interim	
E_x^{oht} (Msi)	Mean	6.58	6.96	6.64	7.02	6.01	6.28
	Minimum	6.42	6.70	6.52	6.74	5.85	6.08
	Maximum	6.78	7.20	6.87	7.12	6.33	6.52
	C.V.(%)	2.10	2.82	1.84	2.03	3.14	2.56
	No. Specimens	6		6		6	
	No. Batches	1		1		1	
	Approval Class	Screening		Screening		Screening	
ε_x^{ohtu} (με)	Mean		7500		7200		7300
	Minimum		7000		7000		7000
	Maximum		7800		7300		7700
	C.V.(%)		3.7		1.8		3.6
	B-value		(1)		(1)		(1)
	Distribution		Normal		Normal		Normal
	C_1		7500		7200		7300
	C_2		270		130		260
	No. Specimens	6		6		6	
	No. Batches	1		1		1	
	Approval Class	Screening		Screening		Screening	

(1) B-values are presented only for fully approved data.

MATERIAL:	AS4 6k/PR 500 RTM 5HS			**Table 4.2.26(m)** **C/Ep 370-5HS** **AS4/PR 500** **OHT, x-axis** **[0/45/90/-45]$_s$** **180/W, 240/W, 300/W** **Screening, Interim**
RESIN CONTENT:	28 - 36 wt%	COMP: DENSITY:	1.55 - 1.60 g/cm^3	
FIBER VOLUME:	55.5 - 64.8 vol %	VOID CONTENT:	NA	
PLY THICKNESS:	0.0128 - 0.0149 in.			
TEST METHOD: SRM 5R-94		MODULUS CALCULATION: Chord between 1000 and 3000 με		
NORMALIZED BY:	Specimen thickness and batch FAW to 57% fiber volume (0.0145 in. CPT)			

		180		240		300	
Temperature (°F)		180		240		300	
Moisture Content (%)		(2)		(2)		(2)	
Equilibrium at T, RH		160°F water		160°F water		160°F water	
Source Code		61		61		61	
		Normalized	Measured	Normalized	Measured	Normalized	Measured
	Mean	47.1	49.3	46.4	48.6	46.5	48.6
	Minimum	43.1	44.2	43.7	46.0	44.4	45.7
	Maximum	50.0	53.6	49.4	53.4	50.1	52.3
	C.V.(%)	3.81	5.13	3.57	4.44	3.57	6.05
	B-value	(1)	(1)	(1)	(1)	(1)	(1)
F_x^{ohtu}	Distribution	Weibull	Weibull	Weibull	Nonpara.	Weibull	Weibull
(ksi)	C_1	47.9	50.4	47.2	8	28.1	26.8
	C_2	29.6	22.0	31.0	1.49	47.3	49.6
	No. Specimens	16		16		21	
	No. Batches	3		3		3	
	Approval Class	Interim		Interim		Screening	
	Mean	6.69	7.08	7.00	7.46	6.64	6.96
	Minimum	6.58	6.77	6.78	7.07	5.95	6.15
	Maximum	6.80	7.43	7.24	7.70	7.01	7.54
E_x^{oht}	C.V.(%)	1.63	3.44	2.96	3.74	4.92	5.93
(Msi)	No. Specimens	6		6		16	
	No. Batches	1		1		3	
	Approval Class	Screening		Screening		Interim	
	Mean		7100		6600		6900
	Minimum		6800		6100		6000
	Maximum		7200		7100		7800
	C.V.(%)		2.2		6.5		6.1
	B-value		(1)		(1)		(1)
ε_x^{ohtu}	Distribution		Normal		Normal		Weibull
(με)	C_1		7100		6600		17.
	C_2		150		430		7100
	No. Specimens	6		6		18	
	No. Batches	1		1		3	
	Approval Class	Screening		Screening		Interim	

(1) B-values are presented only for fully approved data.
(2) Held in 160°F water bath until full saturation or 95% of equilibrium once full saturation was established.

MATERIAL:	AS4 6k/PR 500 RTM 5HS			Table 4.2.26(n) C/Ep 370-5HS AS4/PR 500 OHC, x-axis [0/45/90/-45]$_s$ 72/A,180/A,240/A Screening, Interim
RESIN CONTENT:	28 - 36 wt%	COMP: DENSITY:	1.55 - 1.60 g/cm^3	
FIBER VOLUME:	55.5 - 64.8 vol %	VOID CONTENT:	NA	
PLY THICKNESS:	0.0128 - 0.0149 in.			
TEST METHOD: SRM 5R-94		MODULUS CALCULATION: Chord between 1000 and 3000 $\mu\varepsilon$		
NORMALIZED BY:	Specimen thickness and batch FAW to 57% fiber volume (0.0145 in. CPT)			

Temperature (°F)		72		180		240	
Moisture Content (%)		ambient		ambient		ambient	
Equilibrium at T, RH							
Source Code		61		61		61	
		Normalized	Measured	Normalized	Measured	Normalized	Measured
	Mean	45.3	47.2	38.2	40.4	35.6	37.9
	Minimum	42.7	44.7	34.8	37.0	32.2	33.9
	Maximum	48.2	51.4	44.1	47.3	37.9	41.0
	C.V.(%)	3.57	4.17	6.32	6.93	4.22	4.38
	B-value	(1)	(1)	(1)	(1)	(1)	(1)
F_x^{ohcu}	Distribution	Weibull	Weibull	Weibull	Normal	Weibull	Weibull
(ksi)	C_1	46.1	48.1	39.4	40.4	36.2	38.6
	C_2	30.7	24.0	15.1	2.80	29.6	26.7
	No. Specimens	18		16		16	
	No. Batches	3		3		3	
	Approval Class	Interim		Interim		Interim	
	Mean	6.67	7.10	6.48	6.94	6.43	6.85
	Minimum	6.28	6.67	6.44	6.78	6.24	6.34
	Maximum	7.08	7.59	6.52	7.05	6.70	7.32
E_x^{ohc}	C.V.(%)	4.47	5.02	0.549	1.44	1.87	4.35
(Msi)	No. Specimens	8		5		15	
	No. Batches	1		1		3	
	Approval Class	Screening		Screening		Screening	
	Mean		6900		6100		5500
	Minimum		6500		5400		5100
	Maximum		7500		6800		6000
	C.V.(%)		5.7		9.7		4.6
	B-value		(1)		(1)		(1)
ε_x^{ohcu}	Distribution		Normal		Normal		Weibull
($\mu\varepsilon$)	C_1		6900		6100		5700
	C_2		390		590		24
	No. Specimens	5		5		15	
	No. Batches	1		1		3	
	Approval Class	Screening		Screening		Screening	

(1) B-values are presented only for fully approved data.

MATERIAL:	AS4 6k/PR 500 RTM 5HS			**Table 4.2.26(o)** **C/Ep 370-5HS** **AS4/PR 500** **OHC, x-axis** **[0/45/90/-45]$_s$** **300/A** **Interim**
RESIN CONTENT:	28 - 36 wt%	COMP: DENSITY:	1.55 - 1.60 g/cm^3	
FIBER VOLUME:	55.5 - 64.8 vol %	VOID CONTENT:	NA	
PLY THICKNESS:	0.0128 - 0.0149 in.			
TEST METHOD: SRM 5R-94		MODULUS CALCULATION: Chord between 1000 and 3000 $\mu\varepsilon$		
NORMALIZED BY:	Specimen thickness and batch FAW to 57% fiber volume (0.0145 in. CPT)			

Temperature (°F) Moisture Content (%) Equilibrium at T, RH Source Code		300 ambient 61					
		Normalized	Measured	Normalized	Measured	Normalized	Measured
	Mean	32.1	34.0				
	Minimum	26.2	28.9				
	Maximum	36.6	38.6				
	C.V.(%)	7.92	7.41				
	B-value	(1)	(1)				
F_x^{ohcu}	Distribution	Weibull	Weibull				
(ksi)	C_1	33.2	35.1				
	C_2	15.7	14.9				
	No. Specimens	17					
	No. Batches	3					
	Approval Class	Interim					
	Mean	6.24	6.60				
	Minimum	6.02	6.19				
	Maximum	6.38	7.24				
E_x^{ohc}	C.V.(%)	1.73	4.13				
(Msi)	No. Specimens	17					
	No. Batches	3					
	Approval Class	Interim					
	Mean		5100				
	Minimum		4300				
	Maximum		5700				
	C.V.(%)		7.6				
	B-value		(1)				
ε_x^{ohcu}	Distribution		Weibull				
($\mu\varepsilon$)	C_1		5300				
	C_2		17				
	No. Specimens	17					
	No. Batches	3					
	Approval Class	Interim					

(1) B-values are presented only for fully approved data.

MATERIAL:	AS4 6k/PR 500 RTM 5HS			Table 4.2.26(p) C/Ep 370-5HS AS4/PR 500 OHC, x-axis [0/45/90/-45]$_s$ 180/W, 240/W, 300/W Screening, Interim
RESIN CONTENT:	28 - 36 wt%	COMP: DENSITY:	1.55 - 1.60 g/cm^3	
FIBER VOLUME:	55.5 - 64.8 vol %	VOID CONTENT:	NA	
PLY THICKNESS:	0.0128 - 0.0149 in.			
TEST METHOD: SRM 5R-94		MODULUS CALCULATION: Chord between 1000 and 3000 $\mu\varepsilon$		
NORMALIZED BY:	Specimen thickness and batch FAW to 57% fiber volume (0.0145 in. CPT)			

		180		240		300	
Temperature (°F)		180		240		300	
Moisture Content (%)		(2)		(2)		(2)	
Equilibrium at T, RH		160°F water		160°F water		160°F water	
Source Code		61		61		61	
		Normalized	Measured	Normalized	Measured	Normalized	Measured
F_x^{ohcu} (ksi)	Mean	36.3	38.5	32.8	34.6	27.1	28.4
	Minimum	32.2	34.5	30.3	31.8	25.0	26.1
	Maximum	40.9	44.2	36.5	38.4	30.2	32.1
	C.V.(%)	7.01	7.02	5.76	6.39	6.35	6.52
	B-value	(1)	(1)	(1)	(1)	(1)	(1)
	Distribution	Weibull	ANOVA	Weibull	Weibull	Nonpara.	Weibull
	C_1	37.5	2.90	33.7	35.7	9	29.3
	C_2	16.1	3.97	18.2	17.2	1.35	16.4
	No. Specimens	16		17		18	
	No. Batches	3		3		3	
	Approval Class	Interim		Interim		Interim	
E_x^{ohc} (Msi)	Mean	6.39	6.90	6.45	6.83	6.10	6.40
	Minimum	6.29	6.56	6.22	6.49	5.84	5.78
	Maximum	6.53	7.13	7.05	7.46	6.45	6.87
	C.V.(%)	1.69	2.89	3.54	4.03	2.64	4.57
	No. Specimens	6		15		15	
	No. Batches	1		3		3	
	Approval Class	Screening		Interim		Interim	
ε_x^{ohcu} ($\mu\varepsilon$)	Mean		5800		5100		4500
	Minimum		5400		4500		4100
	Maximum		6500		5800		4900
	C.V.(%)		7.0		7.2		5.4
	B-value		(1)		(1)		(1)
	Distribution		Normal		Weibull		Weibull
	C_1		5800		5300		4600
	C_2		410		15		20
	No. Specimens	6		15		15	
	No. Batches	1		3		3	
	Approval Class	Screening		Interim		Interim	

(1) B-values are presented only for fully approved data.
(2) Held in 160°F water bath until full saturation or 95% of equilibrium once full saturation was established.

MATERIAL:	AS4 6k/PR 500 RTM 5HS			**Table 4.2.26(q)** **C/Ep 370-5HS** **AS4/PR 500** **CAI, x-axis** **[0/45/90/-45]$_{2s}$** **72/A, Impact** **Interim**
RESIN CONTENT:	30 - 33 wt%	COMP: DENSITY:	1.56 - 1.59 g/cm^3	
FIBER VOLUME:	58.5 - 62.4 vol %	VOID CONTENT:	NA	
PLY THICKNESS:	0.0133 - 0.0141 in.			
TEST METHOD: SRM 2-94, Impact energy (see footnotes)		MODULUS CALCULATION:		
NORMALIZED BY:	Specimen thickness and batch FAW to 57% fiber volume (0.0145 in. CPT)			

		Temperature (°F) Moisture Content (%) Equilibrium at T, RH Source Code	72 ambient (2) 61		72 ambient (3) 61		72 ambient (4) 61	
			Normalized	Measured	Normalized	Measured	Normalized	Measured
F_x^{cai} (ksi)	Mean		60.5	64.3	43.1	45.8	39.5	41.9
	Minimum		55.6	59.1	40.6	42.4	35.5	39.0
	Maximum		67.2	71.7	45.3	48.6	45.7	47.6
	C.V.(%)		5.33	5.42	3.31	4.23	6.32	5.47
	B-value		(1)	(1)	(1)	(1)	(1)	(1)
	Distribution		Weibull	Weibull	ANOVA	ANOVA	ANOVA	ANOVA
	C_1		62.0	66.0	1.58	2.17	2.64	2.45
	C_2		19.6	18.9	4.98	5.26	3.99	4.18
	No. Specimens		15		15		15	
	No. Batches		3		3		3	
	Approval Class		Interim		Interim		Interim	

(1) B-values are presented only for fully approved data.
(2) Impact energy: 135 in-lbs.
(3) Impact energy: 270 in-lbs.
(4) Impact energy: 360 in-lbs.

MATERIAL:	AS4 6k/PR 500 RTM 5HS			**Table 4.2.26(r)** **C/Ep 370-5HS** **AS4/PR 500** **CAI, x-axis** **$[0/45/90/-45]_{2s}$** **72/A, Impact** **Interim**
RESIN CONTENT:	30 - 33 wt%	COMP: DENSITY:	1.56 - 1.59 g/cm^3	
FIBER VOLUME:	58.5 - 62.4 vol %	VOID CONTENT:	NA	
PLY THICKNESS:	0.0133 - 0.0141 in.			
TEST METHOD: SRM 2R-94, Impact energy (see footnotes)		MODULUS CALCULATION:		
NORMALIZED BY:	Specimen thickness and batch FAW to 57% fiber volume (0.0145 in. CPT)			

		72		72			
Temperature (°F) Moisture Content (%) Equilibrium at T, RH Source Code		72 ambient (2) 61		72 ambient (3) 61			
		Normalized	Measured	Normalized	Measured	Normalized	Measured
	Mean	37.2	39.4	35.1	37.4		
	Minimum	34.8	36.1	33.0	34.5		
	Maximum	40.9	43.7	37.5	39.8		
	C.V.(%)	4.61	4.91	4.15	4.26		
	B-value	(1)	(1)	(1)	(1)		
F_x^{cai}	Distribution	ANOVA	ANOVA	ANOVA	ANOVA		
(ksi)	C_1	1.91	2.11	1.59	1.74		
	C_2	5.12	4.73	4.65	4.75		
	No. Specimens	15		15			
	No. Batches	3		3			
	Approval Class	Interim		Interim			

(1) B-values are presented only for fully approved data.
(2) Impact energy: 450 in-lbs.
(3) Impact energy: 545 in-lbs.

MATERIAL:	AS4 6k/PR 500 RTM 5HS			**Table 4.2.26(s)** **C/Ep 370-5HS** **AS4/PR 500** **G_{Ic}, x-axis** **$[0_f]_{6s}$** **72/A** **Screening**
RESIN CONTENT:	33 - 34 wt%	COMP: DENSITY:	1.56 g/cm^3	
FIBER VOLUME:	57.3 - 58.3 vol %	VOID CONTENT:	NA	
PLY THICKNESS:	0.0142 - 0.0144 in.			
TEST METHOD: BMS 8-276, Section 8.5.7 Double Cantilever beam (2)		MODULUS CALCULATION:		
NORMALIZED BY:	Not normalized			

Temperature (°F)		72				
Moisture Content (%)		ambient				
Equilibrium at T, RH						
Source Code		61				
	Mean	2.63				
	Minimum	1.64				
	Maximum	3.88				
	C.V.(%)	20.1				
	B-value	(1)				
	Distribution	ANOVA				
G_{Ic}	C_1	0.642				
(in-lbs/in^2)	C_2	8.30				
	No. Specimens	56				
	No. Batches	2				
	Approval Class	Screening				

(1) B-values are presented only for fully approved data.
(2) Equivalent to ASTM D5528-94 with 0.5 inch specimen width.

MATERIAL:	AS4 6k/PR 500 RTM 5HS			Table 4.2.26(t) C/Ep 370-5HS AS4/PR 500 G_{IIc}, x-axis $[0_f]_{6s}$ 72/A Interim
RESIN CONTENT:	33 - 34 wt%	COMP: DENSITY:	1.56 g/cm^3	
FIBER VOLUME:	57.3 - 58.3 vol %	VOID CONTENT:	NA	
PLY THICKNESS:	0.0142 - 0.0144 in.			
TEST METHOD: BMS 8-276, Section 8.5.9 End Notched Flexure		MODULUS CALCULATION:		
NORMALIZED BY:	Not normalized			

Temperature (°F)		72				
Moisture Content (%)		ambient				
Equilibrium at T, RH						
Source Code		61				
	Mean	7.88				
	Minimum	6.21				
	Maximum	10.8				
	C.V.(%)	13.1				
	B-value	(1)				
	Distribution	ANOVA				
G_{II_C}	C_1	1.20				
(in-lbs/in^2)	C_2	5.02				
	No. Specimens	47				
	No. Batches	3				
	Approval Class	Interim				

(1) B-values are presented only for fully approved data.

4.2.27

4.2.28

4.2.29

4.2.30 IM-7 12k/PR381 unitape

These data are presented in the MIL-HDBK-17-2E Annex A.

4.3 CARBON - POLYESTER COMPOSITES

4.4 CARBON - BISMALEIMIDE COMPOSITES

4.4.1 T-300 3k/F650 unitape data set description

Material Description:

Material: T300 3k/F650 tape

Form: Unidirectional tape, fiber areal weight of 189 g/m^2, typical cured resin content of 32%, typical cured ply thickness of 0.0070 inches.

Processing: Autoclave cure; 375°F, 85 psi for 4 hours. Postcure at 475°F for 4 hours

General Supplier Information:

Fiber: T-300 fibers are continuous, no twist carbon filaments made from PAN precursor, surface treated to improve handling characteristics and structural properties. Filament count is 3,000 filaments/tow. Typical tensile modulus is 33×10^6 psi. Typical tensile strength is 530,000 psi.

Matrix: F650 is a 350°F curing bismaleimide resin. It will retain light tack for several weeks at 70°F.

Maximum Short Term Service Temperature: 500°F (dry), 350°F (wet)

Typical applications: Primary and secondary structural applications.

4.4.1 T-300 3k/F650 unidirectional tape*

MATERIAL:	T-300 3k/F650 unidirectional tape			**C/BMI T-300/F650 Summary**
PREPREG:	Hexcel T3T190/F652 unidirectional tape			
FIBER:	Toray T-300 3k	MATRIX:	Hexcel F650	
T_g(dry):	600°F T_g(wet):	T_g METHOD:		
PROCESSING:	Autoclave cure: 375°F, 4 hours, 85 psig; Postcure: 475°F, 4 hours, free-standing oven			

** DATA WERE SUBMITTED BEFORE THE ESTABLISHMENT OF DATA DOCUMENTATION REQUIREMENTS (JUNE 1989). ALL DOCUMENTATION PRESENTLY REQUIRED WERE NOT SUPPLIED FOR THIS MATERIAL.

Date of fiber manufacture		Date of testing	
Date of resin manufacture		Date of data submittal	4/89
Date of prepreg manufacture		Date of analysis	1/93
Date of composite manufacture			

LAMINA PROPERTY SUMMARY

	75°F/A		-67°F/A	400°F/A				
Tension, 1-axis	SS--		S---	SS--				
Tension, 2-axis								
Tension, 3-axis								
Compression, 1-axis								
Compression, 2-axis								
Compression, 3-axis								
Shear, 12-plane								
Shear, 23-plane								
SB Strength, 31-plane	S---			S---				

Classes of data: F - Fully approved, I - Interim, S - Screening in Strength/Modulus/Poisson's ratio/Strain-to-failure order.

** DATA WERE SUBMITTED BEFORE THE ESTABLISHMENT OF DATA DOCUMENTATION REQUIREMENTS (JUNE 1989). ALL DOCUMENTATION PRESENTLY REQUIRED WERE NOT SUPPLIED FOR THIS MATERIAL.

		Nominal	As Submitted	Test Method
Fiber Density	(g/cm^3)	1.76		
Resin Density	(g/cm^3)	1.27		
Composite Density	(g/cm^3)	1.56	1.57	
Fiber Areal Weight	(g/m^2)	189		
Ply Thickness	(in)	0.0070		

LAMINATE PROPERTY SUMMARY

Classes of data: F - Fully approved, I - Interim, S - Screening in Strength/Modulus/Poisson's ratio/Strain-to-failure order.

** DATA WERE SUBMITTED BEFORE THE ESTABLISHMENT OF DATA DOCUMENTATION REQUIREMENTS (JUNE 1989). ALL DOCUMENTATION PRESENTLY REQUIRED WAS NOT SUPPLIED FOR THIS MATERIAL.

MATERIAL: T-300 3k/F650

RESIN CONTENT: 32 wt% | COMP: DENSITY: 1.57 g/cm³
FIBER VOLUME: 61 % | VOID CONTENT:
PLY THICKNESS: 0.0070 in.

TEST METHOD: ASTM D3039-76 | MODULUS CALCULATION:

NORMALIZED BY: Batch fiber volume fraction to 60%

Table 4.4.1(a)
C/BMI 189-UT
T-300/F650
Tension, 1-axis
$[0]_6$
75/A, -67/A, 400/A
Screening

Temperature (°F)		75		-67		400	
Moisture Content (%)		ambient		ambient		ambient	
Equilibrium at T, RH							
Source Code		21		21		21	
		Normalized	Measured	Normalized	Measured	Normalized	Measured
F_1^{tu} (ksi)	Mean	248	252	194	197	229	233
	Minimum	216	220	167	170	216	220
	Maximum	293	298	212	216	243	247
	C.V.(%)	7.14	7.15	8.68	8.68	3.97	3.97
	B-value	(1)	(1)	(1)	(1)	(1)	(1)
	Distribution	Normal	Normal	Normal	Normal	Normal	Normal
	C_1	248	252	194	197	229	233
	C_2	17.7	18.0	16.8	17.1	11.1	9.24
	No. Specimens	15		15		7	
	No. Batches	1		1		1	
	Approval Class	Screening		Screening		Screening	
E_1^t (Msi)	Mean	18.9	19.2			19.1	19.4
	Minimum	16.5	16.8			16.8	17.1
	Maximum	20.3	20.6			21.0	21.4
	C.V.(%)	5.58	5.49			7.26	7.23
	No. Specimens	15				9	
	No. Batches	1				1	
	Approval Class	Screening				Screening	
ν_{12}^t	Mean						
	No. Specimens						
	No. Batches						
	Approval Class						
ε_1^{tu} ($\mu\varepsilon$)	Mean						
	Minimum						
	Maximum						
	C.V.(%)						
	B-value						
	Distribution						
	C_1						
	C_2						
	No. Specimens						
	No. Batches						
	Approval Class						

(1) B-values are presented only for fully approved data.

** DATA WERE SUBMITTED BEFORE THE ESTABLISHMENT OF DATA DOCUMENTATION REQUIREMENTS (JUNE 1989). ALL DOCUMENTATION PRESENTLY REQUIRED WAS NOT SUPPLIED FOR THIS MATERIAL.

MATERIAL:	T-300 3k/F650			**Table 4.4.1(b)** **C/BMI 189-UT** **T-300/F650** **SBS, 31-plane** **$[0]_{34}$** **75/A, 400/A** **Screening**
RESIN CONTENT:	32 wt%	COMP: DENSITY:	1.57 g/cm^3	
FIBER VOLUME:	61 %	VOID CONTENT:		
PLY THICKNESS:	0.0070 in.			
TEST METHOD: ASTM D2344		MODULUS CALCULATION:		
NORMALIZED BY:	Not normalized			

Temperature (°F) Moisture Content (%) Equilibrium at T, RH Source Code		75 ambient 21	400 ambient 21			
	Mean	14.1	9.39			
	Minimum	13.5	8.77			
	Maximum	15.0	10.1			
	C.V.(%)	3.04	4.25			
	B-value	(1)	(1)			
	Distribution	Weibull	Weibull			
F_{31}^{sbs}	C_1	14.3	9.59			
(ksi)	C_2	32.3	24.6			
	No. Specimens	15	15			
	No. Batches	1	1			
	Approval Class	Screening	Screening			

(1) B-values are presented only for fully approved data.

4.4.2 T-300 3k/F650 8HS data set description

Material Description:

Material: T300 3k/F650

Form: 8 harness satin weave fabric, fiber areal weight of 370 g/m^2, typical cured resin content of 40%, typical cured ply thickness of 0.015 inches.

Processing: Autoclave cure; 375°F, 85 psi for 4 hours. Postcure at 475°F for 4 hours

General Supplier Information:

Fiber: T-300 fibers are continuous, no twist carbon filaments made from PAN precursor, surface treated to improve handling characteristics and structural properties. Filament count is 3,000 filaments/tow. Typical tensile modulus is 33 x 10^6 psi. Typical tensile strength is 530,000 psi.

Matrix: F650 is a 350°F curing bismaleimide resin. It will retain light tack for several weeks at 70°F.

Maximum Short Term Service Temperature: 500°F (dry), 350°F (wet)

Typical applications: Primary and secondary structural applications.

4.4.2 T-300 3k/F650 8-harness satin weave*

MATERIAL:	T-300 3k/F650 8-harness satin weave		**C/BMI T-300/F650 Summary**
PREPREG:	Hexcel F3T584/F650 8-harness satin weave		
FIBER:	Toray T-300 3k	MATRIX:	Hexcel F650
T_g(dry):	600°F T_g(wet):	T_g METHOD:	
PROCESSING:	Autoclave cure: 375°F, 4 hours, 85 psig; Postcure: 475°F, 4 hours, free-standing oven		

** DATA WERE SUBMITTED BEFORE THE ESTABLISHMENT OF DATA DOCUMENTATION REQUIREMENTS (JUNE 1989). ALL DOCUMENTATION PRESENTLY REQUIRED WERE NOT SUPPLIED FOR THIS MATERIAL.

Date of fiber manufacture		Date of testing	
Date of resin manufacture		Date of data submittal	4/89
Date of prepreg manufacture		Date of analysis	1.93
Date of composite manufacture			

LAMINA PROPERTY SUMMARY

	75°F/A		350°F/A	450°F/A				
Tension, 1-axis								
Tension, 2-axis								
Tension, 3-axis								
Compression, 1-axis								
Compression, 2-axis								
Compression, 3-axis								
Shear, 12-plane	SS--							
Shear, 23-plane								
SB Strength, 31-plane	S---		S---	S---				

Classes of data: F - Fully approved, I - Interim, S - Screening in Strength/Modulus/Poisson's ratio/Strain-to-failure order.

** DATA WERE SUBMITTED BEFORE THE ESTABLISHMENT OF DATA DOCUMENTATION REQUIREMENTS (JUNE 1989). ALL DOCUMENTATION PRESENTLY REQUIRED WERE NOT SUPPLIED FOR THIS MATERIAL.

		Nominal	As Submitted	Test Method
Fiber Density	(g/cm^3)	1.75		
Resin Density	(g/cm^3)	1.27		
Composite Density	(g/cm^3)	1.54		
Fiber Areal Weight	(g/m^2)	370		
Ply Thickness	(in)	0.015		

LAMINATE PROPERTY SUMMARY

Classes of data: F - Fully approved, I - Interim, S - Screening in Strength/Modulus/Poisson's ratio/Strain-to-failure order.

** DATA WERE SUBMITTED BEFORE THE ESTABLISHMENT OF DATA DOCUMENTATION REQUIREMENTS (JUNE 1989). ALL DOCUMENTATION PRESENTLY REQUIRED WAS NOT SUPPLIED FOR THIS MATERIAL.

MATERIAL: T-300 3k/F650		Table 4.4.2(a) C/BMI 370-8HS T-300/F650 Shear, 12-plane $[\pm 45_f]_{4s}$ 75/A Screening
RESIN CONTENT: 40 wt%	COMP: DENSITY: 1.51 g/cm^3	
FIBER VOLUME: 52 %	VOID CONTENT:	
PLY THICKNESS: 0.015 in.		
TEST METHOD: ASTM D3518-76	MODULUS CALCULATION:	
NORMALIZED BY: Not normalized		

Temperature (°F) Moisture Content (%) Equilibrium at T, RH Source Code		75 ambient 21				
F_{12}^{su} (ksi)	Mean	9.77				
	Minimum	8.57				
	Maximum	11.1				
	C.V.(%)	8.78				
	B-value	(1)				
	Distribution	Weibull				
	C_1	10.2				
	C_2	12.9				
	No. Specimens	15				
	No. Batches	1				
	Approval Class	Screening				
G_{12}^{s} (Msi)	Mean	0.69				
	Minimum	0.59				
	Maximum	0.81				
	C.V.(%)	10				
	No. Specimens	14				
	No. Batches	1				
	Approval Class	Screening				
γ_{12}^{su} (με)	Mean					
	Minimum					
	Maximum					
	C.V.(%)					
	B-value					
	Distribution					
	C_1					
	C_2					
	No. Specimens					
	No. Batches					
	Approval Class					

(1) B-values are presented only for fully approved data.

** DATA WERE SUBMITTED BEFORE THE ESTABLISHMENT OF DATA DOCUMENTATION REQUIREMENTS (JUNE 1989). ALL DOCUMENTATION PRESENTLY REQUIRED WAS NOT SUPPLIED FOR THIS MATERIAL.

Table 4.4.2(b)
C/BMI 370-8HS
T-300/F650
SBS, 31-plane
$[0_f]_8$
75/A, 350/A, 450/A
Screening

MATERIAL: T-300 3k/F650

RESIN CONTENT: 40 wt%
FIBER VOLUME: 52 %
PLY THICKNESS: 0.015 in.

COMP: DENSITY: 1.51 g/cm^3
VOID CONTENT:

TEST METHOD: ASTM D2344

MODULUS CALCULATION:

NORMALIZED BY: Not normalized

Temperature (°F) Moisture Content (%) Equilibrium at T, RH Source Code		75 ambient 21	350 ambient 21	450 ambient 21		
	Mean	5.83	5.59	5.80		
	Minimum	4.75	4.93	5.23		
	Maximum	8.06	6.44	6.57		
	C.V.(%)	15.0	10.9	6.81		
	B-value	(1)	(1)	(1)		
	Distribution	Nonpara.	Weibull	Weibull		
F_{31}^{sbs}	C_1	8	5.86	5.98		
(ksi)	C_2	1.54	11.0	15.5		
	No. Specimens	15	10	10		
	No. Batches	1	1	1		
	Approval Class	Screening	Screening	Screening		

(1) B-values are presented only for fully approved data.

4.4.3 T-300 3k/F652 8HS fabric data set description

Material Description:

Material: T300 3k/F652

Form: 8 harness satin weave fabric, fiber areal weight of 367 g/m^2, typical cured resin content of 27%, typical cured ply thickness of 0.0124 inches.

Processing: Press cure, 400°F, 2.5 hours, 125 psi. Post cure at 550°F, 4 hours

General Supplier Information:

Fiber: T-300 3K fibers are continuous, no twist carbon filaments made from PAN precursor, surface treated to improve handling characteristics and structural properties. Filament count is 3,000 filaments/tow. Typical tensile modulus is 33 x 10^6 psi. Typical tensile strength is 530,000 psi.

Matrix: F652 is a bismaleimide resin that has been modified from F650 to reduce the flow of the resin. The lower flow allows the resin to be used in press forming operations and also for high temperature honeycomb. The properties are equivalent to F650.

Maximum Short Term Service Temperature: 500°F (dry), 350°F (wet)

Typical applications: Primary and secondary structural applications.

4.4.3 T-300 3k/F652 8-harness satin weave*

				C/BMI T-300/F652 Summary
MATERIAL:	T-300 3k/F652 8-harness satin weave			
PREPREG:	Hexcel F3G584/F652 8-harness satin weave			
FIBER:	Amoco Thornel T-300	MATRIX:	Hexcel F652	
T_g(dry):	600°F T_g(wet):	T_g METHOD:		
PROCESSING:	Press cured: 400°F, 2.5 hours, 125 psig; Postcure: 550°F, 4 hours			

** DATA WERE SUBMITTED BEFORE THE ESTABLISHMENT OF DATA DOCUMENTATION REQUIREMENTS (JUNE 1989). ALL DOCUMENTATION PRESENTLY REQUIRED WERE NOT SUPPLIED FOR THIS MATERIAL.

Date of fiber manufacture		Date of testing	
Date of resin manufacture		Date of data submittal	4/89
Date of prepreg manufacture		Date of analysis	1/93
Date of composite manufacture			

LAMINA PROPERTY SUMMARY

	70°F/A		600°F/A					
Tension, 1-axis	SS--							
Tension, 2-axis								
Tension, 3-axis								
Compression, 1-axis								
Compression, 2-axis								
Compression, 3-axis								
Shear, 12-plane								
Shear, 23-plane								
SB Strength, 31-plane	S---		S---					

Classes of data: F - Fully approved, I - Interim, S - Screening in Strength/Modulus/Poisson's ratio/Strain-to-failure order.

** DATA WERE SUBMITTED BEFORE THE ESTABLISHMENT OF DATA DOCUMENTATION REQUIREMENTS (JUNE 1989). ALL DOCUMENTATION PRESENTLY REQUIRED WERE NOT SUPPLIED FOR THIS MATERIAL.

	Nominal	As Submitted	Test Method
Fiber Density (g/cm^3)	1.76		
Resin Density (g/cm^3)	1.26		
Composite Density (g/cm^3)	1.55	1.57	
Fiber Areal Weight (g/m^2)	367		
Ply Thickness (in)	.00124		

LAMINATE PROPERTY SUMMARY

Classes of data: F - Fully approved, I - Interim, S - Screening in Strength/Modulus/Poisson's ratio/Strain-to-failure order.

** DATA WERE SUBMITTED BEFORE THE ESTABLISHMENT OF DATA DOCUMENTATION REQUIREMENTS (JUNE 1989). ALL DOCUMENTATION PRESENTLY REQUIRED WAS NOT SUPPLIED FOR THIS MATERIAL.

MATERIAL:	T-300 3k/F652 8-harness satin weave			**Table 4.4.3(a)** **C/BMI 367-8HS** **T-300/F652** **Tension, 1-axis** **$[0_f]_{10}$** **70/A** **Screening**
RESIN CONTENT:	27.2 wt%	COMP: DENSITY:	1.57 g/cm³	
FIBER VOLUME:	64.8 %	VOID CONTENT:		
PLY THICKNESS:	0.012 in.			
TEST METHOD: ASTM D3039-76		MODULUS CALCULATION:		
NORMALIZED BY:	Batch fiber volume to 57%			

Temperature (°F) Moisture Content (%) Equilibrium at T, RH Source Code		70 ambient 21					
		Normalized	Measured	Normalized	Measured	Normalized	Measured
F_1^{tu} (ksi)	Mean	73.6	84.0				
	Minimum	58.8	67.1				
	Maximum	84.3	96.1				
	C.V.(%)	10.1	10.0				
	B-value	(1)	(1)				
	Distribution	Weibull	Weibull				
	C_1	76.8	87.6				
	C_2	12.3	12.4				
	No. Specimens	15					
	No. Batches	1					
	Approval Class	Screening					
E_1^t (Msi)	Mean	9.71	11.1				
	Minimum	8.94	10.2				
	Maximum	10.2	11.6				
	C.V.(%)	4.36	4.28				
	No. Specimens	15					
	No. Batches	1					
	Approval Class	Screening					
ν_{12}^t	Mean						
	No. Specimens						
	No. Batches						
	Approval Class						
ε_1^{tu} (με)	Mean						
	Minimum						
	Maximum						
	C.V.(%)						
	B-value						
	Distribution						
	C_1						
	C_2						
	No. Specimens						
	No. Batches						
	Approval Class						

(1) B-values are presented only for fully approved data.

** DATA WERE SUBMITTED BEFORE THE ESTABLISHMENT OF DATA DOCUMENTATION REQUIREMENTS (JUNE 1989). ALL DOCUMENTATION PRESENTLY REQUIRED WAS NOT SUPPLIED FOR THIS MATERIAL.

MATERIAL: T-300 3k/F652 8-harness satin weave

Table 4.4.3(b)
C/BMI 367-8HS
T-300/F652
SBS, 31-plane
$[0_f]_{10}$
70/A, 600/A
Screening

RESIN CONTENT: 27.2 wt% COMP: DENSITY: 1.57 g/cm^3
FIBER VOLUME: 64.8 % VOID CONTENT:
PLY THICKNESS: 0.0012 in.

TEST METHOD: MODULUS CALCULATION:
ASTM D2344

NORMALIZED BY: Not normalized

Temperature (°F) Moisture Content (%) Equilibrium at T, RH Source Code		70 ambient 21	600 ambient 21			
	Mean	5.97	4.59			
	Minimum	5.13	4.29			
	Maximum	6.64	4.82			
	C.V.(%)	8.17	3.60			
	B-value	(1)	(1)			
	Distribution	Weibull	Weibull			
F_{31}^{sbs}	C_1	6.18	4.66			
(ksi)	C_2	14.8	36.8			
	No. Specimens	15	15			
	No. Batches	1	1			
	Approval Class	Screening	Screening			

(1) B-values are presented only for fully approved data.

4.4.4 AS4/5250-3 unitape data set description

Material Description:

Material: AS4/5250-3

Form: Unidirectional tape, fiber areal weight of 147 g/m^2, typical cured resin content of 26-38%, typical cured ply thickness of 0.0055 inches.

Processing: Autoclave cure; 250°F, 85 psi, 1 hour; 350°F, 85 psi, 6 hours. Postcure; 475°F, 6 hours.

General Supplier Information:

Fiber: AS4 fibers are continuous carbon filaments made from PAN precursor, surface treated to improve handling characteristics and structural properties. Typical tensile modulus is 34 x 10^6 psi. Typical tensile strength is 550,000 psi.

Matrix: 5250-3 is a modified bismaleimide resin possessing good hot/wet strength and improved toughness over standard bismaleimides. Good high temperature resistance.

Maximum Short Term Service Temperature: 450°F (dry), 350°F (wet)

Typical applications: Primary and secondary structural applications on commercial and military aircraft.

4.4.4 AS4/5250-3 unidirectional tape*

MATERIAL:	AS4/5250-3 unidirectional tape			**C/BMI AS4/5250-3 Summary**
PREPREG:	Narmco AS4/5250-3 unidirectional tape, grade 147			
FIBER:	Hercules AS4	MATRIX:	Narmco 5250-3	
T_g(dry):	642°F T_g(wet): 561°F	T_g METHOD:	DMA	
PROCESSING:	Autoclave cure: 250°F, 60 minutes; 350°F, 360 minutes, 85 psi; Postcure: 475°F, 6 hours			

** DATA WERE SUBMITTED BEFORE THE ESTABLISHMENT OF DATA DOCUMENTATION REQUIREMENTS (JUNE 1989). ALL DOCUMENTATION PRESENTLY REQUIRED WERE NOT SUPPLIED FOR THIS MATERIAL.

Date of fiber manufacture		Date of testing	
Date of resin manufacture		Date of data submittal	12/88
Date of prepreg manufacture		Date of analysis	1/93
Date of composite manufacture			

LAMINA PROPERTY SUMMARY

	72°F/A		-67°F/A	350°F/A	450°F/A		74°F/W	350°F/W
Tension, 1-axis	SSSS		SSSS	SSSS	SSSS		SSSS	SSSS
Tension, 2-axis	SS-S		SS-S	SS-S	SS-S			
Tension, 3-axis								
Compression, 1-axis	SS-S		SS-S	SS-S	SS-S		SS-S	SS-S
Compression, 2-axis								
Compression, 3-axis								
Shear, 12-plane	SS--		SS--	SS--	SS--		SS--	SS--
Shear, 23-plane								
SB Strength, 31-plane								

Classes of data: F - Fully approved, I - Interim, S - Screening in Strength/Modulus/Poisson's ratio/Strain-to-failure order.

** DATA WERE SUBMITTED BEFORE THE ESTABLISHMENT OF DATA DOCUMENTATION REQUIREMENTS (JUNE 1989). ALL DOCUMENTATION PRESENTLY REQUIRED WERE NOT SUPPLIED FOR THIS MATERIAL.

		Nominal	As Submitted	Test Method
Fiber Density	(g/cm^3)	1.80		
Resin Density	(g/cm^3)	1.25		
Composite Density	(g/cm^3)	1.58	1.52 - 1.63	
Fiber Areal Weight	(g/m^2)	147	132 - 165	ASTM D3529
Ply Thickness	(in)	0.0051 - 0.0059	0.0050 - 0.0062	

LAMINATE PROPERTY SUMMARY

Classes of data: F - Fully approved, I - Interim, S - Screening in Strength/Modulus/Poisson's ratio/Strain-to-failure order.

** DATA WERE SUBMITTED BEFORE THE ESTABLISHMENT OF DATA DOCUMENTATION REQUIREMENTS (JUNE 1989). ALL DOCUMENTATION PRESENTLY REQUIRED WAS NOT SUPPLIED FOR THIS MATERIAL.

Table 4.4.4(a)
C/BMI 147-UT
AS4/5250-3
Tension, 1-axis
$[0]_8$
72/A, -67/A, 350/A
Screening

MATERIAL: AS4/5250-3 unidirectional material

RESIN CONTENT: 26-28 wt% COMP: DENSITY: 1.58-1.61 g/cm^3
FIBER VOLUME: 63-66 % VOID CONTENT: 0.1-0.9%
PLY THICKNESS: 0.0050-0.0053 in.

TEST METHOD: ASTM D3039-76 MODULUS CALCULATION:

NORMALIZED BY: Ply thickness to 0.0055 in. and batch fiber volume to 60%

Temperature (°F) Moisture Content (%) Equilibrium at T, RH Source Code		72 ambient (1)		-67 ambient (1)		350 ambient (1)	
		Normalized	Measured	Normalized	Measured	Normalized	Measured
F_1^{tu} (ksi)	Mean	252	291	270	311	266	308
	Minimum	223	255	249	285	241	276
	Maximum	275	322	288	332	283	325
	C.V.(%)	7.63	8.48	6.12	6.48	6.87	7.54
	B-value	(2)	(2)	(2)	(2)	(2)	(2)
	Distribution	Normal	Normal	Normal	Normal	Normal	Nonpara.
	C_1	252	291	270	312	266	5
	C_2	19.2	24.7	16.5	20.2	18.3	3.06
	No. Specimens	6		6		6	
	No. Batches	1		1		1	
	Approval Class	Screening		Screening		Screening	
E_1^t (Msi)	Mean	15.9	18.3	16.4	18.9	16.4	19.0
	Minimum	15.3	17.7	15.9	18.5	15.8	18.2
	Maximum	16.4	18.9	16.8	19.4	16.7	19.5
	C.V.(%)	3.04	2.51	2.23	1.91	2.07	2.85
	No. Specimens	6		6		6	
	No. Batches	1		1		1	
	Approval Class	Screening		Screening		Screening	
ν_{12}^t	Mean		0.300		0.295		0.302
	No. Specimens	6		6		6	
	No. Batches	1		1		1	
	Approval Class	Screening		Screening		Screening	
ε_1^{tu} ($\mu\varepsilon$)	Mean		17100		15800		15900
	Minimum		14900		14100		14800
	Maximum		20000		18000		17100
	C.V.(%)		13.3		9.6		4.98
	B-value		(2)		(2)		(2)
	Distribution		Normal		Normal		Normal
	C_1		17100		15800		15900
	C_2		2270		1520		789
	No. Specimens	6		6		6	
	No. Batches	1		1		1	
	Approval Class	Screening		Screening		Screening	

(1) Reference 4.4.4.
(2) B-values are presented only for fully approved data.

** DATA WERE SUBMITTED BEFORE THE ESTABLISHMENT OF DATA DOCUMENTATION REQUIREMENTS (JUNE 1989). ALL DOCUMENTATION PRESENTLY REQUIRED WAS NOT SUPPLIED FOR THIS MATERIAL.

MATERIAL:	AS4/5250-3 unidirectional material			**Table 4.4.4(b)**
RESIN CONTENT:	26-28 wt%	COMP: DENSITY:	1.61-1.63 g/cm^3	**C/BMI 147-UT**
FIBER VOLUME:	63-67 %	VOID CONTENT:	0.0-0.9%	**AS4/5250-3**
PLY THICKNESS:	0.0050-0.0053 in.			**Tension, 1-axis**
				$[0]_8$
TEST METHOD: ASTM D3039-76		MODULUS CALCULATION:		**450/A, 74/0.70%, 350/0.73%**
NORMALIZED BY:	Ply thickness to 0.0055 in. and batch fiber volume to 60%			**Screening**

Temperature (°F)		450		74		350	
Moisture Content (%)		ambient		0.70		0.73	
Equilibrium at T, RH				160°F, 95%		(1)	
Source Code		25 (2)		25 (2)		25 (2)	
		Normalized	Measured	Normalized	Measured	Normalized	Measured
	Mean	253	292	268	312	249	287
	Minimum	208	237	235	268	232	264
	Maximum	269	314	293	347	261	305
	C.V.(%)	8.87	9.64	7.74	8.99	4.50	5.42
	B-value	(3)	(3)	(3)	(3)	(3)	(3)
F_1^{tu}	Distribution	Nonpara.	Normal	Normal	Normal	Normal	Normal
(ksi)	C_1	5	292	268	312	249	288
	C_2	3.06	28.1	20.7	28.1	11.2	15.6
	No. Specimens	6		6		5	
	No. Batches	1		1		1	
	Approval Class	Screening		Screening		Screening	
	Mean	16.5	19.0	16.6	19.3	15.9	18.4
	Minimum	15.7	18.1	16.2	18.9	15.4	17.8
	Maximum	16.9	19.7	17.3	19.9	16.4	19.1
E_1^t	C.V.(%)	3.43	3.56	2.36	1.82	2.41	2.71
(Msi)	No. Specimens	6		6		5	
	No. Batches	1		1		1	
	Approval Class	Screening		Screening		Screening	
	Mean		0.295		0.335		0.368
	No. Specimens	6		6		5	
ν_{12}^t	No. Batches	1		1		1	
	Approval Class	Screening		Screening		Screening	
	Mean		13900		15200		14900
	Minimum		11700		13500		13200
	Maximum		15000		16600		15500
	C.V.(%)		8.14		7.14		6.46
	B-value		(3)		(3)		(3)
ε_1^{tu}	Distribution		Normal		Normal		Normal
($\mu\varepsilon$)	C_1		13900		15200		14900
	C_2		1130		1080		961
	No. Specimens	6		6		6	
	No. Batches	1		1		1	
	Approval Class	Screening		Screening		Screening	

(1) Conditioned at 160°F, 95% relative humidity for 29 days (75% saturation).

(2) Reference 4.4.4.

(3) B-values are presented only for fully approved data.

** DATA WERE SUBMITTED BEFORE THE ESTABLISHMENT OF DATA DOCUMENTATION REQUIREMENTS (JUNE 1989). ALL DOCUMENTATION PRESENTLY REQUIRED WAS NOT SUPPLIED FOR THIS MATERIAL.

MATERIAL:	AS4/5250-3 unidirectional material			**Table 4.4.4(c)** **C/BMI 147-UT** **AS4/5250-3** **Tension, 1-axis** **$[0]_8$** **350/1.0%** **Screening**
RESIN CONTENT:	26-28 wt%	COMP: DENSITY:	1.61 g/cm^3	
FIBER VOLUME:	63-66 %	VOID CONTENT:	0.1-0.9%	
PLY THICKNESS:	0.0050-0.0053 in.			
TEST METHOD: ASTM D3039-76		MODULUS CALCULATION:		
NORMALIZED BY:	Ply thickness to 0.0055 in. and batch fiber volume to 60%			

Temperature (°F)		350					
Moisture Content (%)		1.0					
Equilibrium at T, RH		160°F, 95%					
Source Code		(1)					
		Normalized	Measured	Normalized	Measured	Normalized	Measured
	Mean	235	270				
	Minimum	176	202				
	Maximum	259	296				
	C.V.(%)	12.8	13.0				
	B-value	(2)	(2)				
F_1^{tu}	Distribution	Normal	Normal				
(ksi)	C_1	235	270				
	C_2	29.9	35.1				
	No. Specimens	6					
	No. Batches	1					
	Approval Class	Screening					
	Mean	16.7	19.2				
	Minimum	15.5	17.7				
	Maximum	18.4	21.2				
E_1^t	C.V.(%)	6.43	6.26				
(Msi)	No. Specimens	6					
	No. Batches	1					
	Approval Class	Screening					
	Mean		0.363				
	No. Specimens	4					
ν_{12}^t	No. Batches	1					
	Approval Class	Screening					
	Mean		14400				
	Minimum		9950				
	Maximum		16200				
	C.V.(%)		16.0				
	B-value		(2)				
ε_1^{tu}	Distribution		Normal				
(με)	C_1		14400				
	C_2		2300				
	No. Specimens	6					
	No. Batches	1					
	Approval Class	Screening					

(1) Reference 4.4.4.

(2) B-values are presented only for fully approved data.

** DATA WERE SUBMITTED BEFORE THE ESTABLISHMENT OF DATA DOCUMENTATION REQUIREMENTS (JUNE 1989). ALL DOCUMENTATION PRESENTLY REQUIRED WAS NOT SUPPLIED FOR THIS MATERIAL.

MATERIAL: AS4/5250-3 unidirectional material

Table 4.4.4(d)
C/BMI 147-UT
AS4/5250-3
Tension, 2-axis
$[90]_8$
72/A, -67/A, 350/A, 450/A
Screening

RESIN CONTENT: 27-40 wt% | COMP: DENSITY: 1.52-1.61 g/cm^3
FIBER VOLUME: 51-65 % | VOID CONTENT: 0.1-0.8%
PLY THICKNESS: 0.0051-0.0059 in.

TEST METHOD: ASTM D3039-76 | MODULUS CALCULATION:

NORMALIZED BY: Not normalized

Temperature (°F)		72	-67	350	450		
Moisture Content (%)		ambient	ambient	ambient	ambient		
Equilibrium at T, RH							
Source Code		25	25	25	25		
	Mean	4.61	4.98	4.63	4.54		
	Minimum	3.52	4.68	3.43	4.13		
	Maximum	5.65	5.94	5.33	5.19		
	C.V.(%)	18.4	9.69	13.7	9.20		
	B-value	(1)	(1)	(1)	(1)		
F_2^{tu}	Distribution	Normal	Nonpara.	Normal	Normal		
(ksi)	C_1	4.61	5	4.63	4.54		
	C_2	0.847	3.06	0.637	0.417		
	No. Specimens	6	6	6	6		
	No. Batches	1	1	1	1		
	Approval Class	Screening	Screening	Screening	Screening		
	Mean	1.24	1.40	1.04	1.08		
	Minimum	1.17	1.26	0.940	0.930		
	Maximum	1.35	1.47	1.16	1.26		
E_2^t	C.V.(%)	5.90	5.50	8.50	10.3		
(Msi)	No. Specimens	6	6	5	6		
	No. Batches	1	1	1	1		
	Approval Class	Screening	Screening	Screening	Screening		
	Mean						
	No. Specimens						
ν_{21}^t	No. Batches						
	Approval Class						
	Mean	3540	3580	4680	4330		
	Minimum	2000	3180	3300	3600		
	Maximum	4900	4740	6000	5600		
	C.V.(%)	26.9	16.5	19.0	18.0		
	B-value	(1)	(1)	(1)	(1)		
ε_2^{tu}	Distribution	Normal	Lognormal	Normal	Normal		
($\mu\varepsilon$)	C_1	3540	8.17	4680	4330		
	C_2	955	0.149	889	782		
	No. Specimens	6	6	6	6		
	No. Batches	1	1	1	1		
	Approval Class	Screening	Screening	Screening	Screening		

(1) B-values are presented only for fully approved data.

** DATA WERE SUBMITTED BEFORE THE ESTABLISHMENT OF DATA DOCUMENTATION REQUIREMENTS (JUNE 1989). ALL DOCUMENTATION PRESENTLY REQUIRED WAS NOT SUPPLIED FOR THIS MATERIAL.

MATERIAL: AS4/5250-3 unidirectional material RESIN CONTENT: 36-38 wt% COMP: DENSITY: 1.55 g/cm³ FIBER VOLUME: 53-56 % VOID CONTENT: 0.1-0.9% PLY THICKNESS: 0.0057-0.0062 in. TEST METHOD: ASTM D3410A-87 MODULUS CALCULATION: NORMALIZED BY: Ply thickness to 0.0055 in. and batch fiber volume to 60%	**Table 4.4.4(e)** **C/BMI 147-UT** **AS4/5250-3** **Compression, 1-axis** **$[0]_6$** **72/A, -67/A, 350/A** **Screening**

		72		-67		350	
Temperature (°F)		72		-67		350	
Moisture Content (%)		ambient		ambient		ambient	
Equilibrium at T, RH							
Source Code		(1)		(1)		(1)	
		Normalized	Measured	Normalized	Measured	Normalized	Measured
F_1^{cu} (ksi)	Mean	175	158	198	179	174	148
	Minimum	122	110	176	160	141	127
	Maximum	203	184	222	201	235	185
	C.V.(%)	15.9	15.9	8.0	8.0	23.6	15.9
	B-value	(2)	(2)	(2)	(2)	(2)	(2)
	Distribution	Normal	Normal	Normal	Normal	Normal	Normal
	C_1	175	158	198	179	174	149
	C_2	27.7	25.1	15.8	14.3	41.1	23.6
	No. Specimens	6		6		6	
	No. Batches	1		1		1	
	Approval Class	Screening		Screening		Screening	
E_1^c (Msi)	Mean	17.0	15.4	15.5	14.0	17.4	14.9
	Minimum	14.1	12.8	13.9	12.6	15.2	13.8
	Maximum	22.7	20.5	18.5	16.7	21.9	17.2
	C.V.(%)	20.1	20.0	10.7	10.6	14.7	8.55
	No. Specimens	6		6		6	
	No. Batches	1		1		1	
	Approval Class	Screening		Screening		Screening	
ν_{12}^c	Mean						
	No. Specimens						
	No. Batches						
	Approval Class						
ε_1^{cu} (με)	Mean		12100		19800		15300
	Minimum		8000		8360		10200
	Maximum		22700		26700		18400
	C.V.(%)		46.2		43.9		18.1
	B-value		(2)		(2)		(2)
	Distribution		Normal		Normal		Normal
	C_1		12100		19800		15300
	C_2		5570		8710		2770
	No. Specimens	6		6		6	
	No. Batches	1		1		1	
	Approval Class	Screening		Screening		Screening	

(1) Reference 4.4.4.
(2) B-values are presented only for fully approved data.

** DATA WERE SUBMITTED BEFORE THE ESTABLISHMENT OF DATA DOCUMENTATION REQUIREMENTS (JUNE 1989). ALL DOCUMENTATION PRESENTLY REQUIRED WAS NOT SUPPLIED FOR THIS MATERIAL.

Table 4.4.4(f)
C/BMI 147-UT
AS4/5250-3
Compression, 1-axis
$[0]_8$
450/A, 74/0.82%, 350/0.79%
Screening

MATERIAL: AS4/5250-3 unidirectional material

RESIN CONTENT: 36-38 wt% COMP: DENSITY: 1.55 g/cm^3
FIBER VOLUME: 53-56 % VOID CONTENT: 0.1-0.9%
PLY THICKNESS: 0.0057-0.0062 in.

TEST METHOD: ASTM D3410A-87

MODULUS CALCULATION:

NORMALIZED BY: Ply thickness to 0.0055 in. and batch fiber volume to 60%

		450		74		350	
Temperature (°F)		450		74		350	
Moisture Content (%)		ambient		0.82		0.79	
Equilibrium at T, RH				160°F, 95%		(1)	
Source Code		(2)		(2)		(2)	
		Normalized	Measured	Normalized	Measured	Normalized	Measured
F_1^{cu} (ksi)	Mean	153	131	194	176	153	139
	Minimum	119	108	175	159	113	102
	Maximum	207	163	216	195	173	157
	C.V.(%)	21.2	15.1	8.6	8.63	15.5	15.5
	B-value	(3)	(3)	(3)	(3)	(3)	(3)
	Distribution	Normal	Normal	Normal	Normal	Normal	Normal
	C_1	153	131	194	176	153	139
	C_2	32.4	19.7	16.7	15.2	23.8	21.5
	No. Specimens	6		6		5	
	No. Batches	1		1		1	
	Approval Class	Screening		Screening		Screening	
E_1^c (Msi)	Mean	18.2	15.6	18.5	16.8	16.1	14.6
	Minimum	14.0	12.6	16.4	14.9	14.3	12.9
	Maximum	21.7	17.1	21.5	19.5	18.2	16.5
	C.V.(%)	16.0	10.4	9.42	9.39	9.78	9.75
	No. Specimens	6		6		5	
	No. Batches	1		1		1	
	Approval Class	Screening		Screening		Screening	
ν_{12}^c	Mean						
	No. Specimens						
	No. Batches						
	Approval Class						
ε_1^{cu} ($\mu\varepsilon$)	Mean		8480		15900		12600
	Minimum		2900		10600		6400
	Maximum		14600		22900		16000
	C.V.(%)		44.7		32.5		30.2
	B-value		(3)		(3)		(3)
	Distribution		Normal		Normal		Normal
	C_1		8480		15900		12600
	C_2		3790		5170		3810
	No. Specimens	6		6		5	
	No. Batches	1		1		1	
	Approval Class	Screening		Screening		Screening	

(1) Conditioned at 160°F, 95% relative humidity for 7 days (75% saturation).
(2) Reference 4.4.4.
(3) B-values are presented only for fully approved data.

** DATA WERE SUBMITTED BEFORE THE ESTABLISHMENT OF DATA DOCUMENTATION REQUIREMENTS (JUNE 1989). ALL DOCUMENTATION PRESENTLY REQUIRED WAS NOT SUPPLIED FOR THIS MATERIAL.

MATERIAL:	AS4/5250-3 unidirectional material			**Table 4.4.4(g)** **C/BMI 147-UT** **AS4/5250-3** **Compression, 1-axis** **$[0]_8$** **350/1.0%** **Screening**
RESIN CONTENT:	36 wt%	COMP: DENSITY:	1.55 g/cm³	
FIBER VOLUME:	56 %	VOID CONTENT:	0.0%	
PLY THICKNESS:	0.0050-0.0053 in.			
TEST METHOD: ASTM D3410A-87		MODULUS CALCULATION:		
NORMALIZED BY:	Ply thickness to 0.0055 in. and batch fiber volume to 60%			

Temperature (°F) Moisture Content (%) Equilibrium at T, RH Source Code		350 1.0 160°F, 95% (1)					
		Normalized	Measured	Normalized	Measured	Normalized	Measured
F_1^{cu} (ksi)	Mean	127	115				
	Minimum	108	97.9				
	Maximum	152	138				
	C.V.(%)	11.4	11.4				
	B-value	(2)	(2)				
	Distribution	Normal	Normal				
	C_1	127	115				
	C_2	14.4	13.0				
	No. Specimens	6					
	No. Batches	1					
	Approval Class	Screening					
E_1^c (Msi)	Mean	18.1	16.4				
	Minimum	16.6	15.0				
	Maximum	20.7	18.7				
	C.V.(%)	7.93	7.89				
	No. Specimens	6					
	No. Batches	1					
	Approval Class	Screening					
ν_{12}^c	Mean						
	No. Specimens						
	No. Batches						
	Approval Class						
ε_1^{cu} ($\mu\varepsilon$)	Mean		8120				
	Minimum		6600				
	Maximum		9180				
	C.V.(%)		11.5				
	B-value		(2)				
	Distribution		Normal				
	C_1		8120				
	C_2		934				
	No. Specimens	6					
	No. Batches	1					
	Approval Class	Screening					

(1) Reference 4.4.4.
(2) B-values are presented only for fully approved data.

** DATA WERE SUBMITTED BEFORE THE ESTABLISHMENT OF DATA DOCUMENTATION REQUIREMENTS (JUNE 1989). ALL DOCUMENTATION PRESENTLY REQUIRED WAS NOT SUPPLIED FOR THIS MATERIAL.

MATERIAL:	AS4/5250-3 unidirectional material			**Table 4.4.4(h)** **C/BMI 147-UT** **AS4/5250-3** **Shear, 12-plane** **$[\pm 45]_{4s}$** **72/A, -67/A, 350/A, 450/A** **Screening**
RESIN CONTENT:	28-32 wt%	COMP: DENSITY:	1.58-1.61 g/cm^3	
FIBER VOLUME:	59-63 %	VOID CONTENT:	0.0-1.2%	
PLY THICKNESS:	0.0055-0.0058 in.			
TEST METHOD: ASTM D3518-76		MODULUS CALCULATION:		
NORMALIZED BY:	Not normalized			

Temperature (°F) Moisture Content (%) Equilibrium at T, RH Source Code		72 ambient (1)	-67 ambient (1)	350 ambient (1)	450 ambient (1)	
	Mean	9.61	10.1	10.4	9.01	
	Minimum	8.49	9.67	9.55	8.44	
	Maximum	10.4	10.5	11.0	9.47	
	C.V.(%)	6.95	3.50	5.31	4.87	
	B-value	(2)	(2)	(2)	(2)	
	Distribution	Normal	Normal	Normal	Normal	
F_{12}^{su}	C_1	9.61	10.1	10.4	9.01	
(ksi)	C_2	0.668	0.352	0.553	0.439	
	No. Specimens	6	6	6	6	
	No. Batches	1	1	1	1	
	Approval Class	Screening	Screening	Screening	Screening	
	Mean	0.77	0.84	0.66	0.62	
	Minimum	0.71	0.78	0.62	0.50	
	Maximum	0.83	0.86	0.72	0.69	
G_{12}^{s}	C.V.(%)	5.6	3.6	5.3	12	
(Msi)	No. Specimens	6	6	6	6	
	No. Batches	1	1	1	1	
	Approval Class	Screening	Screening	Screening	Screening	
	Mean					
	Minimum					
	Maximum					
	C.V.(%)					
	B-value					
γ_{12}^{su}	Distribution					
($\mu\varepsilon$)	C_1					
	C_2					
	No. Specimens					
	No. Batches					
	Approval Class					

(1) Reference 4.4.4.

(2) B-values are presented only for fully approved data.

** DATA WERE SUBMITTED BEFORE THE ESTABLISHMENT OF DATA DOCUMENTATION REQUIREMENTS (JUNE 1989). ALL DOCUMENTATION PRESENTLY REQUIRED WAS NOT SUPPLIED FOR THIS MATERIAL.

MATERIAL:	AS4/5250-3 unidirectional material			**Table 4.4.4(i) C/BMI 147-UT AS4/5250-3 Shear, 12-plane $[\pm 45]_{4s}$ 74/0.55%, 350/0.55%, 350/1.1% Screening**
RESIN CONTENT:	28-32 wt%	COMP: DENSITY:	1.58-1.61 g/cm^3	
FIBER VOLUME:	59-63 %	VOID CONTENT:	0.0-1.2%	
PLY THICKNESS:	0.0055-0.0058 in.			
TEST METHOD: ASTM D3518-76		MODULUS CALCULATION:		
NORMALIZED BY:	Not normalized			

Temperature (°F) Moisture Content (%) Equilibrium at T, RH Source Code		74 0.55 160°F, 95% (2)	350 0.55 (1) (2)	350 1.1 160°F, 95% (2)		
F_{12}^{su} (ksi)	Mean	12.5	8.70	9.81		
	Minimum	11.3	8.24	8.13		
	Maximum	13.2	8.95	10.6		
	C.V.(%)	5.26	3.42	9.27		
	B-value	(3)	(3)	(3)		
	Distribution	Normal	Normal	Normal		
	C_1	12.5	8.70	9.81		
	C_2	0.656	0.298	0.909		
	No. Specimens	6	5	6		
	No. Batches	1	1	1		
	Approval Class	Screening	Screening	Screening		
G_{12}^{s} (Msi)	Mean	0.79	0.46	0.49		
	Minimum	0.77	0.43	0.40		
	Maximum	0.81	0.48	0.56		
	C.V.(%)	1.9	4.0	14		
	No. Specimens	6	6	4		
	No. Batches	1	1	1		
	Approval Class	Screening	Screening	Screening		
γ_{12}^{su} ($\mu\varepsilon$)	Mean					
	Minimum					
	Maximum					
	C.V.(%)					
	B-value					
	Distribution					
	C_1					
	C_2					
	No. Specimens					
	No. Batches					
	Approval Class					

(1) Conditioned at 160°F, 95% relative humidity for 3 days (75% saturation).
(2) Reference 4.4.4.
(3) B-values are presented only for fully approved data.

4.5 CARBON - POLYIMIDE COMPOSITES

4.5.1 Celion 3000/F670 8HS fabric data set description

Material Description:

Material: Celion 3000/F670

Form: 8 harness satin fabric, areal weight of 384 g/m^2, typical cured resin content of 30-34%, typical cured ply thickness of 0.0132-0.0144 inches.

Processing: Autoclave cure; 440°F for 2 hours, 600°F for 3 hours, 200 psi. Postcure to achieve high temperature service.

General Supplier Information:

Fiber: Celion 3000 fibers are continuous carbon filaments made from PAN precursor. Filament count is 3000 filaments/tow. Typical tensile modulus is 34 x 106 psi. Typical tensile strength is 515,000 psi.

Matrix: F670 is a polyimide resin (PMR 15) with good high temperature performance.

Maximum Short Term Service Temperature: 575°F (dry)

Typical applications: Commercial and military aircraft applications where high temperature resistance is a requirement.

4.5.1 Celion 3000/F670 8-harness satin weave*

MATERIAL:	Celion 3000/F670 8-harness satin weave		**C/PI Celion 3000/F670 Summary**
PREPREG:	Hexcel F3L584/F670 8-harness satin weave		
FIBER:	Celanese Celion 3000	MATRIX:	Hexcel F670 (PMR-15)
T_g(dry):	635°F T_g(wet):	T_g METHOD:	
PROCESSING:	Autoclave cure: 440°F, 2 hours; 600°F, 3 Hours, 200 psig; Postcure		

** DATA WERE SUBMITTED BEFORE THE ESTABLISHMENT OF DATA DOCUMENTATION REQUIREMENTS (JUNE 1989). ALL DOCUMENTATION PRESENTLY REQUIRED WERE NOT SUPPLIED FOR THIS MATERIAL.

Date of fiber manufacture		Date of testing	8/87
Date of resin manufacture		Date of data submittal	4/89
Date of prepreg manufacture	2/87-5/87	Date of analysis	1/93
Date of composite manufacture			

LAMINA PROPERTY SUMMARY

	75°F/A		550°F/A					
Tension, 1-axis	SS--		SS--					
Tension, 2-axis	SS--		SS--					
Tension, 3-axis								
Compression, 1-axis	SS--		SS--					
Compression, 2-axis	SS--		SS--					
Compression, 3-axis								
Shear, 12-plane								
SB Strength, 23-plane	S---							
SB Strength, 31-plane	S---							

Classes of data: F - Fully approved, I - Interim, S - Screening in Strength/Modulus/Poisson's ratio/Strain-to-failure order.

** DATA WERE SUBMITTED BEFORE THE ESTABLISHMENT OF DATA DOCUMENTATION REQUIREMENTS (JUNE 1989). ALL DOCUMENTATION PRESENTLY REQUIRED WERE NOT SUPPLIED FOR THIS MATERIAL.

		Nominal	As Submitted	Test Method
Fiber Density	(g/cm^3)	1.8		
Resin Density	(g/cm^3)	1.32		
Composite Density	(g/cm^3)	1.59	1.59 - 1.63	
Fiber Areal Weight	(g/m^2)	384		
Ply Thickness	(in)		0.0132 - 0.0144	

LAMINATE PROPERTY SUMMARY

Classes of data: F - Fully approved, I - Interim, S - Screening in Strength/Modulus/Poisson's ratio/Strain-to-failure order.

** DATA WERE SUBMITTED BEFORE THE ESTABLISHMENT OF DATA DOCUMENTATION REQUIREMENTS (JUNE 1989). ALL DOCUMENTATION PRESENTLY REQUIRED WAS NOT SUPPLIED FOR THIS MATERIAL.

MATERIAL:	Celion 3000/F670 8-harness satin weave			**Table 4.5.1(a)** **C/PI 384-8HS** **Celion 3000/F670** **Tension, 1-axis** **$[0_f]_8$** **75/A, 550/A** **Screening**
RESIN CONTENT:	30-34 wt%	COMP: DENSITY:	1.59-1.63 g/cm³	
FIBER VOLUME:	57-64 %	VOID CONTENT:	0.0-0.62%	
PLY THICKNESS:	0.0132-0.0144 in.			
TEST METHOD: ASTM D3039-76		MODULUS CALCULATION:		
NORMALIZED BY:	Fiber volume to 57%			

Temperature (°F) Moisture Content (%) Equilibrium at T, RH Source Code		75 ambient 22		550 ambient 22			
		Normalized	Measured	Normalized	Measured	Normalized	Measured
F_1^{tu} (ksi)	Mean	132	136	116	120		
	Minimum	127	131	95.4	98.7		
	Maximum	140	144	129	134		
	C.V.(%)	2.75	2.76	7.94	7.95		
	B-value	(1)	(1)	(1)	(1)		
	Distribution	Normal	Normal	Normal	Normal		
	C_1	132	136	116	120		
	C_2	3.63	3.76	9.18	9.52		
	No. Specimens	9		9			
	No. Batches	3		3			
	Approval Class	Screening		Screening			
E_1^t (Msi)	Mean	9.03	9.35	8.67	8.98		
	Minimum	8.66	8.96	8.50	8.80		
	Maximum	9.35	9.68	9.07	9.39		
	C.V.(%)	3.22	3.23	2.54	2.55		
	No. Specimens	9		9			
	No. Batches	3		3			
	Approval Class	Screening		Screening			
ν_{12}^t	Mean						
	No. Specimens						
	No. Batches						
	Approval Class						
ε_1^{tu} (με)	Mean						
	Minimum						
	Maximum						
	C.V.(%)						
	B-value						
	Distribution						
	C_1						
	C_2						
	No. Specimens						
	No. Batches						
	Approval Class						

(1) B-values are presented only for fully approved data.

** DATA WERE SUBMITTED BEFORE THE ESTABLISHMENT OF DATA DOCUMENTATION REQUIREMENTS (JUNE 1989). ALL DOCUMENTATION PRESENTLY REQUIRED WAS NOT SUPPLIED FOR THIS MATERIAL.

MATERIAL:	Celion 3000/F670 8-harness satin weave			**Table 4.5.1(b)** **C/PI 384-8HS** **Celion 3000/F670** **Tension, 2-axis** **$[90_f]_8$** **75/A, 550/A** **Screening**
RESIN CONTENT:	30-34 wt%	COMP: DENSITY:	1.59-1.63 g/cm³	
FIBER VOLUME:	57-64 %	VOID CONTENT:	0.0-0.62%	
PLY THICKNESS:	0.0132-0.0144 in.			
TEST METHOD: ASTM D3039-76		MODULUS CALCULATION:		
NORMALIZED BY:	Fiber volume fraction to 57%			

Temperature (°F) Moisture Content (%) Equilibrium at T, RH Source Code		75 ambient 22		550 ambient 22			
		Normalized	Measured	Normalized	Measured	Normalized	Measured
F_2^{tu} (ksi)	Mean	107	111	90.4	93.5		
	Minimum	85.6	88.6	61.9	64.1		
	Maximum	129	133	123	127		
	C.V.(%)	15.7	15.7	23.8	23.8		
	B-value	(1)	(1)	(1)	(1)		
	Distribution	ANOVA	ANOVA	ANOVA	ANOVA		
	C_1	19.3	20.0	24.7	25.5		
	C_2	6.09	6.09	6.02	6.02		
	No. Specimens	9		9			
	No. Batches	3		3			
	Approval Class	Screening		Screening			
E_2^t (Msi)	Mean	8.43	8.73	8.23	8.52		
	Minimum	7.43	7.69	7.58	7.85		
	Maximum	9.33	9.66	8.84	9.15		
	C.V.(%)	7.45	7.46	5.49	5.48		
	No. Specimens	9		9			
	No. Batches	3		3			
	Approval Class	Screening		Screening			
ν_{21}^t	Mean						
	No. Specimens						
	No. Batches						
	Approval Class						
ε_2^{tu} (με)	Mean						
	Minimum						
	Maximum						
	C.V.(%)						
	B-value						
	Distribution						
	C_1						
	C_2						
	No. Specimens						
	No. Batches						
	Approval Class						

(1) B-values are presented only for fully approved data.

** DATA WERE SUBMITTED BEFORE THE ESTABLISHMENT OF DATA DOCUMENTATION REQUIREMENTS (JUNE 1989). ALL DOCUMENTATION PRESENTLY REQUIRED WAS NOT SUPPLIED FOR THIS MATERIAL.

MATERIAL:	Celion 3000/F670 8-harness satin weave			**Table 4.5.1(c)** **C/PI 384-8HS** **Celion 3000/F670** **Compression, 1-axis** **$[0_f]_8$** **75/A, 550/A** **Screening**
RESIN CONTENT:	30-34 wt%	COMP: DENSITY:	1.59-1.63 g/cm³	
FIBER VOLUME:	57-64 %	VOID CONTENT:	0.0-0.62%	
PLY THICKNESS:	0.0132-0.0144 in.			
TEST METHOD: SACMA SRM 1-88		MODULUS CALCULATION:		
NORMALIZED BY:	Fiber volume fraction to 57%			

Temperature (°F) Moisture Content (%) Equilibrium at T, RH Source Code		75 ambient 22		550 ambient 22			
		Normalized	Measured	Normalized	Measured	Normalized	Measured
F_1^{cu} (ksi)	Mean	99.4	103	66.0	68.3		
	Minimum	87.9	91.3	59.0	61.1		
	Maximum	118	122	71.7	74.2		
	C.V.(%)	9.33	9.33	6.60	6.59		
	B-value	(1)	(1)	(1)	(1)		
	Distribution	ANOVA	ANOVA	Normal	Normal		
	C_1	10.2	10.6	66.0	68.3		
	C_2	5.28	5.28	4.36	4.51		
	No. Specimens	9		9			
	No. Batches	3		3			
	Approval Class	Screening		Screening			
E_1^c (Msi)	Mean	8.61	8.92	8.09	8.38		
	Minimum	8.40	8.69	7.26	7.51		
	Maximum	9.09	9.41	8.78	9.09		
	C.V.(%)	2.54	2.54	5.19	5.21		
	No. Specimens	9		9			
	No. Batches	3		3			
	Approval Class	Screening		Screening			
ν_{12}^c	Mean						
	No. Specimens						
	No. Batches						
	Approval Class						
ε_1^{cu} ($\mu\varepsilon$)	Mean						
	Minimum						
	Maximum						
	C.V.(%)						
	B-value						
	Distribution						
	C_1						
	C_2						
	No. Specimens						
	No. Batches						
	Approval Class						

(1) B-values are presented only for fully approved data.

** DATA WERE SUBMITTED BEFORE THE ESTABLISHMENT OF DATA DOCUMENTATION REQUIREMENTS (JUNE 1989). ALL DOCUMENTATION PRESENTLY REQUIRED WAS NOT SUPPLIED FOR THIS MATERIAL.

MATERIAL:	Celion 3000/F670 8-harness satin weave			**Table 4.5.1(d)** **C/PI 384-8HS** **Celion 3000/F670** **Compression, 2-axis** **$[90_f]_8$** **75/A, 550/A** **Screening**
RESIN CONTENT:	30-34 wt%	COMP: DENSITY:	1.59-1.63 g/cm^3	
FIBER VOLUME:	57-64 %	VOID CONTENT:	0.0-0.62%	
PLY THICKNESS:	0.0132-0.0144 in.			
TEST METHOD: SACMA SRM 1-88		MODULUS CALCULATION:		
NORMALIZED BY:	Fiber volume fraction to 57%			

Temperature (°F) Moisture Content (%) Equilibrium at T, RH Source Code		75 ambient 22		550 ambient 22			
		Normalized	Measured	Normalized	Measured	Normalized	Measured
F_2^{cu} (ksi)	Mean	78.9	81.7	54.2	56.1		
	Minimum	76.1	78.8	52.4	54.2		
	Maximum	80.7	83.5	56.6	58.6		
	C.V.(%)	3.10	3.10	4.02	4.03		
	B-value	(1)					
	Distribution						
	C_1						
	C_2						
	No. Specimens	3		3			
	No. Batches	1		1			
	Approval Class	Screening		Screening			
E_2^c (Msi)	Mean	8.08	8.37	7.67	7.94		
	Minimum	8.03	8.31	7.59	7.86		
	Maximum	8.14	8.43	7.77	8.04		
	C.V.(%)	0.681	0.720	1.19	1.15		
	No. Specimens	3		3			
	No. Batches	1		1			
	Approval Class	Screening		Screening			
ν_{12}^c	Mean						
	No. Specimens						
	No. Batches						
	Approval Class						
ε_2^{cu} ($\mu\varepsilon$)	Mean						
	Minimum						
	Maximum						
	C.V.(%)						
	B-value						
	Distribution						
	C_1						
	C_2						
	No. Specimens						
	No. Batches						
	Approval Class						

(1) Insufficient observations to complete the statistical evaluations.

** DATA WERE SUBMITTED BEFORE THE ESTABLISHMENT OF DATA DOCUMENTATION REQUIREMENTS (JUNE 1989). ALL DOCUMENTATION PRESENTLY REQUIRED WAS NOT SUPPLIED FOR THIS MATERIAL.

MATERIAL:	Celion 3000/F670 8-harness satin weave			**Table 4.5.1(e)** **C/PI 384-8HS** **Celion 3000/F670** **SBS, 23-plane** **$[0_f]_8$** **75/A** **Screening**
RESIN CONTENT:	30-34 wt%	COMP: DENSITY:	1.59-1.63 g/cm³	
FIBER VOLUME:	57-64 %	VOID CONTENT:	0.0-0.62%	
PLY THICKNESS:	0.0132-0.0144 in.			
TEST METHOD: ASTM D2344-84		MODULUS CALCULATION:		
NORMALIZED BY:	Not normalized			

Temperature (°F)		75				
Moisture Content (%)		ambient				
Equilibrium at T, RH						
Source Code		22				
	Mean	11.1				
	Minimum	10.4				
	Maximum	11.7				
	C.V.(%)	5.88				
	B-value	(1)				
	Distribution					
F_{23}^{sbs}	C_1					
(ksi)	C_2					
	No. Specimens	3				
	No. Batches	1				
	Approval Class	Screening				

(1) Insufficient observations to complete the statistical evaluations.

** DATA WERE SUBMITTED BEFORE THE ESTABLISHMENT OF DATA DOCUMENTATION REQUIREMENTS (JUNE 1989). ALL DOCUMENTATION PRESENTLY REQUIRED WAS NOT SUPPLIED FOR THIS MATERIAL.

Table 4.5.1(f)
C/PI 384-8HS
Celion 3000/F670
SBS, 31-plane
$[0_f]_8$
75/A
Screening

MATERIAL: Celion 3000/F670 8-harness satin weave

RESIN CONTENT: 30-34 wt% | COMP: DENSITY: 1.59-1.63 g/cm^3
FIBER VOLUME: 57-64 % | VOID CONTENT: 0.0-0.62%
PLY THICKNESS: 0.0132-0.0144 in.

TEST METHOD: ASTM D2344-84 | MODULUS CALCULATION:

NORMALIZED BY: Not normalized

Temperature (°F) Moisture Content (%) Equilibrium at T, RH Source Code		75 ambient 22				
	Mean	10.9				
	Minimum	9.70				
	Maximum	12.0				
	C.V.(%)	6.15				
	B-value	(1)				
	Distribution	ANOVA				
F_{31}^{sbs}	C_1	0.722				
(ksi)	C_2	4.78				
	No. Specimens	9				
	No. Batches	3				
	Approval Class	Screening				

(1) B-values are presented only for fully approved data.

4.6 CARBON - PHENOLIC COMPOSITES

4.7 CARBON - SILICONE COMPOSITES

4.8 CARBON - POLYBENZIMIDAZOLE COMPOSITES

4.9 CARBON - PEEK COMPOSITES

4.9.1 IM-6 12k/APC-2 unitape data set description

Material Description:

Material: IM-6 12k/APC-2

Form: Unidirectional tape, fiber areal weight of 150 g/m^2, typical cured resin content of 32%, typical cured ply thickness of 0.0053 inches.

Processing: Autoclave cure; 720°F, 30-45 mins., 60 psi.

General Supplier Information:

Fiber: IM6 fibers are continuous, intermediate modulus carbon filaments made from PAN precursor, surface treated to improve handling characteristics and structural properties. Filament count is 12,000 filaments per tow. Typical tensile modulus is 40 x 10^6 psi. Typical tensile strength is 635,000 psi.

Matrix: APC-2 is a semi-crystalline thermoplastic (polyetheretherketone, PEEK) resin that has high toughness and damage tolerance. It can be stored indefinitely at ambient conditions.

Maximum Short Term Service Temperature: 250°F (dry), 250°F (wet)

Typical applications: Primary and secondary structural applications on commercial and military aircraft, space components.

4.9.1 IM-6 12k/APC-2 unidirectional tape*

MATERIAL:	IM-6 12k/APC-2 unidirectional tape			**C/PEEK IM-6/APC-2 Summary**
PREPREG:	Fiberite IM-6/APC-2 unidirectional tape			
FIBER:	Hercules IM-6 12k	MATRIX:	Fiberite APC-2	
T_g(dry):	291°F T_g(wet): 309°F	T_g METHOD:	DMA	
PROCESSING:	Autoclave cure: 720°F, 30 - 45 minutes, 60 psig			

** DATA WERE SUBMITTED BEFORE THE ESTABLISHMENT OF DATA DOCUMENTATION REQUIREMENTS (JUNE 1989). ALL DOCUMENTATION PRESENTLY REQUIRED WERE NOT SUPPLIED FOR THIS MATERIAL.

Date of fiber manufacture		Date of testing	
Date of resin manufacture		Date of data submittal	12/88
Date of prepreg manufacture		Date of analysis	1/93
Date of composite manufacture			

LAMINA PROPERTY SUMMARY

	74°F/A		-67°F/A	180°F/A	250°F/A	180°F/O	74°F/W	180°F/W
Tension, 1-axis	SSSS		SSSS	SSSS	SSSS	SSSS	SSSS	SSSS
Tension, 2-axis	SS-S		SS-S	SS-S	SS-S			
Tension, 3-axis								
Compression, 1-axis	SS-S		SS-S	SS-S	SS-S	SS-S	SS-S	SS-S
Compression, 2-axis								
Compression, 3-axis								
Shear, 12-plane	SS--		SS--	SS--	SS--	SS--	SS--	SS--
Shear, 23-plane								
Shear, 31-plane								

Classes of data: F - Fully approved, I - Interim, S - Screening in Strength/Modulus/Poisson's ratio/Strain-to-failure order.

** DATA WERE SUBMITTED BEFORE THE ESTABLISHMENT OF DATA DOCUMENTATION REQUIREMENTS (JUNE 1989). ALL DOCUMENTATION PRESENTLY REQUIRED WERE NOT SUPPLIED FOR THIS MATERIAL.

		Nominal	As Submitted	Test Method
Fiber Density	(g/cm^3)	1.73		
Resin Density	(g/cm^3)	1.28		
Composite Density	(g/cm^3)	1.55	1.54 - 1.58	ASTM D792
Fiber Areal Weight	(g/m^2)			
Ply Thickness	(in)	0.0054	0.0052 - 0.0058	

LAMINATE PROPERTY SUMMARY

Classes of data: F - Fully approved, I - Interim, S - Screening in Strength/Modulus/Poisson's ratio/Strain-to-failure order.

** DATA WERE SUBMITTED BEFORE THE ESTABLISHMENT OF DATA DOCUMENTATION REQUIREMENTS (JUNE 1989). ALL DOCUMENTATION PRESENTLY REQUIRED WAS NOT SUPPLIED FOR THIS MATERIAL.

MATERIAL:	IM-6 12k/APC-2 unidirectional tape			**Table 4.9.1(a) C/PEEK - UT IM-6/APC-2 Tension, 1-axis $[0]_8$ 74/A, -67/A, 180/A Screening**
RESIN CONTENT:	32 wt%	COMP: DENSITY:	1.55 g/cm^3	
FIBER VOLUME:	61-62 %	VOID CONTENT:	0.0-0.2%	
PLY THICKNESS:	0.0053-0.0054 in.			
TEST METHOD: ASTM D3039-76		MODULUS CALCULATION:		
NORMALIZED BY:	Ply thickness to 0.0055 in. and batch fiber volume to 60%			

		74		-67		180	
Temperature (°F) Moisture Content (%) Equilibrium at T, RH Source Code		74 ambient (1)		-67 ambient (1)		180 ambient (1)	
		Normalized	Measured	Normalized	Measured	Normalized	Measured
F_1^{tu} (ksi)	Mean	350	370	376	398	327	344
	Minimum	266	282	326	345	234	248
	Maximum	426	455	412	439	402	421
	C.V.(%)	15.9	16.0	8.69	8.93	17.3	16.8
	B-value	(2)	(2)	(2)	(2)	(2)	(2)
	Distribution	Normal	Normal	Normal	Normal	Normal	Normal
	C_1	350	370	376	398	327	344
	C_2	55.5	59.3	32.7	35.6	56.4	58.0
	No. Specimens	6		6		6	
	No. Batches	1		1		1	
	Approval Class	Screening		Screening		Screening	
E_1^t (Msi)	Mean	21.6	22.9	22.0	23.3	23.2	24.4
	Minimum	21.3	22.4	20.9	22.2	22.3	23.6
	Maximum	22.0	23.3	23.2	24.5	23.7	25.0
	C.V.(%)	1.41	1.58	3.35	3.26	2.24	2.17
	No. Specimens	6		6		6	
	No. Batches	1		1		1	
	Approval Class	Screening		Screening		Screening	
ν_{12}^t	Mean		0.342		0.357		0.355
	No. Specimens	6		6		6	
	No. Batches	1		1		1	
	Approval Class	Screening		Screening		Screening	
ε_1^{tu} ($\mu\varepsilon$)	Mean		13600		15900		14100
	Minimum		8100		13500		10400
	Maximum		17500		17200		16800
	C.V.(%)		24.6		9.23		14.9
	B-value		(2)		(2)		(2)
	Distribution		Normal		Normal		Normal
	C_1		13600		15900		14100
	C_2		3350		1470		2100
	No. Specimens	6		6		6	
	No. Batches	1		1		1	
	Approval Class	Screening		Screening		Screening	

(1) Reference 4.9.1.
(2) B-values are presented only for fully approved data.

** DATA WERE SUBMITTED BEFORE THE ESTABLISHMENT OF DATA DOCUMENTATION REQUIREMENTS (JUNE 1989). ALL DOCUMENTATION PRESENTLY REQUIRED WAS NOT SUPPLIED FOR THIS MATERIAL.

MATERIAL:	IM-6 12k/APC-2 unidirectional tape		
RESIN CONTENT:	32 wt%	COMP: DENSITY:	1.55 g/cm^3
FIBER VOLUME:	61-62 %	VOID CONTENT:	0.0-0.2%
PLY THICKNESS:	0.0053-0.0054 in.		
TEST METHOD: ASTM D3039-76		MODULUS CALCULATION:	
NORMALIZED BY:	Ply thickness to 0.0055 in. and batch fiber volume to 60%		

Table 4.9.1(b)
C/PEEK - UT
IM-6/APC-2
Tension, 1-axis
$[0]_s$
250/A, 74/0.13%, 180/0.11%
Screening

		250		180		74	
Temperature (°F)		250		180		74	
Moisture Content (%)		ambient		0.11		0.13	
Equilibrium at T, RH				(1)		160°F, 95%	
Source Code		(2)		(2)		(2)	
		Normalized	Measured	Normalized	Measured	Normalized	Measured
	Mean	304	322	369	390	352	371
	Minimum	253	269	303	320	271	286
	Maximum	341	363	403	425	415	434
	C.V.(%)	11.4	11.4	12.3	12.2	14.6	14.2
	B-value	(3)	(3)	(3)	(3)	(3)	(3)
F_1^{tu}	Distribution	Normal	Normal	Normal	Normal	Normal	Normal
(ksi)	C_1	304	322	369	390	352	371
	C_2	34.7	36.6	45.3	47.6	51.4	52.6
	No. Specimens	6		5		6	
	No. Batches	1		1		1	
	Approval Class	Screening		Screening		Screening	
	Mean	21.4	22.7	21.8	23.0	21.2	22.3
	Minimum	20.5	21.9	20.9	22.1	20.4	21.6
	Maximum	22.1	23.4	22.2	23.5	22.0	23.0
E_1^t	C.V.(%)	2.70	2.42	2.42	2.42	3.15	3.04
(Msi)	No. Specimens	6		5		6	
	No. Batches	1		1		1	
	Approval Class	Screening		Screening		Screening	
	Mean		0.338		0.366		0.372
	No. Specimens	6		5		6	
ν_{12}^t	No. Batches	1		1		1	
	Approval Class	Screening		Screening		Screening	
	Mean		14800		16300		18100
	Minimum		12500		14400		15700
	Maximum		16400		17200		20800
	C.V.(%)		11.8		6.70		10.8
	B-value		(3)		(3)		(3)
ε_1^{tu}	Distribution		Normal		Normal		Normal
($\mu\varepsilon$)	C_1		14800		16300		18100
	C_2		1760		1090		1960
	No. Specimens	6		5		6	
	No. Batches	1		1		1	
	Approval Class	Screening		Screening		Screening	

(1) Conditioned at 160°F, 96% relative humidity for 3 days (75% saturation).
(2) Reference 4.9.1.
(3) B-values are presented only for fully approved data.

** DATA WERE SUBMITTED BEFORE THE ESTABLISHMENT OF DATA DOCUMENTATION REQUIREMENTS (JUNE 1989). ALL DOCUMENTATION PRESENTLY REQUIRED WAS NOT SUPPLIED FOR THIS MATERIAL.

MATERIAL:	IM-6 12k/APC-2 unidirectional tape		
RESIN CONTENT:	32 wt%	COMP: DENSITY:	1.55 g/cm^3
FIBER VOLUME:	61-62 %	VOID CONTENT:	0.0-0.2%
PLY THICKNESS:	0.0053-0.0054 in.		
TEST METHOD: ASTM D3039-76		MODULUS CALCULATION:	
NORMALIZED BY:	Ply thickness to 0.0055 in. and batch fiber volume to 60%		

Table 4.9.1(c)
C/PEEK - UT
IM-6/APC-2
Tension, 1-axis
$[0]_8$
180/0.14%
Screening

Temperature (°F)		180					
Moisture Content (%)		0.14					
Equilibrium at T, RH		160°F, 95%					
Source Code		(1)					
		Normalized	Measured	Normalized	Measured	Normalized	Measured
F_1^{tu} (ksi)	Mean	364	385				
	Minimum	325	344				
	Maximum	411	436				
	C.V.(%)	10.2	10.1				
	B-value	(2)	(2)				
	Distribution	Normal	Normal				
	C_1	364	385				
	C_2	37.2	38.8				
	No. Specimens	6					
	No. Batches	1					
	Approval Class	Screening					
E_1^t (Msi)	Mean	21.2	22.4				
	Minimum	20.5	21.8				
	Maximum	22.2	23.2				
	C.V.(%)	3.14	2.77				
	No. Specimens	6					
	No. Batches	1					
	Approval Class	Screening					
ν_{12}^t	Mean		0.332				
	No. Specimens	6					
	No. Batches	1					
	Approval Class	Screening					
ε_1^{tu} (με)	Mean		15400				
	Minimum		13600				
	Maximum		17200				
	C.V.(%)		9.24				
	B-value		(2)				
	Distribution		Normal				
	C_1		15400				
	C_2		1420				
	No. Specimens	6					
	No. Batches	1					
	Approval Class	Screening					

(1) Reference 4.9.1.
(2) B-values are presented only for fully approved data.

** DATA WERE SUBMITTED BEFORE THE ESTABLISHMENT OF DATA DOCUMENTATION REQUIREMENTS (JUNE 1989). ALL DOCUMENTATION PRESENTLY REQUIRED WAS NOT SUPPLIED FOR THIS MATERIAL.

Table 4.9.1(d)
C/PEEK-UT
IM-6/APC-2
Tension, 2-axis
$[90]_{16}$
74/A, -67/A, 180/A, 250/A
Screening

MATERIAL: IM-6 12k/APC-2 unidirectional tape

RESIN CONTENT: 31-34 wt% COMP: DENSITY: 1.55 g/cm^3
FIBER VOLUME: 60-62 % VOID CONTENT: 0.0%
PLY THICKNESS: 0.0054-0.0058 in.

TEST METHOD: ASTM D3039-76 MODULUS CALCULATION:

NORMALIZED BY: Not normalized

Temperature (°F) Moisture Content (%) Equilibrium at T, RH Source Code		74 ambient (1)	-67 ambient (1)	180 ambient (1)	250 ambient (1)		
F_2^{tu} (ksi)	Mean	9.41	9.67	11.1	9.07		
	Minimum	8.53	8.72	10.0	7.30		
	Maximum	10.6	10.7	12.2	9.72		
	C.V.(%)	9.35	6.52	8.87	10.1		
	B-value	(2)	(2)	(2)	(2)		
	Distribution	Normal	Normal	Normal	Normal		
	C_1	9.41	9.67	11.1	9.07		
	C_2	0.880	0.631	0.985	0.916		
	No. Specimens	6	6	6	6		
	No. Batches	1	1	1	1		
	Approval Class	Screening	Screening	Screening	Screening		
E_2^t (Msi)	Mean	1.28	1.41	1.22	1.32		
	Minimum	1.24	1.35	1.17	1.27		
	Maximum	1.36	1.46	1.25	1.38		
	C.V.(%)	3.33	3.32	2.13	3.44		
	No. Specimens	6	6	6	6		
	No. Batches	1	1	1	1		
	Approval Class	Screening	Screening	Screening	Screening		
ν_{21}^t	Mean						
	No. Specimens						
	No. Batches						
	Approval Class						
ε_2^{tu} (με)	Mean	7610	7120	10900	12300		
	Minimum	6650	6450	8850	8510		
	Maximum	8830	8180	14900	23600		
	C.V.(%)	11.2	8.15	20.0	45.5		
	B-value	(2)	(2)	(2)	(2)		
	Distribution	Normal	Normal	Normal	Nonpara.		
	C_1	7610	7120	10900	5		
	C_2	850	581	2180	3.06		
	No. Specimens	6	6	6	6		
	No. Batches	1	1	1	1		
	Approval Class	Screening	Screening	Screening	Screening		

(1) Reference 4.9.1.
(2) B-values are presented only for fully approved data.

** DATA WERE SUBMITTED BEFORE THE ESTABLISHMENT OF DATA DOCUMENTATION REQUIREMENTS (JUNE 1989). ALL DOCUMENTATION PRESENTLY REQUIRED WAS NOT SUPPLIED FOR THIS MATERIAL.

MATERIAL:	IM-6 12k/APC-2 unidirectional tape			**Table 4.9.1(e) C/PEEK - UT IM-6/APC-2 Compression, 1-axis $[0]_{16}$ 74/A, -67/A, 180/A Screening**
RESIN CONTENT:	32 wt%	COMP: DENSITY:	1.55 g/cm³	
FIBER VOLUME:	60-62 %	VOID CONTENT:	0.0%	
PLY THICKNESS:	0.0054-0.0058 in.			
TEST METHOD: ASTM D3410A-87		MODULUS CALCULATION:		
NORMALIZED BY:	Ply thickness to 0.0055 in. and batch fiber volume to 60%			

Temperature (°F) Moisture Content (%) Equilibrium at T, RH Source Code		74 ambient (1)		-67 ambient (1)		180 ambient (1)	
		Normalized	Measured	Normalized	Measured	Normalized	Measured
F_1^{cu} (ksi)	Mean	167	169	156	160	156	155
	Minimum	139	144	115	118	103	96.7
	Maximum	197	200	179	181	195	190
	C.V.(%)	13.3	13.3	16.0	15.6	20.2	20.4
	B-value	(2)	(2)	(2)	(2)	(2)	(2)
	Distribution	Normal	Normal	Normal	Normal	Normal	Normal
	C_1	167	169	156	160	156	155
	C_2	22.1	22.4	25.0	24.9	31.5	31.6
	No. Specimens	6		6		6	
	No. Batches	1		1		1	
	Approval Class	Screening		Screening		Screening	
E_1^c (Msi)	Mean	19.4	19.7	20.4	20.9	21.4	21.2
	Minimum	17.6	18.1	16.9	17.3	17.0	16.0
	Maximum	20.9	21.2	24.0	24.8	27.5	26.7
	C.V.(%)	6.54	7.17	12.2	12.6	16.1	16.1
	No. Specimens	6		6		6	
	No. Batches	1		1		1	
	Approval Class	Screening		Screening		Screening	
ν_{12}^c	Mean						
	No. Specimens						
	No. Batches						
	Approval Class						
ε_1^{cu} (με)	Mean		8790		7910		8010
	Minimum		7780		4510		5950
	Maximum		10500		9630		9350
	C.V.(%)		11.8		24.7		14.9
	B-value		(2)		(2)		(2)
	Distribution		Normal		Normal		Normal
	C_1		8790		7910		8010
	C_2		1040		1950		1200
	No. Specimens	6		6		6	
	No. Batches	1		1		1	
	Approval Class	Screening		Screening		Screening	

(1) Reference 4.9.1.
(2) B-values are presented only for fully approved data.

** DATA WERE SUBMITTED BEFORE THE ESTABLISHMENT OF DATA DOCUMENTATION REQUIREMENTS (JUNE 1989). ALL DOCUMENTATION PRESENTLY REQUIRED WAS NOT SUPPLIED FOR THIS MATERIAL.

MATERIAL:	IM-6 12k/APC-2 unidirectional tape			**Table 4.9.1(f) C/PEEK - UT IM-6/APC-2 Compression, 1-axis $[0]_{16}$ 250/A, 74/0.12%, 180/0.097% Screening**
RESIN CONTENT:	32 wt%	COMP: DENSITY:	1.55 g/cm³	
FIBER VOLUME:	60-62 %	VOID CONTENT:	0.0%	
PLY THICKNESS:	0.0054-0.0058 in.			
TEST METHOD: ASTM D3410A-87		MODULUS CALCULATION:		
NORMALIZED BY:	Ply thickness to 0.0055 in. and batch fiber volume to 60%			

		250		180		74	
Temperature (°F)		250		180		74	
Moisture Content (%)		ambient		0.097		0.12	
Equilibrium at T, RH				(1)		160°F, 95%	
Source Code		(2)		(2)		(2)	
		Normalized	Measured	Normalized	Measured	Normalized	Measured
F_1^{cu} (ksi)	Mean	129	126	162	160	174	176
	Minimum	70.0	71.5	156	146	141	144
	Maximum	154	145	168	169	186	192
	C.V.(%)	23.6	21.8	3.25	5.36	9.6	9.7
	B-value	(3)	(3)	(3)	(3)	(3)	(3)
	Distribution	Normal	Nonpara.	Normal	Normal	Normal	Normal
	C_1	129	5	162	160	174	176
	C_2	30.5	3.06	5.26	8.59	16.7	17.1
	No. Specimens	6		5		6	
	No. Batches	1		1		1	
	Approval Class	Screening		Screening		Screening	
E_1^c (Msi)	Mean	21.2	20.7	19.5	19.3	21.4	21.6
	Minimum	19.6	19.0	18.7	18.6	18.8	19.3
	Maximum	24.7	23.2	20.0	20.7	23.9	23.9
	C.V.(%)	8.47	7.37	2.91	4.42	8.60	7.38
	No. Specimens	6		5		6	
	No. Batches	1		1		1	
	Approval Class	Screening		Screening		Screening	
ν_{12}^c	Mean						
	No. Specimens						
	No. Batches						
	Approval Class						
ε_1^{cu} (με)	Mean		6860		8310		8690
	Minimum		3380		7500		6950
	Maximum		8990		9390		12100
	C.V.(%)		28.7		8.94		23.5
	B-value		(3)		(3)		(3)
	Distribution		Normal		Normal		Normal
	C_1		6860		8310		8690
	C_2		1970		743		2050
	No. Specimens	6		5		6	
	No. Batches	1		1		1	
	Approval Class	Screening		Screening		Screening	

(1) Conditioned at 160°F, 95% relative humidity for 10 days (75% saturation).
(2) Reference 4.9.1.
(3) B-values are presented only for fully approved data.

** DATA WERE SUBMITTED BEFORE THE ESTABLISHMENT OF DATA DOCUMENTATION REQUIREMENTS (JUNE 1989). ALL DOCUMENTATION PRESENTLY REQUIRED WAS NOT SUPPLIED FOR THIS MATERIAL.

MATERIAL:	IM-6 12k/APC-2 unidirectional tape			**Table 4.9.1(g)** **C/PEEK - UT** **IM-6/APC-2** **Compression, 1-axis** **$[0]_{16}$** **180/0.11%** **Screening**
RESIN CONTENT:	32 wt%	COMP: DENSITY:	1.55 g/cm^3	
FIBER VOLUME:	60-62 %	VOID CONTENT:	0.0%	
PLY THICKNESS:	0.0054-0.0058 in.			
TEST METHOD: ASTM D3410A-87		MODULUS CALCULATION:		
NORMALIZED BY:	Ply thickness to 0.0055 in. and batch fiber volume to 60%			

Temperature (°F) Moisture Content (%) Equilibrium at T, RH Source Code		180 0.11 160°F, 95% (1)					
		Normalized	Measured	Normalized	Measured	Normalized	Measured
F_1^{cu} (ksi)	Mean	154	151				
	Minimum	105	98.5				
	Maximum	189	183				
	C.V.(%)	18.2	19.3				
	B-value	(2)	(2)				
	Distribution	Normal	Normal				
	C_1	154	151				
	C_2	28.0	29.3				
	No. Specimens	6					
	No. Batches	1					
	Approval Class	Screening					
E_1^c (Msi)	Mean	20.3	19.8				
	Minimum	15.6	15.7				
	Maximum	25.3	24.6				
	C.V.(%)	18.4	17.6				
	No. Specimens	6					
	No. Batches	1					
	Approval Class	Screening					
ν_{12}^c	Mean						
	No. Specimens						
	No. Batches						
	Approval Class						
ε_1^{cu} (με)	Mean		8180				
	Minimum		6580				
	Maximum		9500				
	C.V.(%)		13.0				
	B-value		(2)				
	Distribution		Normal				
	C_1		8180				
	C_2		1070				
	No. Specimens	6					
	No. Batches	1					
	Approval Class	Screening					

(1) Reference 4.9.1.
(2) B-values are presented only for fully approved data.

** DATA WERE SUBMITTED BEFORE THE ESTABLISHMENT OF DATA DOCUMENTATION REQUIREMENTS (JUNE 1989). ALL DOCUMENTATION PRESENTLY REQUIRED WAS NOT SUPPLIED FOR THIS MATERIAL.

MATERIAL: IM-6 12k/APC-2 unidirectional tape

Table 4.9.1(h)
C/PEEK - UT
IM-6/APC-2
Shear, 12-plane
$[\pm 45]_{4s}$
74/A, -67/A, 180/A, 250/A
Screening

RESIN CONTENT: 31-32 wt% COMP: DENSITY: 1.55 g/cm^3
FIBER VOLUME: 61 % VOID CONTENT: 0.0-0.2%
PLY THICKNESS: 0.0052-0.0056 in.

TEST METHOD: MODULUS CALCULATION:
ASTM D3518-76

NORMALIZED BY: Not normalized

Temperature (°F) Moisture Content (%) Equilibrium at T, RH Source Code		74 ambient (1)	-67 ambient (1)	180 ambient (1)	250 ambient (1)	
	Mean	23.9	25.4	22.4	19.8	
	Minimum	18.9	18.1	17.2	14.2	
	Maximum	27.8	29.0	25.3	23.1	
	C.V.(%)	14.8	14.8	15.6	15.1	
	B-value	(2)	(2)	(2)	(2)	
	Distribution	Normal	Normal	Normal	Normal	
F_{12}^{su}	C_1	23.9	25.4	22.4	19.8	
(ksi)	C_2	3.53	3.77	3.49	2.98	
	No. Specimens	6	6	6	6	
	No. Batches	1	1	1	1	
	Approval Class	Screening	Screening	Screening	Screening	
	Mean	0.78	0.91	0.78	0.71	
	Minimum	0.73	0.83	0.72	0.63	
	Maximum	0.83	0.96	0.86	0.79	
G_{12}^{s}	C.V.(%)	5.5	5.5	6.2	9.3	
(Msi)	No. Specimens	6	6	6	6	
	No. Batches	1	1	1	1	
	Approval Class	Screening	Screening	Screening	Screening	
	Mean					
	Minimum					
	Maximum					
	C.V.(%)					
	B-value					
γ_{12}^{su}	Distribution					
($\mu\varepsilon$)	C_1					
	C_2					
	No. Specimens					
	No. Batches					
	Approval Class					

(1) Reference 4.9.1.
(2) B-values are presented only for fully approved data.

** DATA WERE SUBMITTED BEFORE THE ESTABLISHMENT OF DATA DOCUMENTATION REQUIREMENTS (JUNE 1989). ALL DOCUMENTATION PRESENTLY REQUIRED WAS NOT SUPPLIED FOR THIS MATERIAL.

MATERIAL:	IM-6 12k/APC-2 unidirectional tape			**Table 4.9.1(i)** **C/PEEK - UT** **IM-6/APC-2** **Shear, 12-plane** **$[\pm45]_{4s}$** **74/0.21%, 180/0.17%, 180/0.20%** **Screening**
RESIN CONTENT:	31-32 wt%	COMP: DENSITY:	1.55 g/cm^3	
FIBER VOLUME:	61 %	VOID CONTENT:	0.0-0.2%	
PLY THICKNESS:	0.0052-0.0056 in.			
TEST METHOD: ASTM D3518-76		MODULUS CALCULATION:		
NORMALIZED BY:	Not normalized			

Temperature (°F)		180	74	180		
Moisture Content (%)		0.17	0.21	0.20		
Equilibrium at T, RH		(1)	160°F, 95%	160°F, 95%		
Source Code		(2)	(2)	(2)		
F_{12}^{su} (ksi)	Mean	23.3	23.0	20.0		
	Minimum	21.8	16.2	14.5		
	Maximum	24.0	26.7	26.1		
	C.V.(%)	3.85	15.4	22.4		
	B-value	(3)	(3)	(3)		
	Distribution	Normal	Normal	Normal		
	C_1	23.3	23.0	20.0		
	C_2	0.897	3.55	4.48		
	No. Specimens	5	6	6		
	No. Batches	1	1	1		
	Approval Class	Screening	Screening	Screening		
G_{12}^{s} (Msi)	Mean	0.76	0.79	0.71		
	Minimum	0.74	0.65	0.64		
	Maximum	0.78	0.89	0.78		
	C.V.(%)	2.7	10	9.0		
	No. Specimens	4	6	6		
	No. Batches	1	1	1		
	Approval Class	Screening	Screening	Screening		
γ_{12}^{su} ($\mu\varepsilon$)	Mean					
	Minimum					
	Maximum					
	C.V.(%)					
	B-value					
	Distribution					
	C_1					
	C_2					
	No. Specimens					
	No. Batches					
	Approval Class					

(1) Conditioned at 160°F, 95% relative humidity for 27 days (75% saturation).
(2) Reference 4.9.1.
(3) B-values are presented only for fully approved data.

REFERENCES

4.4.4 Rondeau, R.A., Askins, D. R., and Sjoblom, P., "Development of Engineering Data on New Aerospace Materials," University of Dayton Research Institute, UDR-TR-88-88, AFWAL-TR-88-4217, December 1988, Distribution authorized to DoD and DoD contractors only; critical technology; September 1988. Other requests for this document should be referred to AFWAL/MLSE, OH 45433-6533.

4.9.1 Rondeau, R.A., Askins, D. R., and Sjoblom, P., "Development of Engineering Data on New Aerospace Materials," University of Dayton Research Institute, UDR-TR-88-88, AFWAL-TR-88-4217, December 1988, Distribution authorized to DoD and DoD contractors only; critical technology; September 1988. Other requests for this document should be referred to AFWAL/MLSE, OH 45433-6533.

CHAPTER 5 ARAMID FIBER COMPOSITES

5.1 INTRODUCTION

5.2 ARAMID - EPOXY COMPOSITES

5.3 ARAMID - POLYESTER COMPOSITES

5.4 ARAMID - BISMALEIMIDE COMPOSITES

5.5 ARAMID - POLYIMIDE COMPOSITES

5.6 ARAMID - PHENOLIC COMPOSITES

5.7 ARAMID - SILICON COMPOSITES

5.8 ARAMID - POLYBENZIMIDAZOLE COMPOSITES

5.9 ARAMID - PEEK COMPOSITES

CHAPTER 6 GLASS FIBER COMPOSITES

6.1 INTRODUCTION

6.2 GLASS\EPOXY COMPOSITES

6.2.1 S2-449 43k/SP381 unitape data set description

Material Description:

Material: S2-449 17k/PR381

Form: Unidirectional tape, fiber areal weight of 111 g/m^2, typical cured resin content of 28-33%, typical cured ply thickness of 0.0033 - 0.0037 inches.

Processing: Autoclave cure; 260° F, 50 psi for two hours

General Supplier Information:

Fiber: S2 glass has enhanced properties in strength, modulus, impact resistance and fatigue when compared to conventional E glass roving. The sizing for these fibers is an epoxy compatible 449 finish. Roving of 17,000 filaments. Typical tensile modulus is 12.5 to 13.0 Msi. Typical tensile strength is 665,000 psi.

Matrix: PR381 is a 250°F curing epoxy resin providing properties similar to conventional 350°F curing systems. Light tack for up to 30 days at 75°F.

Maximum Short Term Service Temperature: 220°F (dry), 160°F (wet)

Typical applications: Primary and secondary structural applications where improved fatigue and excellent mechanical strength is important such as helicopters and general aviation.

6.2.1 S2-449 43k/SP381 unidirectional tape

MATERIAL:	S2-449 43.5k/SP 381 unidirectional tape			**SGI/Ep 284-UT** **S2-449/SP 381** **Summary**
PREPREG:	3M Scotchply SP 381 Uni S29 284 BW 33RC Prepreg			
FIBER:	Owens Corning S2-449, 0 twist, no surface treatment, typical 449 glass sizing	MATRIX:	3M PR 381	
T_g(dry):	280°F T_g(wet): 234°F	T_g METHOD:	SRM 18-94, RDA, G' onset	
PROCESSING:	Autoclave cure: 260±10°F, 120±20 min., 50 psi			

Date of fiber manufacture	5/92 - 12/94	Date of testing	5/93 - 4/95
Date of resin manufacture	1/93 - 12/94	Date of data submittal	6/96
Date of prepreg manufacture	4/93 - 3/95	Date of analysis	2/97
Date of composite manufacture	12/91 - 3/96		

LAMINA PROPERTY SUMMARY

	75°F/A		-65°F/A	180°F/A		160°F/W		
Tension, 1-axis	FF-F		SS-S	SS-S		SS-S		
Tension, 2-axis	SS-S		SS-S	SS-S		SS-S		
Tension, 3-axis								
Compression, 1-axis	SS-S		SS-S	SS-S		SS-S		
Compression, 2-axis								
Compression, 3-axis								
Shear, 12-plane	SS--		SS--	SS--		SS--		
Shear, 23-plane								
Shear, 31-plane								
SBS, 31-plane	S---		S---	S---		S---		

Classes of data: F - Fully approved, I - Interim, S - Screening in Strength/Modulus/Poisson's ratio/Strain-to-failure order.
Data are also included for F^{sbs} conditioned in eight fluids.

		Nominal	As Submitted	Test Method
Fiber Density	(g/cm^3)	2.49		ASTM C693
Resin Density	(g/cm^3)	1.216		ASTM D792
Composite Density	(g/cm^3)	1.85	1.84 - 1.97	
Fiber Areal Weight	(g/m^2)	284	283 - 291	SRM 23B
Ply Thickness	(in)	0.009	0.0070 - 0.0097	
Fiber Volume	(%)	50	47.3 - 56.1	

LAMINATE PROPERTY SUMMARY

	73°F/A							
[±45/0/∓ 45]								
Tension, x-axis	SS-S							
Tension, y-axis	SS-S							

Classes of data: F - Fully approved, I - Interim, S - Screening in Strength/Modulus/Poisson's ratio/Strain-to-failure order.

MATERIAL:	S2-449 43.5k/SP 381 unidirectional tape			Table 6.2.1(a) SGI/Ep 284-UT S2-449/SP 381 Tension, 1-axis $[0]_s$ 73/A, -65/A, 180/A Fully Approved, Screening
RESIN CONTENT:	29-34 wt%	COMP: DENSITY:	1.84-1.97 g/cm³	
FIBER VOLUME:	47.3-54.7 %	VOID CONTENT:	0-0.07%	
PLY THICKNESS:	0.0080-0.0096 in.			
TEST METHOD: SRM 4-88		MODULUS CALCULATION: Chord between 1000 and 6000 με		
NORMALIZED BY:	Specimen thickness and batch fiber areal weight to 50% (0.0090 in. CPT)			

		73		-65		180	
Temperature (°F) Moisture Content (%) Equilibrium at T, RH Source Code		73 Ambient 69		-65 Ambient 69		180 Ambient 69	
		Normalized	Measured	Normalized	Measured	Normalized	Measured
F_1^{tu} (ksi)	Mean	246	243	236	246	208	211
	Minimum	217	228	204	218	200	200
	Maximum	287	267	257	261	220	228
	C.V.(%)	6.45	3.89	7.44	5.19	3.62	4.79
	B-value	198	219	(1)	(1)	(1)	(1)
	Distribution	ANOVA	ANOVA	ANOVA	Weibull	ANOVA	ANOVA
	C_1	16.8	9.78	21.4	252	8.15	11.7
	C_2	2.82	2.45	16.6	28.3	9.69	14.1
	No. Specimens	32		11		11	
	No. Batches	6		2		2	
	Approval Class	Fully Approved		Screening		Screening	
E_1^t (Msi)	Mean	6.91	6.83	6.93	7.24	6.62	6.70
	Minimum	6.32	6.47	6.41	6.91	6.42	6.55
	Maximum	7.54	7.22	7.24	7.53	6.78	7.09
	C.V.(%)	4.34	2.68	3.03	3.26	1.62	2.48
	No. Specimens	32		11		11	
	No. Batches	6		2		2	
	Approval Class	Fully Approved		Screening		Screening	
ν_{12}^t	Mean						
	No. Specimens						
	No. Batches						
	Approval Class						
ε_1^{tu} (με)	Mean		35600		34100		31500
	Minimum		33400		29500		30000
	Maximum		38300		36700		33800
	C.V.(%)		3.83		6.23		4.21
	B-value		32400		(1)		(1)
	Distribution		ANOVA		ANOVA		ANOVA
	C_1		1400		2440		1390
	C_2		2.28		13.9		7.11
	No. Specimens	32		11		11	
	No. Batches	6		2		2	
	Approval Class	Fully Approved		Screening		Screening	

(1) B-values are presented only for fully approved data.

MATERIAL:	S2-449 43.5k/SP 381 unidirectional tape		
RESIN CONTENT:	32-33 wt%	COMP: DENSITY:	1.89-1.97 g/cm³
FIBER VOLUME:	49.3-51.1 %	VOID CONTENT:	0-0.07%
PLY THICKNESS:	0.0088-0.0092 in.		
TEST METHOD: SRM 4-88		MODULUS CALCULATION: Chord between 1000 and 6000 με	
NORMALIZED BY:	Specimen thickness and batch fiber areal weight to 50% (0.0090 in. CPT)		

Table 6.2.1(b)
SGI/Ep 284-UT
S2-449/SP 381
Tension, 1-axis
$[0]_6$
160/W
Screening

	Temperature (°F)	160					
	Moisture Content (%)	Wet					
	Equilibrium at T, RH	(2)					
	Source Code	69					
		Normalized	Measured	Normalized	Measured	Normalized	Measured
F_1^{tu} (ksi)	Mean	113	115				
	Minimum	105	106				
	Maximum	119	120				
	C.V.(%)	3.90	3.22				
	B-value	(1)	(1)				
	Distribution	Weibull	Weibull				
	C_1	115	116				
	C_2	32.6	40.5				
	No. Specimens	13					
	No. Batches	2					
	Approval Class	Screening					
E_1^t (Msi)	Mean	6.86	6.95				
	Minimum	6.52	6.71				
	Maximum	7.25	7.16				
	C.V.(%)	3.19	2.06				
	No. Specimens	13					
	No. Batches	2					
	Approval Class	Screening					
ν_{12}^t	Mean						
	No. Specimens						
	No. Batches						
	Approval Class						
ε_1^{tu} (με)	Mean		16500				
	Minimum		15600				
	Maximum		17100				
	C.V.(%)		2.76				
	B-value		(1)				
	Distribution		Weibull				
	C_1		16700				
	C_2		45.9				
	No. Specimens	13					
	No. Batches	2					
	Approval Class	Screening					

(1) B-values are presented only for fully approved data.
(2) Conditioned in 160°F water for 14 days.

MATERIAL:	S2-449 43.5k/SP 381 unidirectional tape			Table 6.2.1(c) SGI/Ep 284-UT S2-449/SP 381 Tension, 2-axis $[90]_{10}$ 73/A, -65A, 180/A, 160/W Screening
RESIN CONTENT:	31-32 wt%	COMP: DENSITY:	1.84-1.86 g/cm^3	
FIBER VOLUME:	51.0-53.2 %	VOID CONTENT:	0-0.99%	
PLY THICKNESS:	0.0081-0.0092 in.			
TEST METHOD:		MODULUS CALCULATION:		
SRM 4-88		Chord between 1000 and 3000 με (2)		
NORMALIZED BY:	Not normalized			

	Temperature (°F) Moisture Content (%) Equilibrium at T, RH Source Code	73 Ambient 69	-65 Ambient 69	180 Ambient 69	160 Wet (3) 69		
	Mean	9.0	9.1	7.5	4.2		
	Minimum	8.7	8.3	7.1	3.8		
	Maximum	9.3	9.8	7.6	4.7		
	C.V.(%)	2.3	4.7	2.7	7.5		
	B-value	(1)	(1)	(1)	(1)		
F_2^{tu}	Distribution	Weibull	Weibull	Normal	Weibull		
(ksi)	C_1	9.1	9.3	7.5	4.3		
	C_2	49	24	0.20	14		
	No. Specimens	10	11	6	10		
	No. Batches	2	2	1	2		
	Approval Class	Screening	Screening	Screening	Screening		
	Mean	1.93	2.10	1.53	1.07		
	Minimum	1.85	1.88	1.47	1.00		
	Maximum	2.07	2.31	1.59	1.12		
E_2^t	C.V.(%)	3.31	5.57	2.58	3.23		
(Msi)	No. Specimens	10	11	6	10		
	No. Batches	2	2	1	2		
	Approval Class	Screening	Screening	Screening	Screening		
	Mean						
	No. Specimens						
ν_{21}^t	No. Batches						
	Approval Class						
	Mean	4700	4300	4900	3900		
	Minimum	4200	3800	4600	3400		
	Maximum	5100	4800	5100	4300		
	C.V.(%)	4.6	7.2	4.6	6.7		
	B-value	(1)	(1)	(1)	(1)		
ε_2^{tu}	Distribution	Nonpara.	Weibull	Normal	Weibull		
(με)	C_1	6	4500	4900	4000		
	C_2	2.1	16	220	17		
	No. Specimens	10	11	6	10		
	No. Batches	2	2	1	2		
	Approval Class	Screening	Screening	Screening	Screening		

(1) B-values are presented only for fully approved data.
(2) Exception to SRM 4-88.
(2) Conditioned in 160°F water for 14 days.

MATERIAL:	S2-449 43.5k/SP 381 unidirectional tape			Table 6.2.1(d) SGI/Ep 284-UT S2-449/SP 381 Compression, 1-axis $[0]_s$ 73/A, -65/A, 180/A Screening
RESIN CONTENT:	28-33 wt%	COMP: DENSITY:	1.90-1.94 g/cm^3	
FIBER VOLUME:	49.3-56.1 %	VOID CONTENT:	0.12-0.50%	
PLY THICKNESS:	0.0080-0.0094 in.			
TEST METHOD: SRM 1-88		MODULUS CALCULATION: Chord between 1000 and 3000 με		
NORMALIZED BY:	Specimen thickness and batch fiber areal weight to 50% (0.0090 in. CPT)			

Temperature (°F) Moisture Content (%) Equilibrium at T, RH Source Code		73 Ambient 69		-65 Ambient 69		180 Ambient 69	
		Normalized	Measured	Normalized	Measured	Normalized	Measured
	Mean	168	182	170	177	150	166
	Minimum	141	149	153	162	137	154
	Maximum	199	215	184	196	166	179
	C.V.(%)	10.4	10.8	5.20	5.59	6.70	4.93
	B-value	(1)	(1)	(1)	(1)	(1)	(1)
F_1^{cu}	Distribution	Weibull	Weibull	Weibull	ANOVA	ANOVA	Weibull
(ksi)	C_1	176	191	174	10.9	12.3	170
	C_2	10.6	10.5	22.0	11.3	16.6	22.2
	No. Specimens	20		14		12	
	No. Batches	2		2		2	
	Approval Class	Screening		Screening		Screening	
	Mean	6.96	7.06	6.87	7.20	6.76	6.95
	Minimum	6.71	6.67	6.75	6.75	6.54	6.75
	Maximum	7.20	7.34	7.01	7.68	6.94	7.16
E_1^c	C.V.(%)	2.43	2.68	1.40	4.16	1.74	2.22
(Msi)	No. Specimens	10		10		10	
	No. Batches	2		2		2	
	Approval Class	Screening		Screening		Screening	
	Mean						
	No. Specimens						
ν_{12}^c	No. Batches						
	Approval Class						
	Mean						
	Minimum						
	Maximum						
	C.V.(%)						
	B-value						
ε_1^{cu}	Distribution						
(με)	C_1						
	C_2						
	No. Specimens						
	No. Batches						
	Approval Class						

(1) B-values are presented only for fully approved data.

MATERIAL:	S2-449 43.5k/SP 381 unidirectional tape			**Table 6.2.1(e)** **SGI/Ep 284-UT** **S2-449/SP 381** **Compression, 1-axis** **$[0]_5$** **160/W** **Screening**
RESIN CONTENT:	28-33 wt%	COMP: DENSITY:	1.90-1.94 g/cm^3	
FIBER VOLUME:	49.3-56.1 %	VOID CONTENT:	0.12-0.50%	
PLY THICKNESS:	0.0082-0.0090 in.			
TEST METHOD: SRM 1-88		MODULUS CALCULATION: Chord between 1000 and 3000 $\mu\varepsilon$		
NORMALIZED BY:	Specimen thickness and batch fiber areal weight to 50% (0.0090 in. CPT)			

Temperature (°F)		160					
Moisture Content (%)		Wet					
Equilibrium at T, RH		(2)					
Source Code		69					
		Normalized	Measured	Normalized	Measured	Normalized	Measured
F_1^{cu} (ksi)	Mean	139	146				
	Minimum	130	131				
	Maximum	146	157				
	C.V.(%)	3.48	5.27				
	B-value	(1)	(1)				
	Distribution	Weibull	Weibull				
	C_1	141	149				
	C_2	37.4	22.6				
	No. Specimens	10					
	No. Batches	2					
	Approval Class	Screening					
E_1^c (Msi)	Mean	6.92	7.16				
	Minimum	6.69	6.85				
	Maximum	7.08	7.43				
	C.V.(%)	2.11	2.83				
	No. Specimens	10					
	No. Batches	2					
	Approval Class	Screening					
ν_{12}^c	Mean						
	No. Specimens						
	No. Batches						
	Approval Class						
ε_1^{cu} ($\mu\varepsilon$)	Mean						
	Minimum						
	Maximum						
	C.V.(%)						
	B-value						
	Distribution						
	C_1						
	C_2						
	No. Specimens						
	No. Batches						
	Approval Class						

(1) B-values are presented only for fully approved data.

(2) Conditioned in 160°F water for 14 days.

MATERIAL:	S2-449 43.5k/SP 381 unidirectional tape			**Table 6.2.1(f)** **SGI/Ep 284-UT** **S2-449/SP 381** **Shear, 12-plane** **$[\pm 45]_{2S}$** **73/A, -65A, 180/A, 160/W** **Screening**
RESIN CONTENT:	29-32 wt%	COMP: DENSITY:	1.88-1.94 g/cm^3	
FIBER VOLUME:	51.1-54.5 %	VOID CONTENT:	0.21-0.60%	
PLY THICKNESS:	0.0081-0.0090 in.			
TEST METHOD:		MODULUS CALCULATION:		
SRM 7-88		Chord between 500 and 3000 $\mu\varepsilon$, axial		
NORMALIZED BY:	Not normalized			

Temperature (°F) Moisture Content (%) Equilibrium at T, RH Source Code		73 Ambient 69	-65 Ambient 69	180 Ambient 69	160 Wet (2) 69		
F_{12}^{su} (ksi)	Mean	14.3	13.6	11.8	9.5		
	Minimum	13.2	12.9	10.8	9.0		
	Maximum	14.7	14.5	12.3	9.8		
	C.V.(%)	3.52	3.77	3.66	2.9		
	B-value	(1)	(1)	(1)	(1)		
	Distribution	Nonpara.	Normal	Weibull	Weibull		
	C_1	6	13.6	12.0	9.6		
	C_2	2.14	0.515	38.4	44		
	No. Specimens	10	9	10	12		
	No. Batches	2	2	2	2		
	Approval Class	Screening	Screening	Screening	Screening		
G_{12} (Msi)	Mean	0.689	0.881	0.555	0.470		
	Minimum	0.648	0.837	0.541	0.455		
	Maximum	0.729	0.952	0.578	0.480		
	C.V.(%)	3.62	5.06	2.26	1.76		
	No. Specimens	9	6	10	10		
	No. Batches	2	2	2	2		
	Approval Class	Screening	Screening	Screening	Screening		

(1) B-values are presented only for fully approved data.
(2) Conditioned in 160°F water for 14 days.

MATERIAL:	S2-449 43.5k/SP 381 unidirectional tape		
RESIN CONTENT:	30-34 wt%	COMP: DENSITY:	1.84-1.94 g/cm^3
FIBER VOLUME:	47.6-53.1 %	VOID CONTENT:	0.0-0.64%
PLY THICKNESS:	0.0070-0.0092 in.		
TEST METHOD: SRM 8-88		MODULUS CALCULATION:	
NORMALIZED BY:	Not normalized		

Table 6.2.1(g)
SGI/Ep 284-UT
S2-449/SP 381
SBS, 31-plane
$[0]_{12}$
73/A, -65A, 180/A, 160/W
Screening

Temperature (°F)		73	-65	180	160		
Moisture Content (%)		Ambient	Ambient	Ambient	Wet		
Equilibrium at T, RH					(2)		
Source Code		69	69	69	69		
	Mean	12.4	14.6	8.7	7.2		
	Minimum	11.6	13.9	8.2	7.0		
	Maximum	13.2	15.6	9.0	7.4		
	C.V.(%)	4.16	3.32	2.9	1.7		
	B-value	(1)	(1)	(1)	(1)		
F_{31}^{sbs}	Distribution	ANOVA	Normal	ANOVA	Weibull		
(ksi)	C_1	0.573	14.6	0.31	7.3		
	C_2	3.85	0.485	18	67		
	No. Specimens	25	14	14	13		
	No. Batches	4	2	2	2		
	Approval Class	Screening	Screening	Screening	Screening		

(1) B-values are presented only for fully approved data.
(2) Conditioned in 160°F water for 14 days.

MATERIAL:	S2-449 43.5k/SP 381 unidirectional tape			**Table 6.2.1(h)** **SGI/Ep 284-UT** **S2-449/SP 381** **SBS, 31-plane** **$[0]_{12}$** **73/Fluids** **Screening**
RESIN CONTENT:	30 wt%	COMP: DENSITY:	1.93-1.94 g/cm³	
FIBER VOLUME:	52.9-53.1 %	VOID CONTENT:	0.0-0.64%	
PLY THICKNESS:	0.00792-0.00925 in.			
TEST METHOD: SRM 8-88		MODULUS CALCULATION:		
NORMALIZED BY:	Not normalized			

Temperature (°F)	73	73	73	73		
Moisture Content (%)	(2)	(3)	(4)	(5)		
Equilibrium at T, RH						
Source Code	69	69	69	69		
Mean	11.8	12.3	11.6	11.9		
Minimum	11.0	11.8	9.40	11.4		
Maximum	12.3	13.0	12.8	12.6		
C.V.(%)	3.49	2.87	8.23	3.17		
B-value	(1)	(1)	(1)	(1)		
F_{31}^{sbs} (ksi) Distribution	Weibull	Normal	ANOVA	Normal		
C_1	11.9	12.4	1.07	11.9		
C_2	34.7	0.355	12.2	0.376		
No. Specimens	14	14	14	14		
No. Batches	2	2	2	2		
Approval Class	Screening	Screening	Screening	Screening		

(1) B-values are presented only for fully approved data.
(2) Conditioned in MIL-A-8243 Anti-Icing Fluid at 32°F for 30 days.
(3) Conditioned in MIL-H-83282 hydraulic Fluid at 160°F for 90 days. MIL-H-83282 was converted to MIL-PRF-83282 on September 30, 1997.
(4) Conditioned in MIL-H-5606 hydraulic fluid at 160°F for 90 days.
(5) Conditioned in MIL-T-5624 fuel at 75°F for 90 days. MIL-T-5624 was converted to MIL-PRF-5624 on November 22, 1996.

MATERIAL:	S2-449 43.5k/SP 381 unidirectional tape			**Table 6.2.1(i)** **SGI/Ep 284-UT** **S2-449/SP 381** **SBS, 31-plane** **$[0]_{12}$** **73/Fluids** **Screening**
RESIN CONTENT:	30 wt%	COMP: DENSITY:	193-1.94 g/cm^3	
FIBER VOLUME:	52.9-53.1 %	VOID CONTENT:	0.0-0.64%	
PLY THICKNESS:	0.00758-0.00933 in.			
TEST METHOD: SRM 8-88		MODULUS CALCULATION:		
NORMALIZED BY:	Not normalized			

Temperature (°F) Moisture Content (%) Equilibrium at T, RH Source Code	73 (2) 69	73 (3) 69	73 (4) 69	73 (5) 69		
Mean	11.8	12.1	11.7	11.8		
Minimum	11.1	10.9	10.6	11.3		
Maximum	12.6	12.6	12.3	12.3		
C.V.(%)	3.47	3.84	4.02	2.91		
F_{31}^{sbs} (ksi) B-value	(1)	(1)	(1)	(1)		
Distribution	Weibull	Weibull	Weibull	ANOVA		
C_1	12.0	12.3	11.9	0.386		
C_2	30.7	39.5	37.2	12.6		
No. Specimens	14	14	13	14		
No. Batches	2	2	2	2		
Approval Class	Screening	Screening	Screening	Screening		

(1) B-values are presented only for fully approved data.
(2) Conditioned in MIL-L-23699 lubricating oil at 160°F for 90 days. MIL-L-23699 was converted to MIL-PRF-23699 on May 21, 1997.
(3) Conditioned in MIL-L-7808 lubricating oil at 160°F for 90 days. MIL-L-7808 was converted to MIL-PRF-7808 on May 2, 1997.
(4) Conditioned in MIL-C-87936 cleaning fluid at 75°F for 7 days. MIL-C-87936 was canceled on March 1, 1995 and replaced with MIL-C-87937. MIL-C-87937 was converted to MIL-PRF-87937 on August 14, 1997.
(5) Conditioned in ASTM D740 methyl ethyl ketone (MEK) at 75°F for 7 days.

MATERIAL:	S2-449 43.5k/SP 381 unidirectional tape		
RESIN CONTENT:	30-31wt%	COMP: DENSITY:	1.92-1.94 g/cm^3
FIBER VOLUME:	51.6-53.5 %	VOID CONTENT:	0-0.50%
PLY THICKNESS:	0.0086-0.0089 in.		
TEST METHOD: SRM 4-88		MODULUS CALCULATION: Chord between 1000 and 3000 με	
NORMALIZED BY:	Specimen thickness and batch fiber areal weight to 50% (0.0090 in. CPT)		

Table 6.2.1(j)
SGI/Ep 284-UT
S2-449/SP 381
Tension, x-axis
$[\pm45/0/\mp45]_s$
73/A
Screening

Temperature (°F) Moisture Content (%) Equilibrium at T, RH Source Code		73 Ambient 69					
		Normalized	Measured	Normalized	Measured	Normalized	Measured
	Mean	69.5	72.9				
	Minimum	66.7	71.4				
	Maximum	71.3	75.6				
	C.V.(%)	2.18	1.67				
	B-value	(1)	(1)				
F_x^{tu}	Distribution	ANOVA	Normal				
(ksi)	C_1	1.74	72.9				
	C_2	13.7	1.22				
	No. Specimens	10					
	No. Batches	2					
	Approval Class	Screening					
	Mean	2.87	3.01				
	Minimum	2.78	2.94				
	Maximum	2.96	3.11				
E_x^t	C.V.(%)	2.21	1.58				
(Msi)	No. Specimens	10					
	No. Batches	2					
	Approval Class	Screening					
	Mean						
	No. Specimens						
ν_{xy}^t	No. Batches						
	Approval Class						
	Mean		24200				
	Minimum		23600				
	Maximum		24900				
	C.V.(%)		1.69				
	B-value		(1)				
ε_x^{tu}	Distribution		Weibull				
(με)	C_1		24400				
	C_2		65.4				
	No. Specimens	10					
	No. Batches	2					
	Approval Class	Screening					

(1) B-values are presented only for fully approved data.

MATERIAL:	S2-449 43.5k/SP 381 unidirectional tape			**Table 6.2.1(k)** **SGI/Ep 284-UT** **S2-449/SP 381** **Tension, y-axis** **[±45/90/∓45]$_s$** **73/A** **Screening**
RESIN CONTENT:	30-31 wt%	COMP: DENSITY:	1.92-1.94 g/cm^3	
FIBER VOLUME:	51.6-53.5 %	VOID CONTENT:	0-0.50%	
PLY THICKNESS:	0.0083-0.0090 in.			
TEST METHOD: SRM 4-88		MODULUS CALCULATION: Chord between 1000 and 3000 με		
NORMALIZED BY:	Specimen thickness and batch fiber areal weight to 50% (0.0090 in. CPT)			

Temperature (°F) Moisture Content (%) Equilibrium at T, RH Source Code		73 Ambient 69					
		Normalized	Measured	Normalized	Measured	Normalized	Measured
F_y^{tu} (ksi)	Mean	24.9	26.2				
	Minimum	23.9	24.7				
	Maximum	25.9	27.3				
	C.V.(%)	2.29	2.94				
	B-value	(1)	(1)				
	Distribution	Weibull	Weibull				
	C_1	25.1	26.5				
	C_2	47.1	42.2				
	No. Specimens	10					
	No. Batches	2					
	Approval Class	Screening					
E_y^t (Msi)	Mean	2.15	2.26				
	Minimum	2.10	2.18				
	Maximum	2.20	2.39				
	C.V.(%)	1.33	3.50				
	No. Specimens	10					
	No. Batches	2					
	Approval Class	Screening					
ν_{yx}^t	Mean						
	No. Specimens						
	No. Batches						
	Approval Class						
ε_y^{tu} (με)	Mean		11600				
	Minimum		10900				
	Maximum		12000				
	C.V.(%)		2.65				
	B-value		(1)				
	Distribution		Weibull				
	C_1		11700				
	C_2		49.8				
	No. Specimens	10					
	No. Batches	2					
	Approval Class	Screening					

(1) B-values are presented only for fully approved data.

6.2.2 S2-449 17k/SP 381 unitape data set description

Material Description:

Material: S2-449 43.5k/3M PR381

Form: Unidirectional tape, fiber areal weight of 284 g/m^2, typical cured resin content of 28-33%, typical cured ply thickness of 0.0081 - 0.009 inches.

Processing: Autoclave cure; 260° F, 50 psi for two hours

General Supplier Information:

Fiber: S2 glass has enhanced properties in strength, modulus impact resistance and fatigue when compared to conventional E glass roving. The sizing for these fibers is an epoxy compatible 449 finish material. Rovings of 43,500 filaments. Typical tensile modulus is 12.5 to 13.0 Msi. Typical tensile strength is 665,000 psi.

Matrix: PR381 is a 250°F curing epoxy resin providing properties similar to conventional 350°F curing systems. Light tack for up to 30 days at 75°F.

Maximum Short Term Service Temperature: 220°F (dry), 160°F (wet)

Typical applications: Primary and secondary structural applications where improved fatigue and excellent mechanical strength is important such as helicopters and general aviation.

6.2.2 S2-449 17k/SP 381 unidirectional tape

MATERIAL:	S2-449 17k/SP 381 unidirectional tape			**SGI/Ep 111-UT S2-449/SP 381 Summary**
PREPREG:	3M Scotchply SP 381 Uni S29 111BW 33 RC			
FIBER:	Owens Corning S2-449, 0 twist, no surface treatment, typical 449 glass sizing	MATRIX:	3M SP 381	
T_g(dry):	291°F T_g(wet): 234°F	T_g METHOD:	SRM 18, RDA, G" peak	
PROCESSING:	Autoclave cure: 260±10°F, 120±20 min., 50 psi			

Date of fiber manufacture	8/91 - 12/94	Date of testing	6/93 - 4/96
Date of resin manufacture	11/91 - 5/95	Date of data submittal	6/96
Date of prepreg manufacture	11/91 - 2/96	Date of analysis	2/97
Date of composite manufacture	12/91 - 3/96		

LAMINA PROPERTY SUMMARY

	73°F/A		-65°F/A	180°F/A		160°F/W		
Tension, 1-axis	II-I		SS-S	SS-S		SS-S		
Tension, 2-axis	SS-S		SS-S	SS-S		SS-S		
Tension, 3-axis								
Compression, 1-axis	SS-S		SS-S	SS-S		SS-S		
Compression, 2-axis								
Compression, 3-axis								
Shear, 12-plane	SS--		SS--	SS--		SS--		
Shear, 23-plane								
Shear, 31-plane								
SBS, 31-plane	S---		S---	S---		S---		

Classes of data: F - Fully approved, I - Interim, S - Screening in Strength/Modulus/Poisson's ratio/Strain-to-failure order.
Data are also included for F^{sbs} conditioned in eight fluids.

		Nominal	As Submitted	Test Method
Fiber Density	(g/cm^3)	2.49		ASTM C693
Resin Density	(g/cm^3)	1.216		ASTM D792
Composite Density	(g/cm^3)	1.85	1.82 - 1.94	
Fiber Areal Weight	(g/m^2)	111	111 - 113	SRM 23B
Ply Thickness	(in)	0.0035	0.00303 - 0.00375	
Fiber Volume	(%)	50	47.6 - 55.2	

LAMINATE PROPERTY SUMMARY

	73°F/A							
[±45/0/∓ 45]								
Tension, x-axis	SS-S							
Tension, y-axis	SS-S							

Classes of data: F - Fully approved, I - Interim, S - Screening in Strength/Modulus/Poisson's ratio/Strain-to-failure order.

MATERIAL:	S2-449 17k/SP 381 unidirectional tape			**Table 6.2.2(a)** **SGI/Ep 111-UT** **S2-449/SP 381** **Tension, 1-axis** **$[0]_{12}$** **73/A, -65/A, 180/A** **Interim, Screening**
RESIN CONTENT:	29-36 wt%	COMP: DENSITY:	1.85-1.93 g/cm³	
FIBER VOLUME:	47.6-54.0 %	VOID CONTENT:	0.0-0.17%	
PLY THICKNESS:	0.0032-0.0038 in.			
TEST METHOD: SRM 4-88		MODULUS CALCULATION: Chord between 1000 and 6000 με		
NORMALIZED BY:	Specimen thickness and batch fiber areal weight to 50% (0.0035 in. CPT)			

		73 / Ambient		-65 / Ambient		180 / Ambient	
	Temperature (°F)	73		-65		180	
	Moisture Content (%)	Ambient		Ambient		Ambient	
	Equilibrium at T, RH						
	Source Code	70		70		70	
		Normalized	Measured	Normalized	Measured	Normalized	Measured
F_1^{tu} (ksi)	Mean	255	248	267	274	225	225
	Minimum	243	228	233	251	218	216
	Maximum	277	274	287	302	237	234
	C.V.(%)	3.40	5.07	6.52	5.96	3.13	2.59
	B-value	(1)	(1)	(1)	(1)	(1)	(1)
	Distribution	Normal	ANOVA	Weibull	Weibull	Weibull	Weibull
	C_1	255	13.6	274	281	228	228
	C_2	8.65	3.53	21.3	18.1	32.9	43.2
	No. Specimens	21		11		11	
	No. Batches	4		2		2	
	Approval Class	Interim		Screening		Screening	
E_1^t (Msi)	Mean	6.93	6.75	7.01	7.19	6.73	6.73
	Minimum	6.61	6.26	6.70	6.98	6.50	6.50
	Maximum	7.18	7.16	7.31	7.49	7.09	7.09
	C.V.(%)	2.29	4.37	2.98	2.19	2.80	2.95
	No. Specimens	21		11		11	
	No. Batches	4		2		2	
	Approval Class	Interim		Screening		Screening	
ν_{12}^t	Mean						
	No. Specimens						
	No. Batches						
	Approval Class						
ε_1^{tu} (με)	Mean		36800		38000		33400
	Minimum		34600		33500		31000
	Maximum		38600		40900		35100
	C.V.(%)		3.09		5.85		3.84
	B-value		(1)		(1)		(1)
	Distribution		Weibull		Weibull		Weibull
	C_1		37300		39000		34000
	C_2		37.9		22.5		34.9
	No. Specimens	21		11		11	
	No. Batches	4		2		2	
	Approval Class	Interim		Screening		Screening	

(1) B-values are presented only for fully approved data.

MATERIAL:	S2-449 17k/SP 381 unidirectional tape			Table 6.2.2(b) SGI/Ep 111-UT S2-449/SP 381 Tension, 1-axis $[0]_{12}$ 160/W Screening
RESIN CONTENT:	29-31 wt%	COMP: DENSITY:	1.90-1.93 g/cm³	
FIBER VOLUME:	49.0-50.1 %	VOID CONTENT:	0.00%	
PLY THICKNESS:	0.0034-0.0038 in.			
TEST METHOD: SRM 4-88		MODULUS CALCULATION: Chord between 1000 and 6000 $\mu\varepsilon$		
NORMALIZED BY:	Specimen thickness and batch fiber areal weight to 50% (0.0035 in. CPT)			

		160 / Wet / (2) / 70					
Temperature (°F) / Moisture Content (%) / Equilibrium at T, RH / Source Code		160					
		Wet					
		(2)					
		70					
		Normalized	Measured	Normalized	Measured	Normalized	Measured
F_1^{tu} (ksi)	Mean	116	113				
	Minimum	107	108				
	Maximum	123	123				
	C.V.(%)	4.34	3.54				
	B-value	(1)	(1)				
	Distribution	Weibull	Normal				
	C_1	118	113				
	C_2	26.8	4.01				
	No. Specimens	13					
	No. Batches	2					
	Approval Class	Screening					
E_1^t (Msi)	Mean	6.84	6.71				
	Minimum	6.50	6.49				
	Maximum	7.12	6.97				
	C.V.(%)	2.57	1.99				
	No. Specimens	13					
	No. Batches	2					
	Approval Class	Screening					
ν_{12}^t	Mean						
	No. Specimens						
	No. Batches						
	Approval Class						
ε_1^{tu} ($\mu\varepsilon$)	Mean		16900				
	Minimum		15800				
	Maximum		18100				
	C.V.(%)		3.90				
	B-value		(1)				
	Distribution		Weibull				
	C_1		17200				
	C_2		28.7				
	No. Specimens	13					
	No. Batches	2					
	Approval Class	Screening					

(1) B-values are presented only for fully approved data.
(2) Conditioned in 160°F water for 14 days.

MATERIAL:	S2-449 17k/SP 381 unidirectional tape		
RESIN CONTENT:	29-31 wt%	COMP: DENSITY:	1.88-1.92 g/cm^3
FIBER VOLUME:	48.8-50.1 %	VOID CONTENT:	0.0%
PLY THICKNESS:	0.0033-0.0036 in.		
TEST METHOD: SRM 4-88		MODULUS CALCULATION: Chord between 1000 and 3000 $\mu\varepsilon$ (2)	
NORMALIZED BY:	Not normalized		

Table 6.2.2(c)
SGI/Ep 111-UT
S2-449/SP 381
Tension, 2-axis
$[90]_{20}$
73/A, -65/A, 180/A, 160/W
Screening

Temperature (°F)		73	-65	180	160		
Moisture Content (%)		Ambient	Ambient	Ambient	Wet		
Equilibrium at T, RH					(3)		
Source Code		70	70	70	70		
F_2^{tu} (ksi)	Mean	8.7	10.0	6.4	3.6		
	Minimum	8.1	9.6	5.9	3.1		
	Maximum	9.0	10.3	6.7	3.9		
	C.V.(%)	3.9	3.6	4.0	9.0		
	B-value	(1)	(4)	(1)	(1)		
	Distribution	Normal		Normal	Normal		
	C_1	8.7		6.4	3.6		
	C_2	0.34		0.26	0.32		
	No. Specimens	5	3	8	5		
	No. Batches	1	1	2	1		
	Approval Class	Screening	Screening	Screening	Screening		
E_2^t (Msi)	Mean	1.84	2.11	1.42	1.10		
	Minimum	1.82	2.06	1.34	1.05		
	Maximum	1.91	2.15	1.55	1.16		
	C.V.(%)	2.05	2.14	6.43	4.59		
	No. Specimens	5	3	4	5		
	No. Batches	1	1	1	1		
	Approval Class	Screening	Screening	Screening	Screening		
ν_{21}^t	Mean						
	No. Specimens						
	No. Batches						
	Approval Class						
ε_2^{tu} ($\mu\varepsilon$)	Mean	4700	4730	4450	3280		
	Minimum	4400	4500	4200	3000		
	Maximum	4900	5000	4800	3600		
	C.V.(%)	4.26	5.32	5.95	8.18		
	B-value	(1)	(4)	(1)	(1)		
	Distribution	Normal		Normal	Normal		
	C_1	4700		4450	3280		
	C_2	200.0		265	268		
	No. Specimens	5	3	4	5		
	No. Batches	1	1	1	1		
	Approval Class	Screening	Screening	Screening	Screening		

(1) B-values are presented only for fully approved data.
(2) Exception to SRM 4-88.
(3) Conditioned in 160°F water for 14 days.
(4) The statistical analysis is not completed for less than four specimens.

MATERIAL:	S2-449 17k/SP 381 unidirectional tape			**Table 6.2.2(d)** **SGI/Ep 111-UT** **S2-449/SP 381** **Compression, 1-axis** **$[0]_{12}$** **73/A, -65/A, 180/A** **Screening**
RESIN CONTENT: FIBER VOLUME: PLY THICKNESS:	28-29 wt% 50.1-54.0 % 0.0032-0.0035 in.	COMP: DENSITY: VOID CONTENT:	1.85-1.92 g/cm^3 0.22-1.53%	
TEST METHOD: SRM 1-88		MODULUS CALCULATION: Chord between 1000 and 3000 $\mu\varepsilon$		
NORMALIZED BY:	Specimen thickness and batch fiber areal weight to 50% (0.0035 in. CPT)			

Temperature (°F) Moisture Content (%) Equilibrium at T, RH Source Code		73 Ambient 70		-65 Ambient 70		180 Ambient 70	
		Normalized	Measured	Normalized	Measured	Normalized	Measured
	Mean	172	178	166	177	165	175
	Minimum	145	142	147	152	146	155
	Maximum	193	198	184	198	185	196
	C.V.(%)	8.09	9.35	6.62	7.46	6.81	7.28
	B-value	(1)	(1)	(1)	(1)	(1)	(1)
F_1^{cu}	Distribution	Weibull	Weibull	Weibull	Weibull	Weibull	Weibull
(ksi)	C_1	178	185	171	183	170	181
	C_2	15.2	14.7	17.7	16.0	16.6	16.4
	No. Specimens	13		13		12	
	No. Batches	2		2		2	
	Approval Class	Screening		Screening		Screening	
	Mean	6.86	7.14	6.91	7.19	6.97	7.47
	Minimum	6.43	6.81	6.63	6.96	6.63	7.19
	Maximum	7.24	7.52	7.10	7.49	7.24	7.59
E_1^c	C.V.(%)	3.79	3.39	2.35	2.22	3.18	1.85
(Msi)	No. Specimens	10		10		10	
	No. Batches	2		2		2	
	Approval Class	Screening		Screening		Screening	
	Mean						
	No. Specimens						
ν_{12}^c	No. Batches						
	Approval Class						
	Mean						
	Minimum						
	Maximum						
	C.V.(%)						
	B-value						
ε_1^{cu}	Distribution						
($\mu\varepsilon$)	C_1						
	C_2						
	No. Specimens						
	No. Batches						
	Approval Class						

(1) B-values are presented only for fully approved data.

MATERIAL:	S2-449 17k/SP 381 unidirectional tape			Table 6.2.2(e) SGI/Ep 111-UT S2-449/SP 381 Compression, 1-axis $[0]_{12}$ 160/W Screening
RESIN CONTENT:	28-29 wt%	COMP: DENSITY:	1.85-1.92 g/cm^3	
FIBER VOLUME:	50.1-54.0 %	VOID CONTENT:	0-1.15%	
PLY THICKNESS:	0.0033-0.0037 in.			
TEST METHOD:		MODULUS CALCULATION:		
SRM 1-88		Chord between 1000 and 3000 με		
NORMALIZED BY:	Specimen thickness and batch fiber areal weight to 50% (0.0035 in. CPT)			

Temperature (°F) Moisture Content (%) Equilibrium at T, RH Source Code		160 Wet (2) 70					
		Normalized	Measured	Normalized	Measured	Normalized	Measured
F_1^{cu} (ksi)	Mean	135	137				
	Minimum	124	123				
	Maximum	143	146				
	C.V.(%)	3.51	4.83				
	B-value	(1)	(1)				
	Distribution	Nonpara.	ANOVA				
	C_1	6	8.02				
	C_2	2.14	16.7				
	No. Specimens	10					
	No. Batches	2					
	Approval Class	Screening					
E_1^c (Msi)	Mean	6.96	6.97				
	Minimum	6.69	6.75				
	Maximum	7.24	7.23				
	C.V.(%)	2.44	2.16				
	No. Specimens	10					
	No. Batches	2					
	Approval Class	Screening					
ν_{12}^c	Mean						
	No. Specimens						
	No. Batches						
	Approval Class						
ε_1^{cu} (με)	Mean						
	Minimum						
	Maximum						
	C.V.(%)						
	B-value						
	Distribution						
	C_1						
	C_2						
	No. Specimens						
	No. Batches						
	Approval Class						

(1) B-values are presented only for fully approved data.
(2) Conditioned in 160°F water for 14 days.

MATERIAL:	S2-449 17k/SP 381 unidirectional tape		
RESIN CONTENT:	29-32 wt%	COMP: DENSITY:	1.85-1.89 g/cm^3
FIBER VOLUME:	48.8-51.6 %	VOID CONTENT:	0-0.74%
PLY THICKNESS:	0.0032-0.0037 in.		
TEST METHOD: SRM 7-88		MODULUS CALCULATION: Chord between 1000 and 3000 $\mu\varepsilon$, axial	
NORMALIZED BY:	Not normalized		

Table 6.2.2(f)
SGI/Ep 111-UT
S2-449/SP 381
Shear, 12-plane
$[\pm 45]_{ss}$
73/A, -65/A,180/A, 160/W
Screening

Temperature (°F)		73	-65	180	160		
Moisture Content (%)		Ambient	Ambient	Ambient	Wet		
Equilibrium at T, RH					(2)		
Source Code		70	70	70	70		
F_{12}^{su} (ksi)	Mean	19.7	25.7	15.0	11.1		
	Minimum	18.9	24.7	14.0	10.7		
	Maximum	20.3	26.2	15.5	11.9		
	C.V.(%)	2.18	1.85	2.67	3.43		
	B-value	(1)	(1)	(1)	(1)		
	Distribution	Weibull	Weibull	ANOVA	ANOVA		
	C_1	20.0	25.9	0.452	0.442		
	C_2	61.1	73.2	4.88	5.83		
	No. Specimens	16	16	16	14		
	No. Batches	3	3	3	3		
	Approval Class	Screening	Screening	Screening	Screening		
G_{12} (Msi)	Mean	0.681	0.808	0.539	0.467		
	Minimum	0.627	0.772	0.513	0.440		
	Maximum	0.745	0.850	0.583	0.490		
	C.V.(%)	5.29	3.32	4.06	2.96		
	No. Specimens	9	9	10	10		
	No. Batches	2	2	2	2		
	Approval Class	Screening	Screening	Screening	Screening		

(1) B-values are presented only for fully approved data.
(2) Conditioned in 160°F water for 14 days.

MATERIAL:	S2-449 17k/SP 381 unidirectional tape			**Table 6.2.2(g)** **SGI/Ep 111-UT** **S2-449/SP 381** **SBS, 31-plane** **$[0]_{30}$** **73/A, -65/A, 180/A, 160/W** **Screening**
RESIN CONTENT:	27-35 wt%	COMP: DENSITY:	1.85-1.94 g/cm^3	
FIBER VOLUME:	48.3-55.2 %	VOID CONTENT:	0.0-0.12%	
PLY THICKNESS:	0.0029-0.0035 in.			
TEST METHOD: SRM 8-88		MODULUS CALCULATION:		
NORMALIZED BY:	Not normalized			

Temperature (°F) Moisture Content (%) Equilibrium at T, RH Source Code		73 Ambient 70	-65 Ambient 70	180 Ambient 70	160 Wet (2) 70		
	Mean	12.6	14.9	9.5	7.6		
	Minimum	11.6	13.1	9.1	7.0		
	Maximum	13.7	16.8	9.8	8.7		
	C.V.(%)	4.64	6.89	2.2	7.1		
	B-value	(1)	(1)	(1)	(1)		
F_{31}^{sbs}	Distribution	ANOVA	Weibull	Normal	ANOVA		
(ksi)	C_1	0.613	15.4	9.5	0.63		
	C_2	2.77	17.1	0.21	5.2		
	No. Specimens	32	14	17	18		
	No. Batches	5	2	3	3		
	Approval Class	Screening	Screening	Screening	Screening		

(1) B-values are presented only for fully approved data.
(2) Conditioned in 160°F water for 14 days.

MATERIAL:	S2-449 17k/SP 381 unidirectional tape			Table 6.2.2(h) SGI/Ep 111-UT S2-449/SP 381 SBS, 31-plane $[0]_{30}$ 73/Fluids Screening
RESIN CONTENT:	27-30 wt%	COMP: DENSITY:	1.92-1.94 g/cm³	
FIBER VOLUME:	50.1-51.6 %	VOID CONTENT:	0.0-0.12%	
PLY THICKNESS:	0.0033-0.0037 in.			
TEST METHOD: SRM 8-88		MODULUS CALCULATION:		
NORMALIZED BY:	Not normalized			

Temperature (°F) Moisture Content (%) Equilibrium at T, RH Source Code	73 (2) 70	73 (3) 70	73 (4) 70	73 (5) 70		
Mean	12.0	12.4	12.6	12.1		
Minimum	10.7	10.9	11.3	10.5		
Maximum	13.0	13.4	13.5	12.8		
C.V.(%)	5.20	5.81	4.44	5.22		
B-value	(1)	(1)	(1)	(1)		
F_{31}^{sb} (ksi) Distribution	Weibull	Weibull	Weibull	ANOVA		
C_1	12.3	12.7	12.9	0.683		
C_2	24.0	21.9	27.8	9.78		
No. Specimens	12	14	14	14		
No. Batches	2	2	2	2		
Approval Class	Screening	Screening	Screening	Screening		

(1) B-values are presented only for fully approved data.
(2) Conditioned in MIL-A-8243 Anti-Icing Fluid at 32°F for 30 days.
(3) Conditioned in MIL-H-83282 hydraulic fluid at 160°F for 90 days. MIL-H-83282 was converted to MIL-PRF-83282 on September 30, 1997.
(4) Conditioned in MIL-H-5606 hydraulic fluid at 160°F for 90 days.
(5) Conditioned in MIL-T-5624 fuel at 75°F for 90 days. MIL-T-5624 was converted to MIL-PRF-5624 on November 22, 1996.

MATERIAL:	S2-449 17k/SP 381 unidirectional tape			Table 6.2.2(i) SGI/Ep 111-UT S2-449/SP 381 SBS, 31-plane $[0]_{30}$ 73/Fluids Screening
RESIN CONTENT:	27-30 wt%	COMP: DENSITY:	1.92-1.94 g/cm³	
FIBER VOLUME:	50.1-51.6 %	VOID CONTENT:	0.0-0.12%	
PLY THICKNESS:	0.0033-0.0037 in.			
TEST METHOD: SRM 8-88		MODULUS CALCULATION:		
NORMALIZED BY:	Not normalized			

Temperature (°F)	73	73	73	73		
Moisture Content (%)	(2)	(3)	(4)	(5)		
Equilibrium at T, RH						
Source Code	70	70	70	70		
F_{31}^{sbs} (ksi) Mean	12.6	12.6	11.8	11.9		
Minimum	10.3	11.6	11.1	10.2		
Maximum	13.5	13.6	12.4	12.9		
C.V.(%)	6.49	3.86	3.79	6.19		
B-value	(1)	(1)	(1)	(1)		
Distribution	Weibull	Weibull	Weibull	Weibull		
C_1	12.9	12.8	12.0	12.2		
C_2	23.1	26.6	32.8	21.5		
No. Specimens	14	14	13	13		
No. Batches	2	2	2	2		
Approval Class	Screening	Screening	Screening	Screening		

(1) B-values are presented only for fully approved data.
(2) Conditioned in MIL-L-23699 lubricating oil at 160°F for 90 days. MIL-L-23699 was converted to MIL-PRF-23699 on May 21, 1997.
(3) Conditioned in MIL-L-7808 lubricating oil at 160°F for 90 days. MIL-L-7808 was converted to MIL-PRF-7808 on May 2, 1997.
(4) Conditioned in MIL-C-87936 cleaning fluid at 75°F for 7 days. MIL-C-87936 was canceled on March 1, 1995 and replaced with MIL-C-87937. MIL-C-87937 was converted to MIL-PRF-87937 on August 14, 1997.
(5) Conditioned in ASTM D740 methyl ethyl ketone (MEK) at 75°F for 7 days.

MATERIAL:	S2-449 17k/SP 381 unidirectional tape			Table 6.2.2(j) SGI/Ep 111-UT S2-449/SP 381 Tension, x-axis $[\pm 45/0/\mp 45]_{2S}$ 73/A Screening
RESIN CONTENT:	29-32 wt%	COMP: DENSITY:	1.88-1.89 g/cm^3	
FIBER VOLUME:	50.1-51.6 %	VOID CONTENT:	0.0-0.74%	
PLY THICKNESS:	0.0034-0.0036 in.			
TEST METHOD: SRM 4-88		MODULUS CALCULATION: Chord between 1000 and 3000 $\mu\varepsilon$		
NORMALIZED BY:	Specimen thickness and batch fiber areal weight to 50% (0.0035 in. CPT)			

Temperature (°F) Moisture Content (%) Equilibrium at T, RH Source Code		73 Ambient 70					
		Normalized	Measured	Normalized	Measured	Normalized	Measured
F_x^{tu} (ksi)	Mean	69.7	71.4				
	Minimum	68.1	69.8				
	Maximum	72.5	73.9				
	C.V.(%)	1.78	1.92				
	B-value	(1)	(1)				
	Distribution	Normal	Weibull				
	C_1	69.7	72.1				
	C_2	1.24	55.0				
	No. Specimens	10					
	No. Batches	2					
	Approval Class	Screening					
E_x^t (Msi)	Mean	2.90	2.97				
	Minimum	2.80	2.85				
	Maximum	2.96	3.08				
	C.V.(%)	1.86	2.30				
	No. Specimens	10					
	No. Batches	2					
	Approval Class	Screening					
ν_{xy}^t	Mean						
	No. Specimens						
	No. Batches						
	Approval Class						
ε_x^{tu} ($\mu\varepsilon$)	Mean		24100				
	Minimum		23300				
	Maximum		25200				
	C.V.(%)		2.49				
	B-value		(1)				
	Distribution		Weibull				
	C_1		24400				
	C_2		40.9				
	No. Specimens	10					
	No. Batches	2					
	Approval Class	Screening					

(1) B-values are presented only for fully approved data.

MATERIAL:	S2-449 17k/SP 381 unidirectional tape		
RESIN CONTENT:	30-32 wt%	COMP: DENSITY:	1.87-1.88 g/cm^3
FIBER VOLUME:	50.1 %	VOID CONTENT:	0.0-0.60%
PLY THICKNESS:	0.0035-0.0036 in.		
TEST METHOD: SRM 4-88		MODULUS CALCULATION: Chord between 1000 and 3000 $\mu\varepsilon$	
NORMALIZED BY:	Specimen thickness and batch fiber areal weight to 50% (0.0035 in. CPT)		

Table 6.2.2(k)
SGl/Ep 111-UT
S2-449/SP 381
Tension, y-axis
[±45/90/∓45]$_{2S}$
73/A
Screening

Temperature (°F)		73					
Moisture Content (%)		Ambient					
Equilibrium at T, RH							
Source Code		70					
		Normalized	Measured	Normalized	Measured	Normalized	Measured
F_y^{tu} (ksi)	Mean	36.2	36.6				
	Minimum	35.3	35.8				
	Maximum	37.1	37.6				
	C.V.(%)	1.77	1.77				
	B-value	(1)	(1)				
	Distribution	ANOVA	ANOVA				
	C_1	0.813	0.755				
	C_2	18.6	14.8				
	No. Specimens	10					
	No. Batches	2					
	Approval Class	Screening					
E_y^t (Msi)	Mean	2.21	2.24				
	Minimum	2.14	2.17				
	Maximum	2.28	2.31				
	C.V.(%)	1.88	2.01				
	No. Specimens	10					
	No. Batches	2					
	Approval Class	Screening					
ν_{yx}^t	Mean						
	No. Specimens						
	No. Batches						
	Approval Class						
ε_y^{tu} ($\mu\varepsilon$)	Mean		16400				
	Minimum		15600				
	Maximum		16800				
	C.V.(%)		2.40				
	B-value		(1)				
	Distribution		Weibull				
	C_1		16500				
	C_2		58.7				
	No. Specimens	10					
	No. Batches	2					
	Approval Class	Screening					

(1) B-values are presented only for fully approved data.

6.2.3 7781G 816/PR381 Weave

Material Description:

Material: 7781 E-glass/3M PR381

Form: 8 Harness satin fabric, fiber areal weight of 300 g/m^2, typical cured resin content of 32-38%, typical cured ply thickness of 0.009 - 0.0105 inches.

Processing: Autoclave cure; 260° F, 50 psi for two hours

General Supplier Information:

Fiber: Continuous, E-glass fiber. Typical tensile modulus is 10 x 10^6 psi. Typical tensile strength is 500,000 psi.

Matrix: PR381 is a 250°F curing epoxy resin providing properties similar to conventional 350°F curing systems. Light tack for up to 30 days at 75°F.

Maximum Short Term Service Temperature: 220°F (dry), 160°F (wet)

Typical applications: Aircraft secondary structure, fuselage skins and general industrial applications where improved fatigue and excellent mechanical strengths are required.

6.2.3 7781 G-816/PR381 Weave

MATERIAL:	7781G 816/PR 381 weave			**EGI/Ep 300-W** **7781G/PR 381** **Summary**
FORM:	3M SP 381/7781 E-Glass Fabric Prepreg, 57 Yarn Count/in. (Warp), 54 Yarn Count/in. (Fill)			
FIBER:	Clark-Schwebel 7781 E-glass Fabric, per MIL-C-9084C Type VIII B, Yarn DE-75 1/0.0 twist, no surface treatment, 558 Finish	MATRIX:	3M PR 381	
T_g(ambient):	282/F T_g(wet): 225 /F	T_g METHOD:	SRM-18, DMA E' knee	
PROCESSING:	Autoclave cure: 260/F, 100 min., 50 psi			

Date of fiber manufacture	11/92 - 7/95	Date of testing	3/93 - 4/96
Date of resin manufacture	12/92 - 3/96	Date of data submittal	6/96
Date of prepreg manufacture	12/92 - 3/96	Date of analysis	8/97
Date of composite manufacture	3/93 - 4/96		

LAMINA PROPERTY SUMMARY

	73/F/A		220/F/A				
Tension, 1-axis	II-I		SS-S				
Tension, 2-axis							
Tension, 3-axis							
Compression, 1-axis							
Compression, 2-axis							
Compression, 3-axis							
Shear, 12-plane							
Shear, 23-plane							
Shear, 31-plane							
SBS, 31-plane	S---						
Flexure	S---		S---				

Classes of data: F - Fully approved, I - Interim, S - Screening in Strength/Modulus/Poisson's ratio/Strain-to-failure order

	Nominal	As Submitted	Test Method
Fiber Density (g/cm^3)	2.6		ASTM C693
Resin Density (g/cm^3)			ASTM D792
Composite Density (g/cm^3)	1.85	1.75 - 2.04	ASTM D792
Fiber Areal Weight (g/m^2)	300	288 - 297	SRM 23B
Ply Thickness (in)	0.0099	0.0087 - 0.0104	
Fiber Volume (%)	48	43.0 - 50.9	SRM 10

LAMINATE PROPERTY SUMMARY

Classes of data: F - Fully approved, I - Interim, S - Screening in Strength/Modulus/Poisson's ratio/Strain-to-failure order

MATERIAL:	7781G 816/PR 381 weave			**Table 6.2.3(a) EGI/Ep 300-W 7781G/PR 381 Tension, 1-axis $[0]_s$ 73/A, 220/A Interim, Screening**
RESIN CONTENT:	34-36 wt%	COMP. DENSITY:	1.75-1.97 g/cm^3	
FIBER VOLUME:	43.0-48.4%	VOID CONTENT:	-	
PLY THICKNESS:	0.0091-0.0104 in.			
TEST METHOD: SRM 4-88 (1)		MODULUS CALCULATION: Chord between 1000 and 6000 $\mu\varepsilon$		
NORMALIZED BY:	Specimen thickness and batch fiber areal weight to 50% (0.0091 in. CPT)			

Temperature(°F)		73		220			
Moisture Content(%)		Ambient		Ambient			
Equilibrium at T, RH							
Source Code		72		72			
		Normalized	Measured	Normalized	Measured		
	Mean	74.9	70.9	71.3	67.5		
	Minimum	70.4	62.9	67.0	60.5		
	Maximum	79.6	77.8	77.4	74.4		
	C.V. (%)	3.66	7.07	4.02	5.89		
F_1^{tu}	B-value	(2)	(2)	(2)	(2)		
	Distribution	ANOVA	ANOVA	Weibull	ANOVA		
(ksi)	C_1	2.90	5.37	72.7	4.22		
	C_2	3.10	3.26	24.9	3.45		
	No. Specimens	16		13			
	No. Batches	5		4			
	Approval Class	Interim		Screening			
	Mean	3.83	3.64	3.64	3.44		
	Minimum	3.70	3.37	3.45	3.24		
	Maximum	3.97	3.96	3.75	3.77		
E_1^t	C.V. (%)	2.63	4.51	2.78	5.40		
(Msi)	No. Specimens	15		13			
	No. Batches	5		4			
	Approval Class	Interim		Screening			
	Mean						
ν_{12}^t	No. Specimens						
	No. Batches						
	Approval Class						
	Mean		17800		19600		
	Minimum		15200		18400		
	Maximum		19600		21100		
	C.V. (%)		6.23		4.01		
ε_1^{tu}	B-value		(2)		(2)		
	Distribution		ANOVA		Weibull		
($\mu\varepsilon$)	C_1		1310		20000		
	C_2		3.32		25.7		
	No. Specimens	15		13			
	No. Batches	5		4			
	Approval Class	Interim		Screening			

(1) Three batches were tested according to SRM 4R-94 with modulus calculated as noted above.
(2) B-values are presented only for fully approved data.

MATERIAL:	7781G 816/PR 381 weave			**Table 6.2.3(b)** **EGl/Ep 300-W** **7781G/PR 381** **SBS, 13-axis** **$[0]_{5s}$** **73/A** **Interim**
RESIN CONTENT:	34-36 wt%	COMP. DENSITY:	1.76-2.04 g/cm^3	
FIBER VOLUME:	43.0-50.9%	VOID CONTENT:	%	
PLY THICKNESS:	0.0088-0.0103 in.			
TEST METHOD:		MODULUS CALCULATION:		
SRM 8-88 (1)		NA		
NORMALIZED BY:	Not normalized			

Temperature(°F)		73					
Moisture Content(%)		Ambient					
Equilibrium at T, RH							
Source Code		72					
	Mean	10.4					
	Minimum	9.6					
	Maximum	11.5					
	C.V. (%)	4.8					
F_{13}^{sbs}	B-value	(2)					
	Distribution	ANOVA					
(ksi)	C_1	0.53					
	C_2	3.2					
	No. Specimens	22					
	No. Batches	5					
	Approval Class	Screening					

(1) Three batches were tested according to SRM 8R-94.
(2) B-values are presented only for fully approved data.

MATERIAL: 7781G 816/PR 381 weave		**Table 6.2.3(c)** **EGl/Ep 300-W** **7781G/PR 381** **Flexure** **$[0]_{5s}$** **73/A, 220/A** **Interim, Screening**
RESIN CONTENT: 34-36 wt% FIBER VOLUME: 43.4-48.7% PLY THICKNESS: 0.0091-0.0103 in.	COMP. DENSITY: 1.76-1.97 g/cm^3 VOID CONTENT: %	
TEST METHOD: ASTM D790 Method 1	MODULUS CALCULATION: NA	
NORMALIZED BY: Not normalized		

Temperature(°F) Moisture Content(%) Equilibrium at T, RH Source Code		73 Ambient 72	220 Ambient 72				
	Mean	109	93.2				
	Minimum	94.2	83.4				
	Maximum	121	104				
	C.V. (%)	7.52	8.15				
F^{flex}	B-value	(1)	(1)				
	Distribution	ANOVA	ANOVA				
(ksi)	C_1	8.92	8.45				
	C_2	3.33	4.13				
	No. Specimens	21	14				
	No. Batches	5	4				
	Approval Class	Interim	Screening				

(1) B-values are presented only for fully approved data.

6.3 GLASS - POLYESTER COMPOSITES

6.4 GLASS - BISMALEIMIDE COMPOSITES

6.5 GLASS - POLYIMIDE COMPOSITES

6.6 GLASS - PHENOLIC COMPOSITES

6.7 GLASS - SILICONE COMPOSITES

6.8 GLASS - POLYBENZIMIDAZOLE COMPOSITES

6.9 GLASS - PEEK COMPOSITES

CHAPTER 7 BORON FIBER COMPOSITES

CHAPTER 8 ALUMINA FIBER COMPOSITES

CHAPTER 9 SILICON CARBIDE FIBER COMPOSITES

CHAPTER 10 QUARTZ FIBER COMPOSITES

10.4.1 Astroquartz II/F650 8HS data set description

Material Description:

Material: Astroquartz II/F650

Form: 8 harness satin weave fabric, fiber areal weight of 285 g/m^2, typical cured resin content of 37%, typical cured ply thickness of 0.010 inches.

Processing: Autoclave cure; 375°F, 85 psi for 4 hours. Postcure at 475°F for 4 hours

General Supplier Information:

Fiber: Astroquartz II fiber is a continuous, high strength, low modulus ceramic fiber made of pure fused silica. Typical tensile modulus is 10 x 10^6 psi. Typical tensile strength is 500,000 psi.

Matrix: F650 is a 350°F curing bismaleimide resin. It will retain light tack for several weeks at 70°F.

Maximum Short Term Service Temperature: 500°F (dry), 350°F (wet)

Typical applications: Primary and secondary structural applications, fire containment structures, radomes or any application where high strength and/or electrical properties are required.

10.4.1 Astroquartz II/F650 8-harness satin weave*

MATERIAL:	Astroquartz II/F650 8-harness satin weave			**Q/BMI Astroquartz II/F650 Summary**
PREPREG:	Hexcel AQII581/F650 8-harness satin weave			
FIBER:	J.P. Stevens Astroquartz II	MATRIX:	Hexcel F650	
T_g(dry):	600°F T_g(wet):	T_g METHOD:		
PROCESSING:	Autoclave cure: 375°F, 4 hours, 85 psig; Postcure: 475°F, 4 hours			

* DATA WERE SUBMITTED BEFORE THE ESTABLISHMENT OF DATA DOCUMENTATION REQUIREMENTS (JUNE 1989). ALL DOCUMENTATION PRESENTLY REQUIRED WERE NOT SUPPLIED FOR THIS MATERIAL.

Date of fiber manufacture		Date of testing	
Date of resin manufacture		Date of data submittal	4/89
Date of prepreg manufacture		Date of analysis	1/93
Date of composite manufacture			

LAMINA PROPERTY SUMMARY

	75°F/A		450°F/A					
Tension, 1-axis								
Tension, 2-axis								
Tension, 3-axis								
Compression, 1-axis								
Compression, 2-axis								
Compression, 3-axis								
Shear, 12-plane								
Shear, 23-plane								
SB strength, 31-plane	S---		S---					

Classes of data: F - Fully approved, I - Interim, S - Screening in Strength/Modulus/Poisson's ratio/Strain-to-failure order.

* DATA WERE SUBMITTED BEFORE THE ESTABLISHMENT OF DATA DOCUMENTATION REQUIREMENTS (JUNE 1989). ALL DOCUMENTATION PRESENTLY REQUIRED WERE NOT SUPPLIED FOR THIS MATERIAL.

		Nominal	As Submitted	Test Method
Fiber Density	(g/cm^3)	2.17		
Resin Density	(g/cm^3)	1.27		
Composite Density	(g/cm^3)	1.78	1.73	
Fiber Areal Weight	(g/m^2)	285		
Ply Thickness	(in)	0.0100	0.010	

LAMINATE PROPERTY SUMMARY

Classes of data: F - Fully approved, I - Interim, S - Screening in Strength/Modulus/Poisson's ratio/Strain-to-failure order.

* DATA WERE SUBMITTED BEFORE THE ESTABLISHMENT OF DATA DOCUMENTATION REQUIREMENTS (JUNE 1989). ALL DOCUMENTATION PRESENTLY REQUIRED WERE NOT SUPPLIED FOR THIS MATERIAL.

MATERIAL:	Astroquartz II/F650 8-harness satin weave			**Table 10.4.1(a)** **Q/BMI 285-8HS** **Astroquartz II/F650** **SB strength, 31-plane** **$[0_f]_{12}$** **75/A, 450/A** **Screening**
RESIN CONTENT:	37 wt%	COMP: DENSITY:	1.73 g/cm^3	
FIBER VOLUME:	51 %	VOID CONTENT:		
PLY THICKNESS:	0.010 in.			
TEST METHOD: ASTM D2344		MODULUS CALCULATION:		
NORMALIZED BY:	Not normalized			

Temperature (°F) Moisture Content (%) Equilibrium at T, RH Source Code	75 ambient 21	450 ambient 21				
Mean	6.41	6.56				
Minimum	6.31	6.43				
Maximum	6.50	6.72				
C.V.(%)	1.06	1.69				
B-value	(1)	(1)				
F_{31}^{sbs} (ksi) Distribution	Normal	Normal				
C_1	6.41	6.56				
C_2	0.068	0.111				
No. Specimens	5	5				
No. Batches	1	1				
Approval Class	Screening	Screening				

(1) B-values are presented only for fully approved data.

10.5 QUARTZ - POLYIMIDE COMPOSITES

10.6 QUARTZ - PHENOLIC COMPOSITES

10.7 QUARTZ - SILICONE COMPOSITES

10.8 QUARTZ - POLYBENZIMIDAZOLE COMPOSITES

10.9 QUARTZ - PEEK COMPOSITE

APPENDIX A1

MIL-HDBK-17A DATA

A1.1 GENERAL INFORMATION

The data on polymer matrix composite materials which were presented in MIL-HDBK-17A, dated January 1971, are presented in this appendix. MIL-HDBK-17A has been superseded so these data are presented here so they can be referenced in a current publication. However, these data do not meet the data requirements in Volume 1. The materials which were included in MIL-HDBK-17A are listed in Table A1. Of the sixteen materials, six are still available, five are no longer available, and the availability of the other five materials could not be determined. The data from the six available materials are provided in this appendix. The data from the remaining materials may be added as availability of the material or usefulness of the data is determined. Note that Narmco 5505 has been licensed to AVCO and those data are presented herein as AVCO 5505.

TABLE A1 *Materials from MIL-HDBK-17A.*

Materials
Available:
U.S. Polymeric E-720E/7781 (ECDE-1/0-550) Fiberglass Epoxy
Hexcel F-161/7743(550) Fiberglass Epoxy
Hexcel F-161/7781(ECDE-1/0-550) Fiberglass Epoxy
Narmco N588/7781 (ECDE-1/0-550) Fiberglass Epoxy
Narmco 506/7781 (ECDE-1/0-A1100) Fiberglass Phenolic
AVCO 5505 Boron Epoxy
Not available:
U.S. Polymeric E-779/7743 (Volan) Fiberglass Epoxy
3M XP251S Fiberglass Epoxy
U.S. Polymeric S-860/1581 (ECG-1/2-112) Neutral pH Fiberglass Silicone
U.S. Polymeric P670A/7781 (ECDE-1/0) Fiberglass Modified DAP Polyester
SP272 Boron Epoxy
Availability unknown:
Bloomingdale BP915/7781 (ECDE-1/0-550) Fiberglass Epoxy
Bloomingdale BP911/7781 (ECDE-1/0 Volan) Fiberglass Epoxy
Cordo E293/7781 (ECDE-1/0-550) Fiberglass Epoxy
Styrene-Alkyd Polyester/7781 Fiberglass
Cordo IFRR/7781 (ECDE-1/0) Fiberglass Modified DAP Polyester

The table and figure numbers used in this appendix are similar to those in MIL-HDBK-17A. The chapter identification has been changed from 4 to A1 but the rest of all figure and table numbers has not been changed. For example, Table A1.40 is the same as Table 4.40 in MIL-HDBK-17A. The MIL-HDBK-17A text describing the test program and methods is reproduced in Sections A1.2 through A1.4.

A1.2 INTRODUCTION

The laminate properties presented in this chapter have been generated in test programs conducted at the U.S. Forest Products Laboratory and elsewhere (Reference A1.2).[1] Properties are given for fiberglass with epoxy, phenolic, silicone and polyester resins and for boron with epoxy. Additional information on these and other material combinations will be issued as supplements or revisions of the present handbook edition.

A1.3 HANDBOOK TEST PROGRAM

A1.3.1 Objectives

The objectives of the handbook test program are to obtain statistically significant data for materials currently in use and to determine the degree of reproducibility attained in their fabrication. A minimum requirement is that test results include data from three sets of panels which are representative of the manufacturing procedures employed by three different fabricators. The properties listed in the charts and tables of this chapter represent test results from only one set of panels for each material system. Properties are therefore not given minimum values and are considered to be "typical" for each material. When the minimum number of tests has been completed for a material, its properties will be assigned values on a B-basis; that is, the value above which 90 percent of the population of values is expected to fall with a confidence of 95 percent.

A1.3.2 Preimpregnated materials

All test panels are fabricated from prepregs. Emphasis is placed on materials for use as facings in sandwich type structures. The prepregs for facings are normally processed to conform with two methods of sandwich fabrication. These are the laminate grades for two-step sandwich constructions and the controlled flow adhesive grades for one-step sandwich constructions. Only laminates simulating precured facings, that is, for use in two-step sandwiches, have been subjected to the narrow coupon tests listed in this chapter. The controlled flow adhesive prepregs are best tested as sandwich panels, and such testing is not at present included in the handbook program.

The prepreg materials comply with the specifications established by the individual fabricators. In general, the materials are autoclave molding grades with flows controlled to attain minimum bleedout and optimum bonding of the plies. When possible handling characteristics are specified consistent with the objectives of collimated plies in the laminate and the retention of fiber orientation during lay-up and cure.

Imposed tolerances on the gravimetric resin content of the prepregs are dependent on the type of reinforcement. For bidirectional woven broadgoods such as style 7781 fabric, the resin fraction is specified as not varying by more than two percent from the assigned devolatilized resin content. For directionally woven broadgoods such as style 7743 fabric, and nonwoven parallel fiber tapes such as XP251S, variation from the assigned devolatilized resin content is not to exceed three percent.

---

[1]Exceptions are the data for fiberglass-polyester laminates, taken from earlier sources, and the data for boron-epoxy panels which were compiled under special contract and published separately (Reference A1.2).

APPENDIX A1

A1.3.3 Test panels

A minimum size of the test panels has been established as two feet parallel to the warp direction by three feet parallel to the width for woven fabrics. For the non-woven laminates, including unidirectional,

crossplied and quasi-isotropic configurations, the three foot dimension is parallel to the fiber direction in the outer plies.

It is desirable that the test laminates be fabricated so that fiber alignment and orthotropy are maintained and that they are symmetrically balanced. Such conditions are generally attained in the test panels and they are designated in the following data summary tables as balanced and parallel. One set of panels (Table A1.1) is not balanced. In this case the laminates are parallel plied.

A1.3.4 Test procedures

Conventional uniaxial tests are conducted at constant crosshead rates. The direction parallel to the warp of woven fabrics is designated as the 0° or 1-direction. The direction perpendicular to the 0° direction is designated as the 90° or 2-direction. For non-woven unidirectional laminates, the 0° direction corresponds to the fiber direction. For crossplied and quasi-isotropic laminates, the 0° direction corresponds to the fiber direction in the outer plies.

A1.3.4.1 Tensile tests

Tensile tests for woven fabric laminates have been conducted initially using the method of ASTM D638 and Type I specimens (Reference A1.3.4.1(a)). Later tests are conducted with a modified specimen (Reference A1.2) and the method is designated as MIL-HDBK-17 tensile test. Tab ended specimens are used to test the 0° tensile properties of the non-woven unidirectional laminates (Reference A1.3.4.1(b)).

A1.3.4.2 Compression tests

Compression tests have been conducted with the end clamped and jig stabilized ASTM D695 specimen (Reference A1.3.4.2) and with the MIL-HDBK-17 compression specimen (Reference A1.2) in which the specimen and fixture have been modified.

A1.3.4.3 Shear tests

The picture frame method (Reference A1.2) has been used to determine the 0° - 90° shear properties of one material system at three resin fractions (Figure A1.6.3). In these tests it is assumed that 88 percent of the load is reacted by the specimen, while the pins in the fixture react the remainder. The other materials are tested by a modified rail shear method (Reference A1.3.4.3).

A1.3.4.4 Interlaminar shear

Interlaminar shear properties are determined by the short beam test method (Reference A1.3.4.1(b)), or by the method of ASTM D2733-68T when indicated (Reference A1.3.4.4).

A1.3.4.5 Flexural tests

Flexural properties are determined by the method of ASTM D790 (Reference A1.3.4.5).

A1.3.4.6 Bearing strength

Bearing strengths are determined by the method of ASTM D953 (Reference A1.3.4.6).

A1.3.5 Dry conditioning

Specimens are dry conditioned by allowing them to attain equilibrium at 70°F to 75°F and 45 percent to 55 percent relative humidity for a minimum of ten days. When tested at other than room temperature, the dry conditioned specimens are soaked at the test temperature for one-half hour prior to applying load.

A1.3.6 Wet conditioning

Specimens are wet conditioned at 125°F and 95 percent to 100 percent relative humidity for 1000 hours (42 days). When tested at temperatures below freezing, the wet conditioned specimens are cycled four times from the wet condition at 125°F to the sub-freezing test temperature; the dwell time at each temperature being one-half hour. Wet specimens tested at 160°F are soaked for one-half hour at this temperature immediately prior to testing. Some materials are shown as being tested at 220°F after wet conditioning. Such testing has been discontinued since these results appear inconclusive.

A1.3.7 Test schedule

The 0° and 90° tension and compression properties are determined at three reference temperatures, 65°F, 70°F - 75°F and 160°F, for both dry and wet conditioned specimens. Dry conditioned specimens are tested at maximum temperature for those materials which are potentially serviceable at elevated temperatures. Ten test results are obtained for the stress-strain relations at each of these conditions. Tests at intermediate temperatures are conducted to verify property changes, in which cases five specimens are tested. Ten test results are also required for the 0° - 90° shear at -65°F, 70°F - 75°F, and 160°F in the dry condition. Five tests are conducted at 70°F - 75°F to determine the stress-strain relations for Poisson's ratio. Flexure, bearing and interlaminar shear are determined in the 0° direction and dry condition at -65°F, 70°F - 75°F and 160°F. Five specimens are tested for each temperature.

A1.4 DATA PRESENTATION

Uniaxial tension, compression and shear are shown as stress-strain relations at each temperature and the properties are summarized in tabular form. Flexural, bearing and interlaminar shear properties are listed in summary tables. Poisson's ratio is shown as the response of the 0° elongation and 90° contraction to the applied tensile stress.

When ten or more results are available at a test condition, average values and the associated standard deviations are given in the tables. Stress-strain relations are plotted as an average curve and a plot of the average minus three times the standard deviation is also shown. When five to nine results are obtained from a test condition, average, maximum, and minimum values and curves are shown.

A1.4.1 Epoxy-fiberglass laminates

All data on fiberglass-epoxy systems are results obtained from the handbook test program. Properties are summarized in Tables A1.1 through A1.8. Detailed data are shown in Figures A1.1.1(a) through A1.8.5. [Four of the nine materials are known to be available.]

A1.4.2 Phenolic-fiberglass laminates

Handbook tested properties are summarized in Table A1.40 and Figures A1.40.1(a) through A1.40.5 for one fiberglass-phenolic system. [This material is available.]

A1.4.3 Silicone-fiberglass laminates

Partial handbook test results were listed in MIL-HDBK-17A for one fiberglass-silicone system. [This material is not available]

A1.4.4 Polyester-fiberglass laminates

Previous data for fiberglass-polyester laminates were listed in MIL-HDBK-17A. [None of these materials are known to be available.]

A1.4.5 Boron-epoxy laminates

Data on two boron-epoxy systems have been abstracted from the literature (Reference A1.4.5) and are presented in Tables A1.110 and A1.111 and in Figures A1.110.1(a) through A1.111.3. [One of these materials is available.]

The laminate thickness is controlled by the number of plies in the construction and the desired resin content. In general, the thickness of woven fabric laminates is maintained at eight plies, except for low resin content laminates which may require as many as ten plies. Nonwoven laminate monolayers are constructed with six plies to reduce the shear lag apparent in testing, and eight plies for the crossplied and quasi-isotropic panels.

APPENDIX A1

TABLE A1.1 *Summary of Mechanical Properties of U.S. Polymeric E-720E/7781 (ECDE-1/0-550) Fiberglass Epoxy*

Fabrication	Lay-up: Parallel	Vacuum: None	Pressure: 55-65 PSI	Bleedout: Edge & Vertical	Cure: 2 hr/350°F	Postcure: 4 hrs/400°F	Plies: 8
Physical Properties	Weight Percent Resin: 34.9	Avg. Specific Gravity: 1.78	Avg. Percent Voids: 2.0	Avg. Thickness: 0.082 inches			
Test Methods	Tension: ASTM D638 TYPE-1	Compression: MIL-HDBK-17	Shear: Rail	Flexure: ASTM D790	Bearing: ASTM D953	Interlaminar Shear: Short Beam	

Temperature		-65°F				75°F				160°F				400°F	
Condition		Dry		Wet		Dry		Wet		Dry		Wet		Dry	
		Avg	SD	Avg	SD	Avg	SD	Avg	SD	Avg	SD	Avg	SD	Avg	SD
Tension															
ultimate stress, ksi	0°	69.2	1.6	69.1	1.7	60.4	1.7	55.7	1.5	52.5	1.0	42.9	0.8	44.8	2.0
	90°	56.0	2.0	56.5	2.0	49.0	1.8	45.9	1.4	42.3	1.2	36.9	1.1	34.9	1.6
ultimate strain, %	0°	2.93	0.08	2.70	0.11	2.43	0.14	2.12	0.08	2.05	0.08	1.61	0.06	1.80	0.20
	90°	2.92	0.22	2.54	0.19	2.33	0.09	2.04	0.09	1.98	0.08	1.70	0.13	1.72	0.22
proportional limit, ksi	0°														
	90°														
initial modulus, 10^6 psi	0°	3.30		3.38		3.12		3.12		2.95		2.76		2.60	
	90°	2.90		3.02		2.82		2.78		2.50		2.65		2.30	
secondary modulus, 10^6 psi	0°	2.30		2.85		2.45		2.50		2.46		2.37			
	90°	1.90		1.74		2.05		2.19		2.01		1.97			
Compression															
ultimate stress, ksi	0°	77.1	4.0	75.0	3.7	64.8	2.9	57.3	3.8	54.0	1.4	46.2	1.4	23.8	2.2
	90°	57.2	2.7	53.9	2.7	50.2	2.9	45.2	2.4	40.8	2.9	36.2	3.1	14.7	1.6
ultimate strain, %	0°	2.48	0.16	2.44	0.15	2.14	0.11	1.99	0.09	1.86	0.08	1.62	0.06	1.12	0.22
	90°	1.93	0.16	1.81	0.19	1.70	0.14	1.58	0.14	1.46	0.17	1.37	0.15	0.91	0.08
proportional limit, ksi	0°														
	90°														
initial modulus, 10^6 psi	0°	3.50		3.45		3.25		3.10		3.15		3.03		2.45	
	90°	3.20		3.26		3.21		3.03		2.99		2.85		1.85	
Shear															
ultimate stress, ksi	0°-90°	17.5				14.3	0.6			11.2					
	±45°														

		-65°F Dry			75°F Dry			160° Dry		
		Avg	Max	Min	Avg	Max	Min	Avg	Max	Min
Flexure										
ultimate stress, ksi	0°	115.6	119.4	111.5	91.7	93.4	90.3	69.4	71.1	67.2
proportional limit, ksi	0°	88.1	100.7	77.5	32.5	36.2	30.8	56.2	62.8	49.4
initial modulus, 10^6 psi	0°	2.87	2.91	2.74	3.21	3.36	3.03	2.81	2.87	2.76
Bearing										
ultimate stress, ksi	0°	74.1	78.4	70.7	60.8	64.4	58.2	50.0	53.0	47.9
stress at 4% elong., ksi	0°	32.1	34.8	29.1	23.9	34.2	20.1	18.1	21.5	15.9
Interlaminar Shear										
ultimate stress, ksi	0°	7.09	7.36	6.80	5.90	6.07	5.72	6.05	6.16	5.91

APPENDIX A1

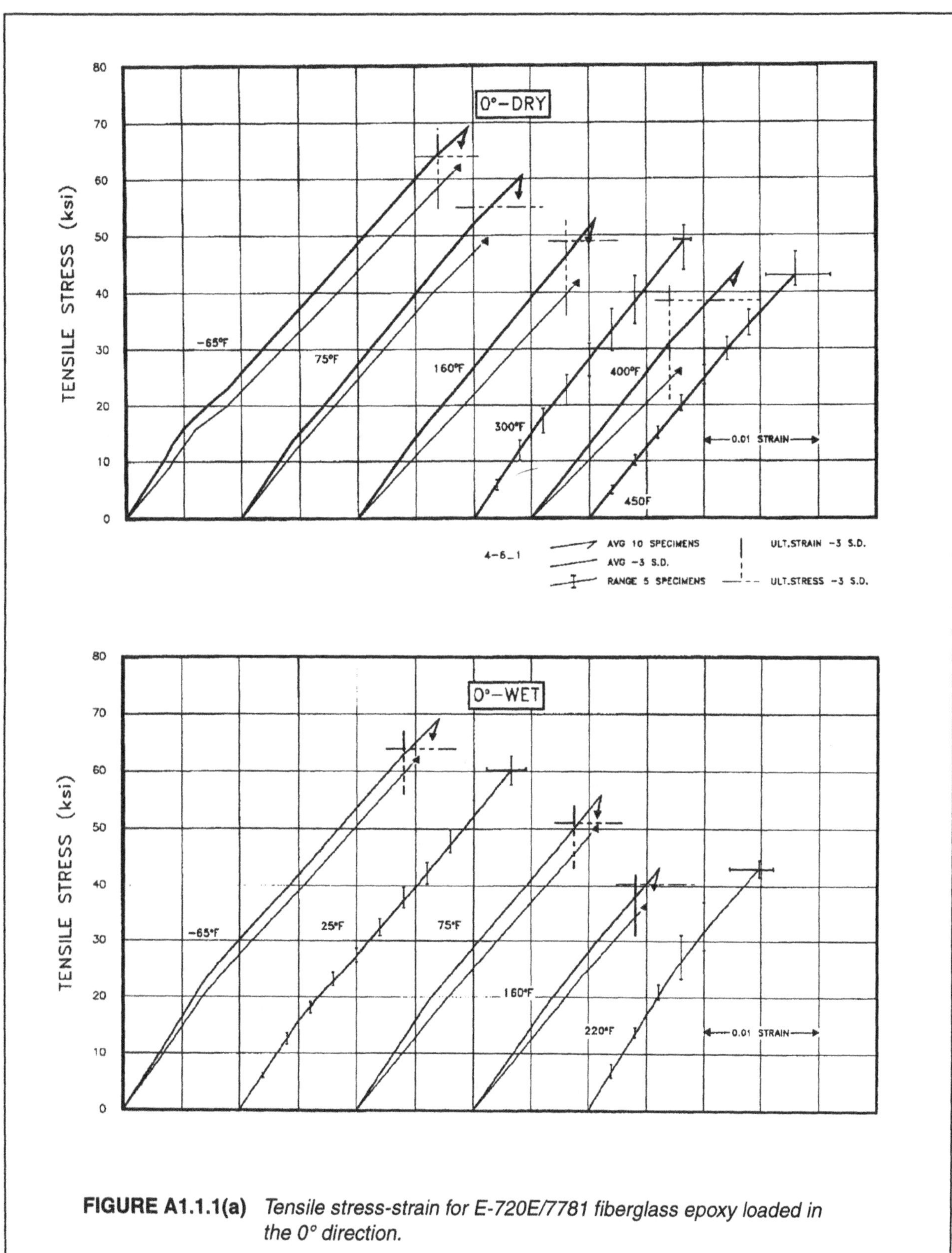

FIGURE A1.1.1(a) *Tensile stress-strain for E-720E/7781 fiberglass epoxy loaded in the 0° direction.*

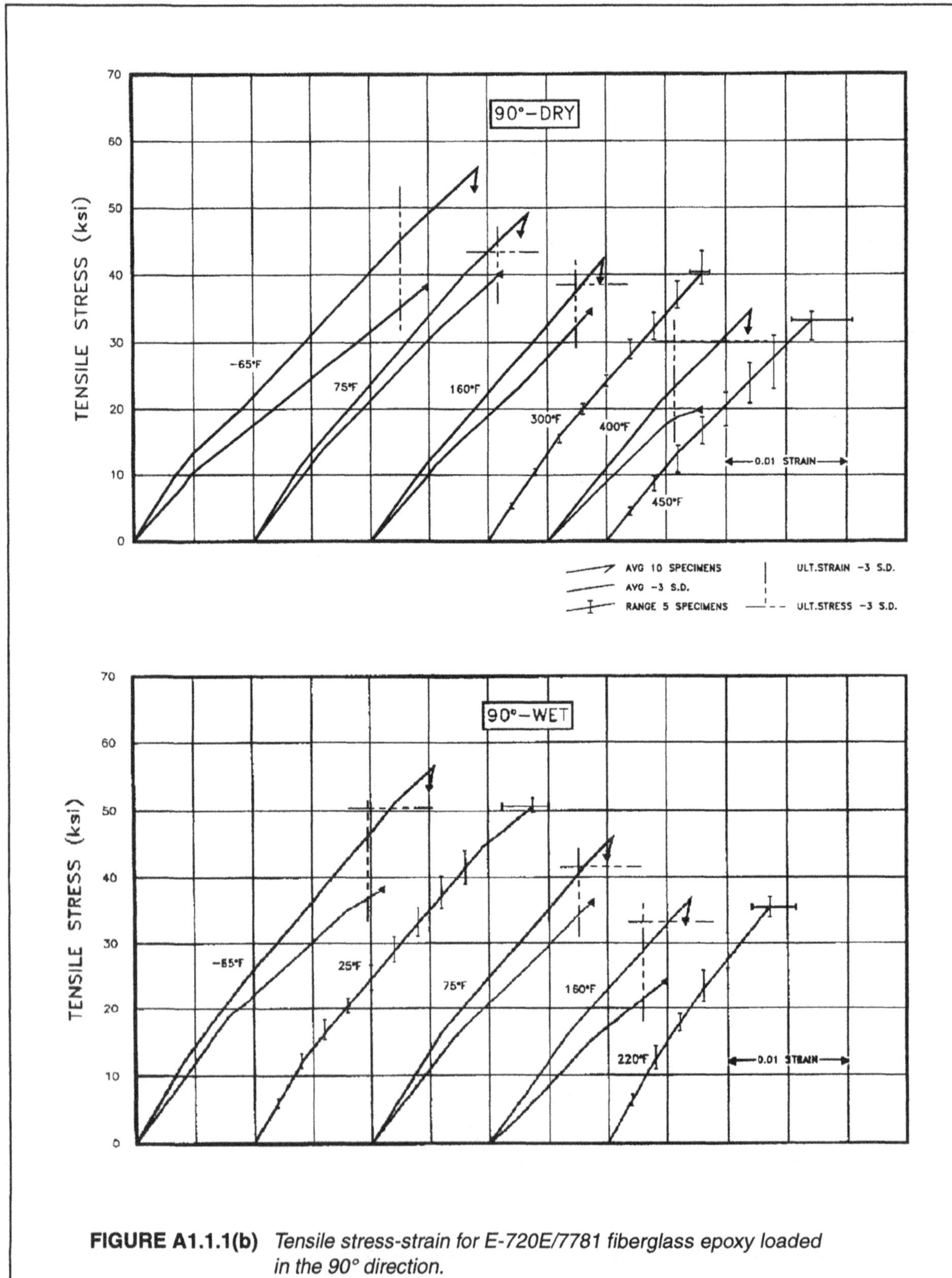

FIGURE A1.1.1(b) *Tensile stress-strain for E-720E/7781 fiberglass epoxy loaded in the 90° direction.*

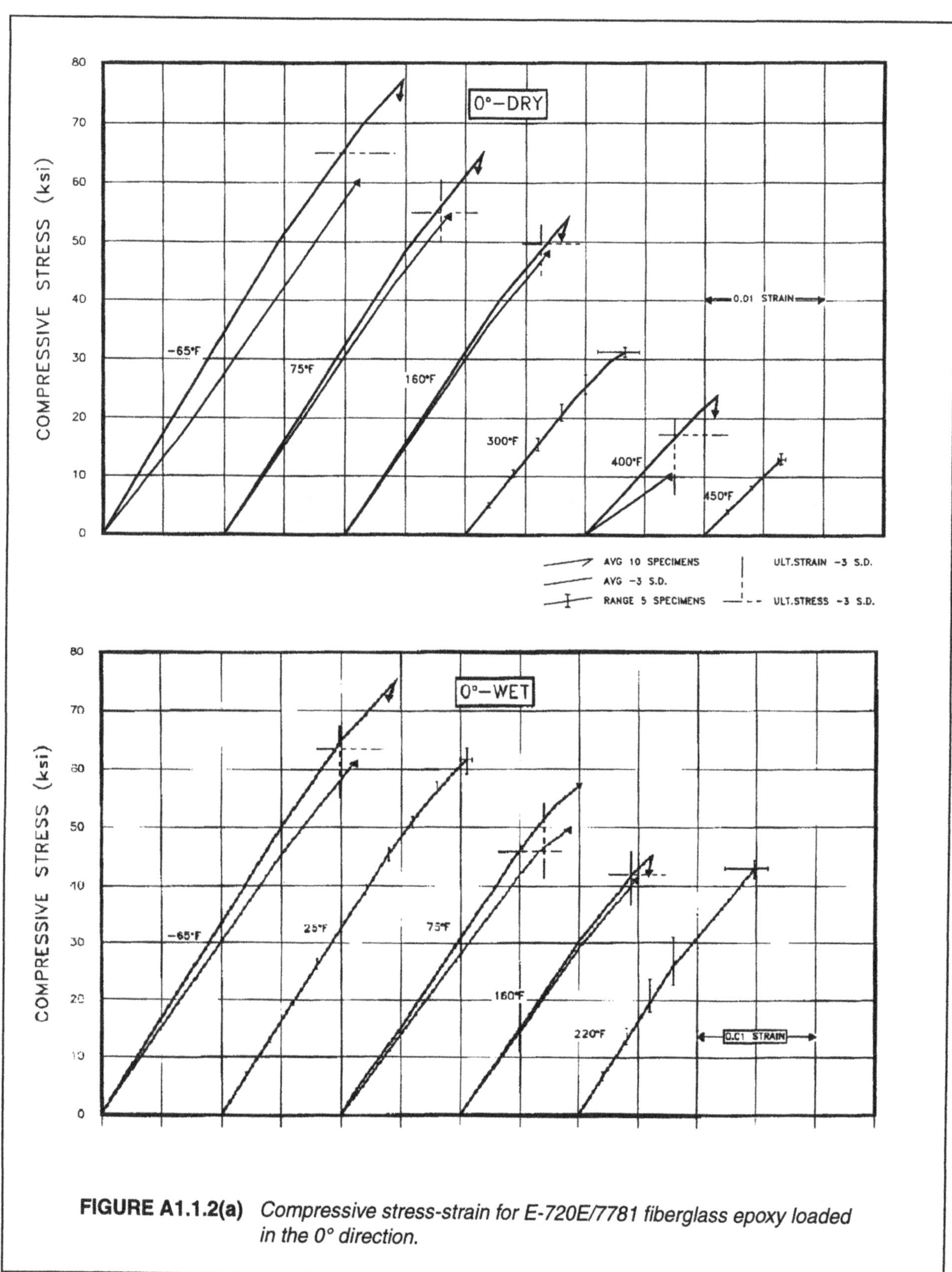

FIGURE A1.1.2(a) *Compressive stress-strain for E-720E/7781 fiberglass epoxy loaded in the 0° direction.*

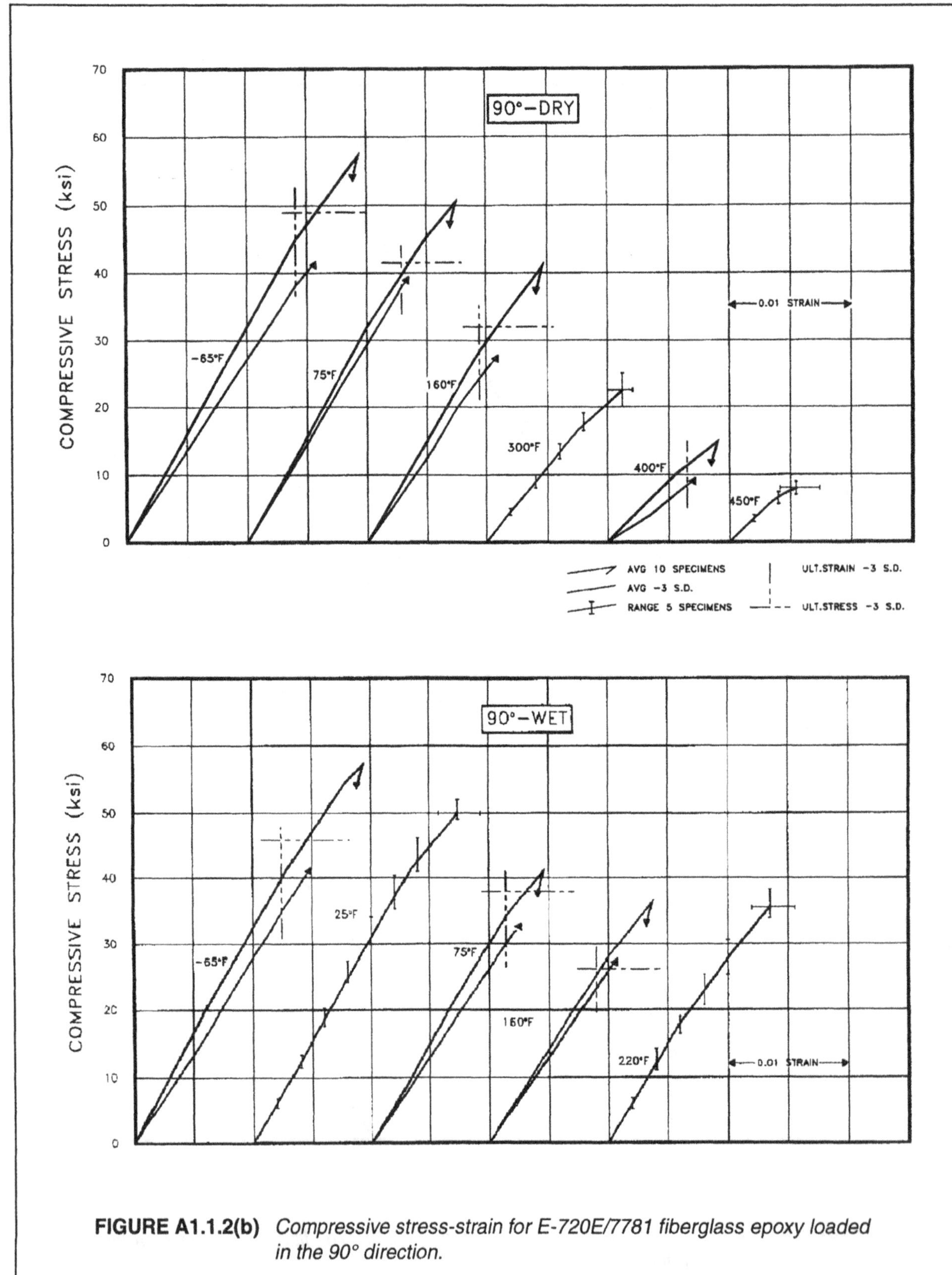

FIGURE A1.1.2(b) *Compressive stress-strain for E-720E/7781 fiberglass epoxy loaded in the 90° direction.*

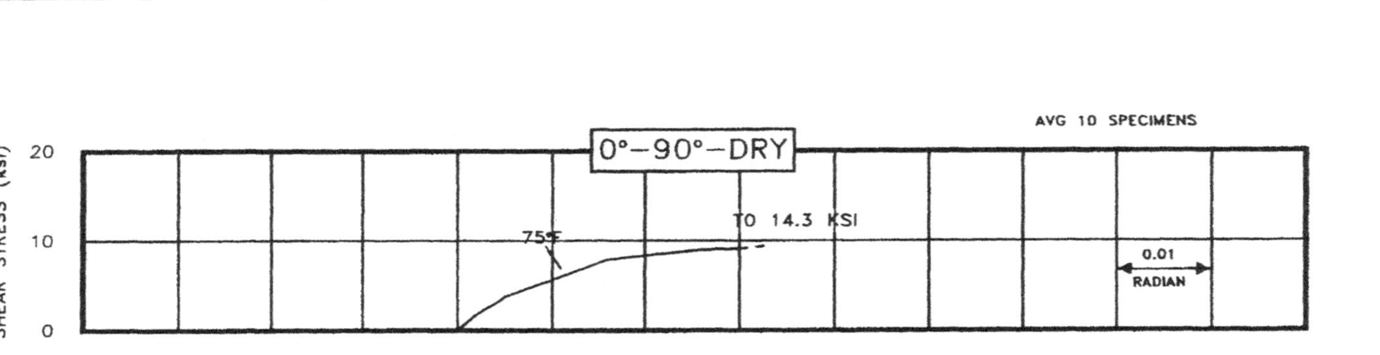

FIGURE A1.1.3 *0° - 90° rail shear for E-720E/7781 fiberglass.*

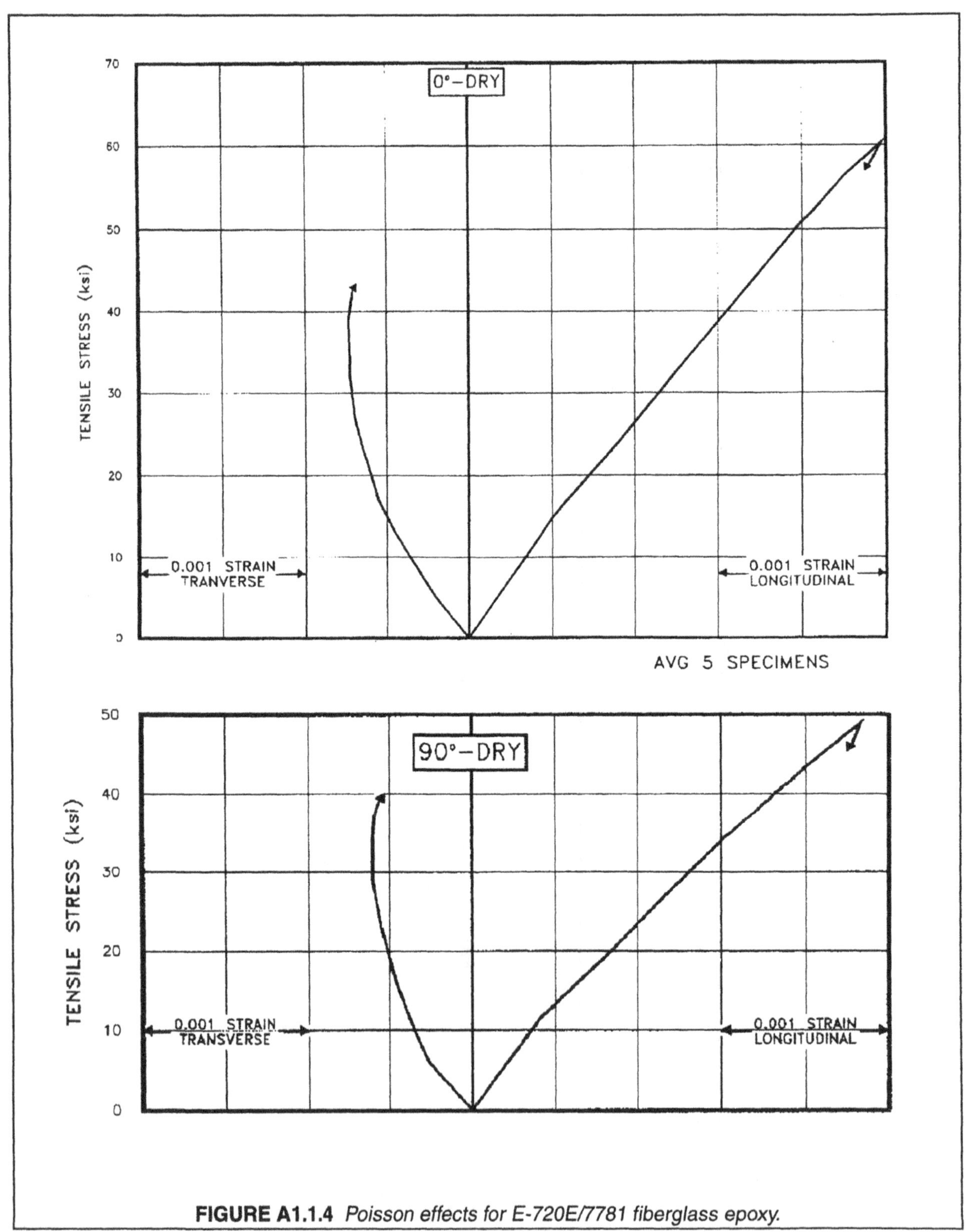

FIGURE A1.1.4 *Poisson effects for E-720E/7781 fiberglass epoxy.*

APPENDIX A1

TABLE A1.3 *Summary of Mechanical Properties of Hexcel F-161/7743(550) Fiberglass Epoxy.*

Fabrication	Lay-up: Balanced	Vacuum: 14 psi	Pressure: 35 psi	Bleedout: Pinched Edge	Cure: 2 hr/350°F	Postcure: 2 hr/350°F	Plies: 8
Physical Properties	Weight Percent Resin: 32.4 $v_f = 0.496$		Avg. Specific Gravity: 1.85		Avg. Percent Voids: 3.0		Avg. Thickness: 0.086 inches
Test Methods	Tension: ASTM-D638 TYPE 1	Compression: MIL-HDBK-17	Shear: Rail	Flexure: ASTM-D790	Bearing: ASTM-D953	Interlaminar Shear: Short Beam	

Temperature		-65°F				75°F				160°F				400°F	
Condition		Dry		Wet		Dry		Wet		Dry		Wet		Dry	
		Avg	SD	Avg	SD	Avg	SD	Avg	SD	Avg	SD	Avg	SD	Avg	SD
Tension															
ultimate stress, ksi	0°	111.3	1.12	107.3	3.60	95.5	7.57	87.3	5.2	80.9	4.05	71.7	2.73	74.5	5.90
	90°	9.84	0.78	9.42	0.59	8.15	0.40	7.27	0.28	6.78	0.18	6.16	0.21	6.59	0.41
ultimate strain, %	0°	2.10	0.31	2.11	0.10	1.88	0.10	1.72	0.17	1.56	0.15	1.35	0.12	1.64	0.09
	90°	2.43	0.25	2.03	0.21	1.82	0.23	1.20	0.28	1.26	0.19	0.61	0.13	1.44	0.19
proportional limit, ksi	0°	86.2		87.8		74.7		81.5		64.0		65.4		61.0	
	90°	5.6		5.0		5.2		4.8		5.0		5.0		3.0	
initial modulus, 10^6 psi	0°	5.42		5.35		5.30		5.55		5.36		5.47		4.52	
	90°	1.61		1.73		1.73		1.41		1.11		1.30		0.74	
secondary modulus, 10^6 psi	0°					5.15									
	90°					0.09									
Compression															
ultimate stress, ksi	0°	95.0	7.42	89.7	7.0	75.9	5.43	67.4	4.43	66.3	5.53	55.0	2.80	26.7	1.93
	90°	40.3	1.93	37.6	2.93	32.1	2.87	30.4	1.27	27.4	1.93	23.0	1.30	8.3	0.90
ultimate strain, %	0°	1.90	0.11	1.83	0.14	1.58	0.11	1.36	0.11	1.47	0.08	1.22	0.06	0.68	0.08
	90°	2.57	0.16	2.46	0.25	2.51	0.19	2.38	1.90	2.58	0.22	2.53	0.30	1.62	0.12
proportional limit, ksi	0°	83.0		70.0		52.2		49.8		55.6		40.8		20.0	
	90°	18.1		15.0		11.9		10.6		9.2		8.2			
initial modulus, 10^6 psi	0°	5.02		4.98		4.96		5.09		4.59		4.66		4.12	
	90°	1.91		1.88		1.65		1.77		1.46		1.37			
Shear															
ultimate stress, ksi	0°-90°	12.5				9.2	0.2			7.7					
	±45°														

		-65°F Dry			75°F Dry			160° Dry		
		Avg	Max	Min	Avg	Max	Min	Avg	Max	Min
Flexure										
ultimate stress, ksi	0°	203.0	210.0	196.0	160.0	163.0	155.0	138.0	142.0	135.0
proportional limit, ksi	0°	153.0	158.0	147.0	127.0	139.0	116.0	116.0	118.0	112.0
initial modulus, 10^6 psi	0°	5.71	5.80	5.63	5.18	5.27	5.10	5.43	5.46	5.32
Bearing										
ultimate stress, ksi	0°	79.4	90.2	64.8	58.8	63.2	52.7	53.7	57.5	50.6
stress at 4% elong., ksi	0°	37.9	45.6	31.5	23.0	27.1	19.5	21.9	23.6	20.5
Interlaminar Shear										
ultimate stress, ksi	0°	9.55	10.15	8.72	9.35	9.55	9.17	8.31	8.65	8.02

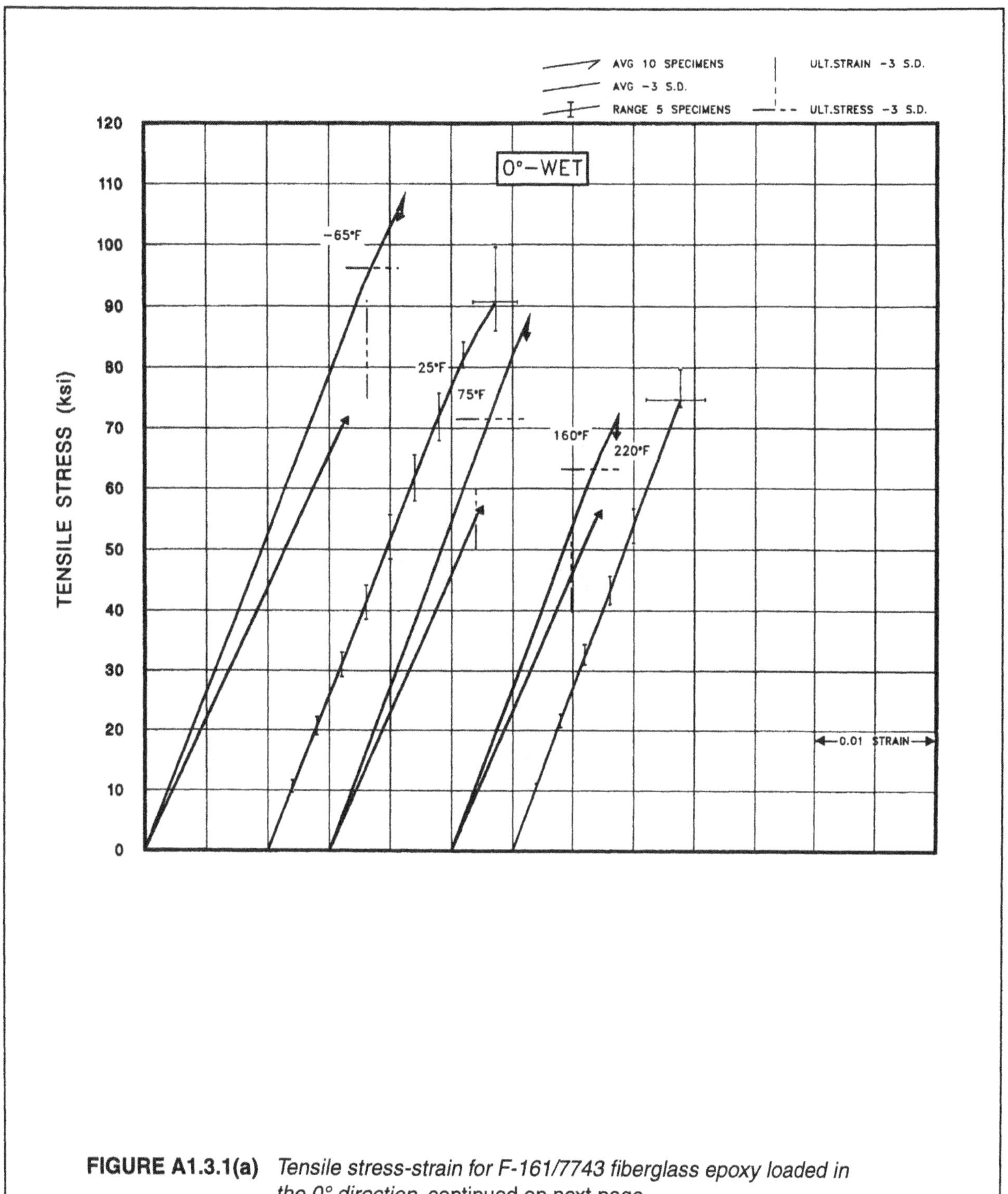

FIGURE A1.3.1(a) *Tensile stress-strain for F-161/7743 fiberglass epoxy loaded in the 0° direction,* continued on next page.

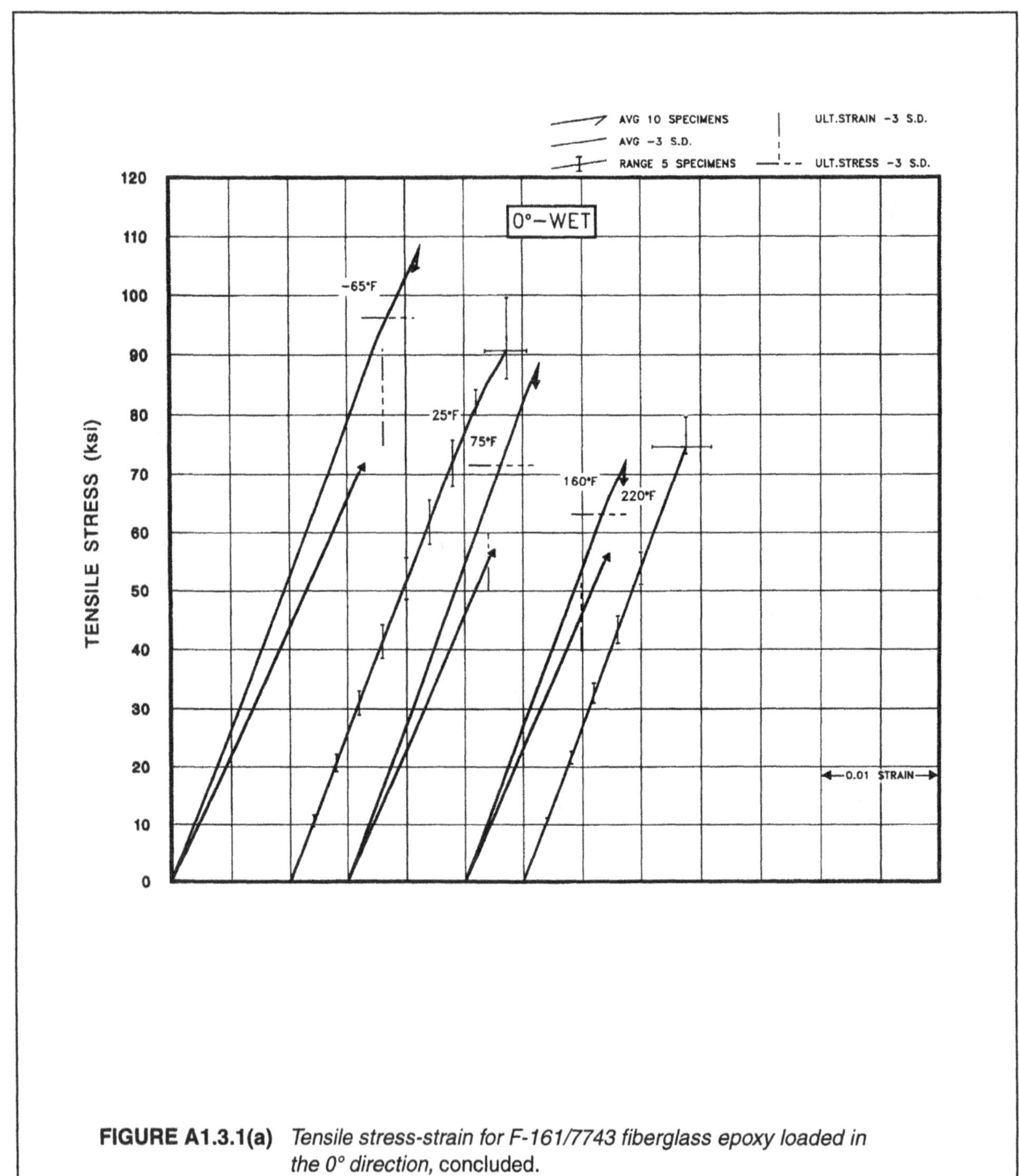

FIGURE A1.3.1(a) *Tensile stress-strain for F-161/7743 fiberglass epoxy loaded in the 0° direction*, concluded.

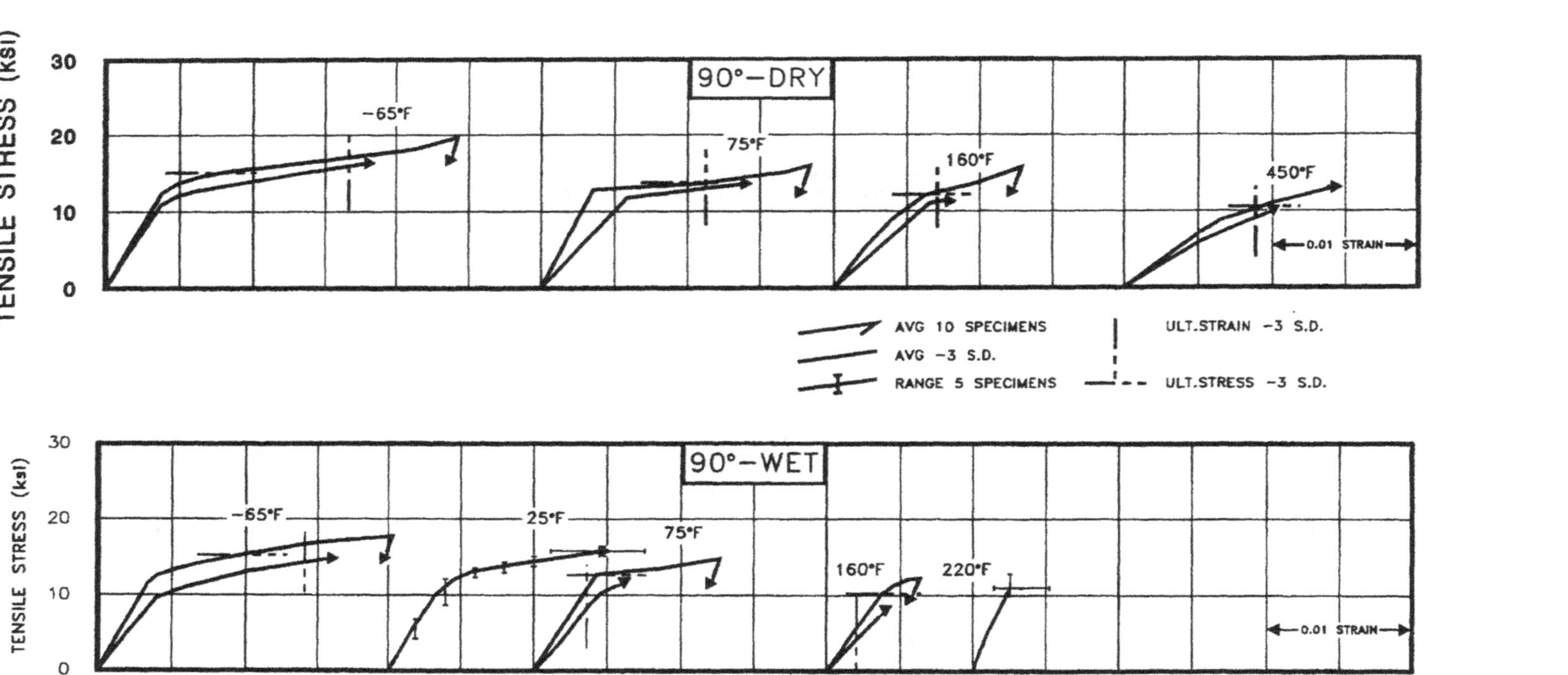

FIGURE A1.3.1(b) *Tensile stress-strain for F-161/7743 fiberglass epoxy loaded in the 90° direction.*

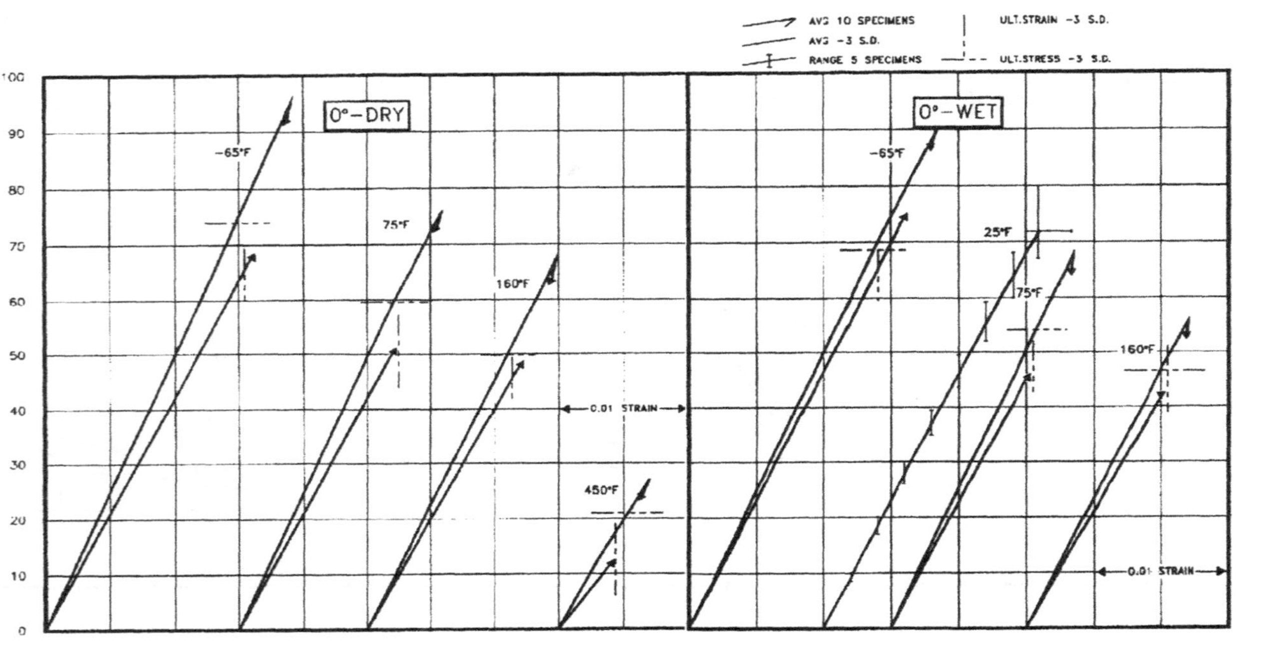

FIGURE A1.3.2(a) *Compressive stress-strain for F-161/7743 fiberglass epoxy loaded in the 0° direction.*

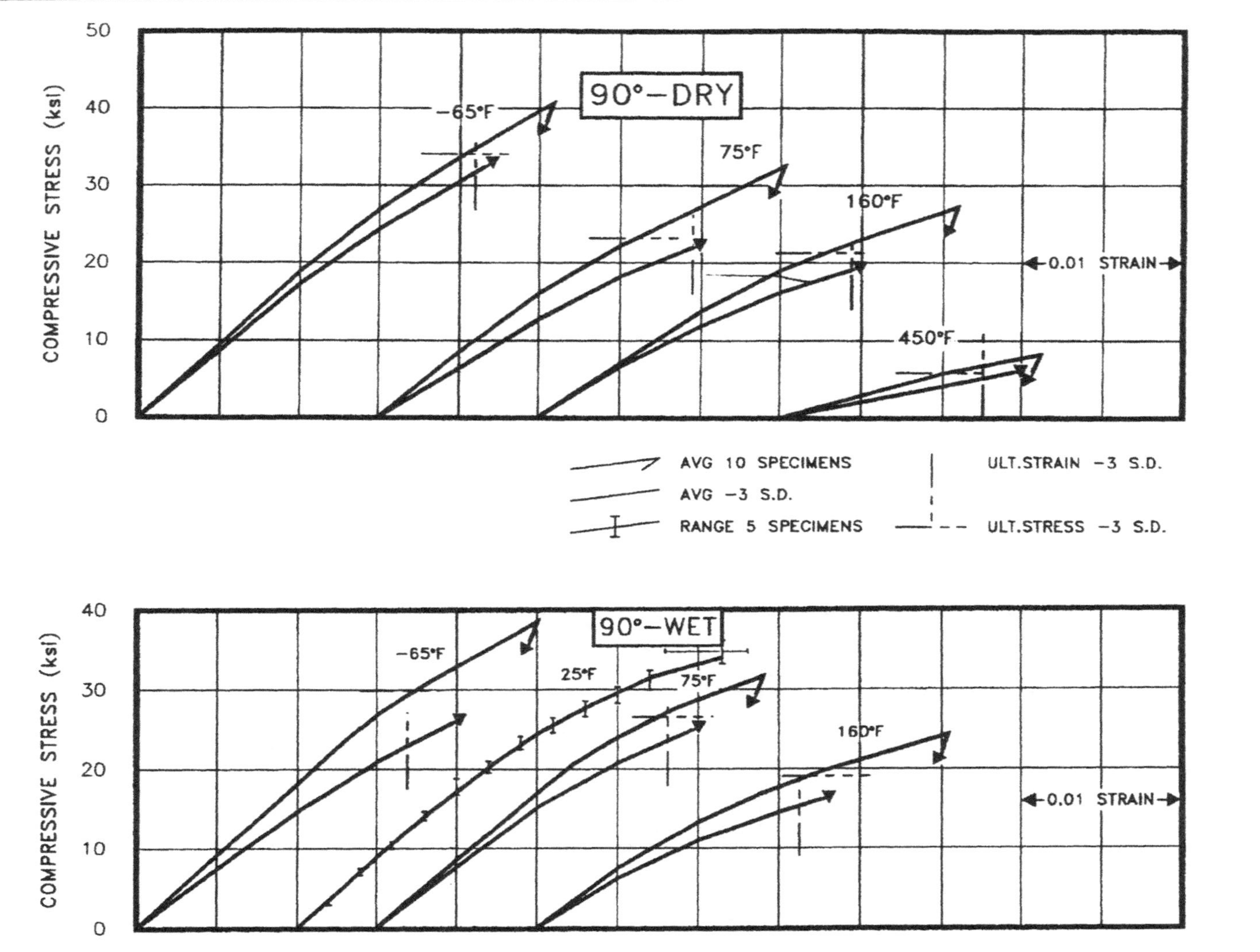

FIGURE A1.3.2(b) *Compressive stress-strain F-161/7743 fiberglass epoxy loaded in the 90° direction.*

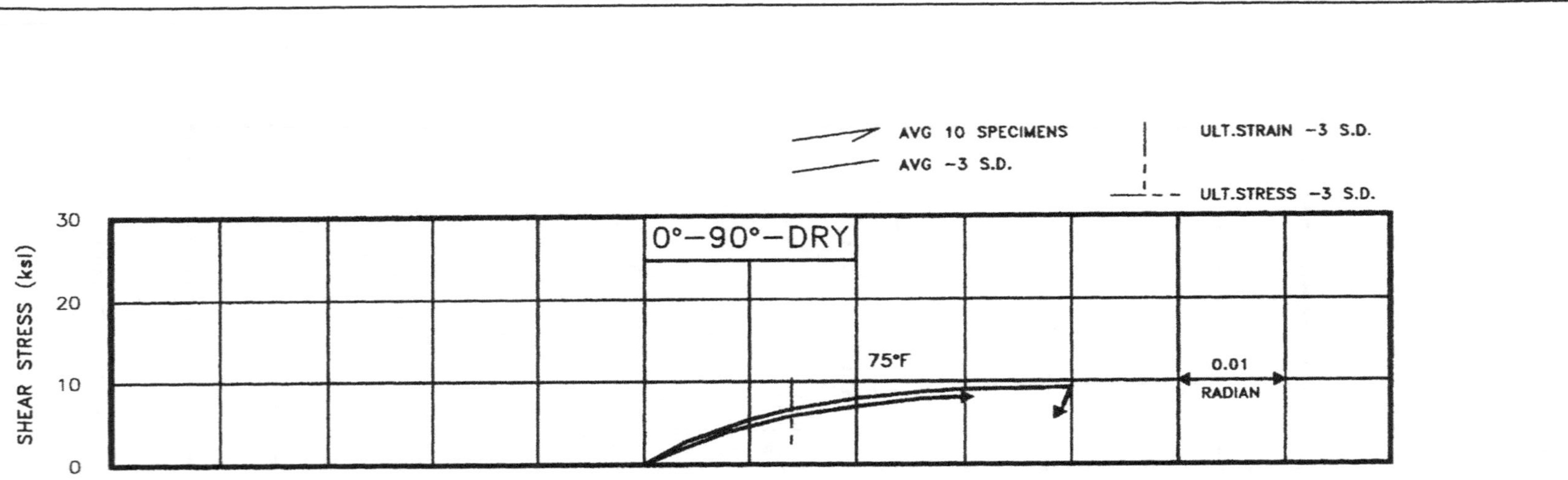

FIGURE A1.3.3 *0° - 90° rail shear for F-161/7743 fiberglass epoxy.*

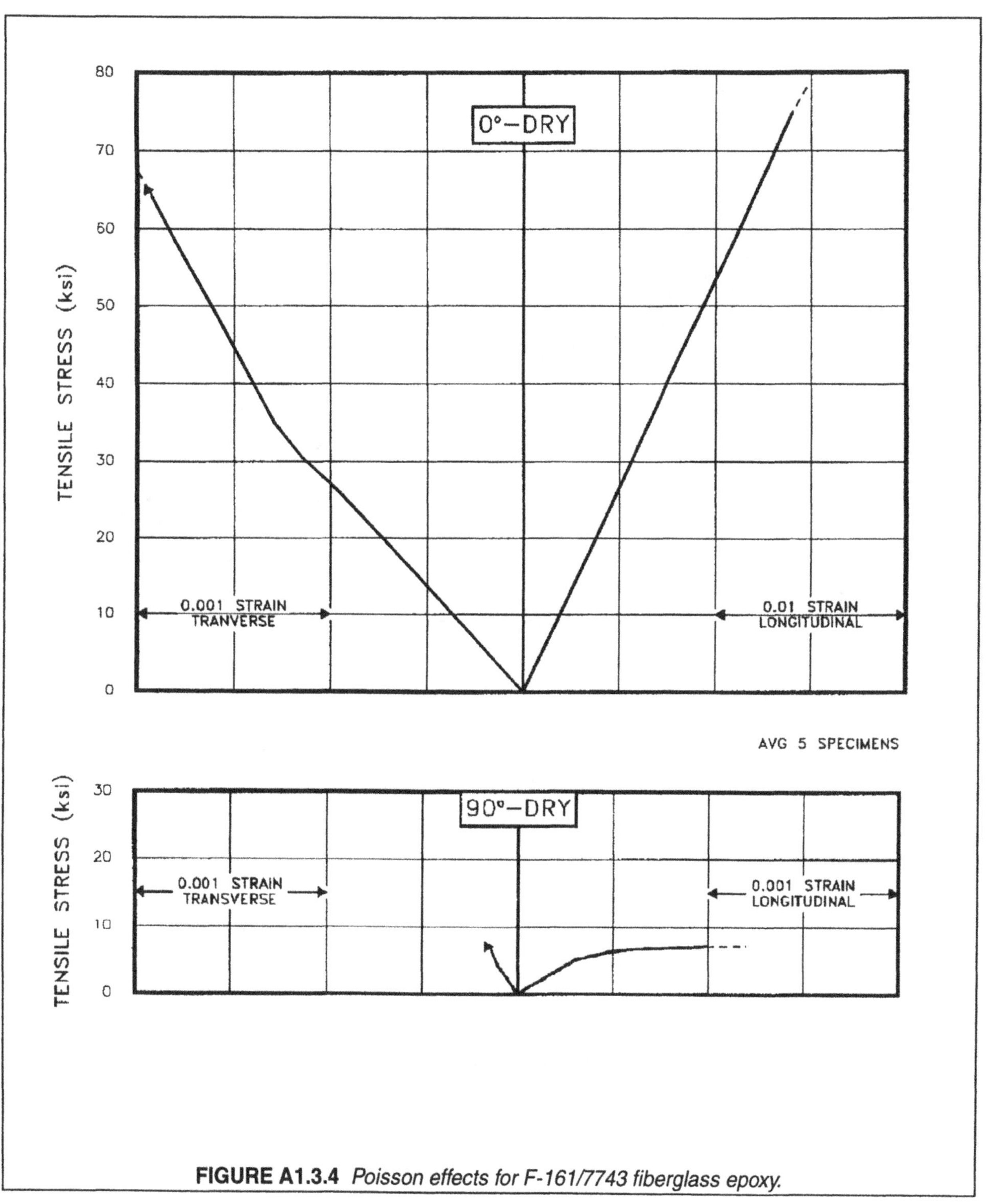

FIGURE A1.3.4 *Poisson effects for F-161/7743 fiberglass epoxy.*

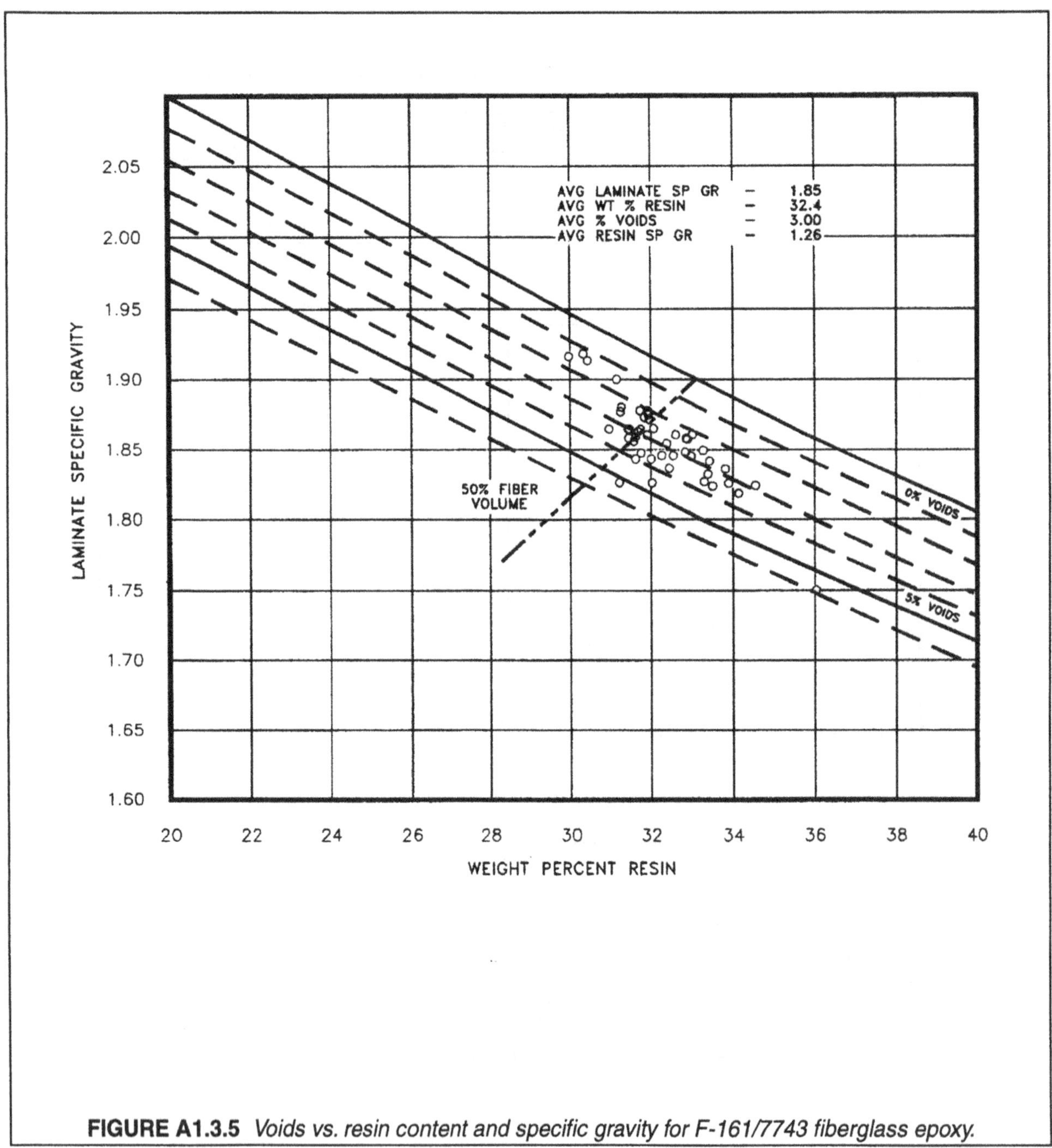

FIGURE A1.3.5 *Voids vs. resin content and specific gravity for F-161/7743 fiberglass epoxy.*

APPENDIX A1

TABLE A1.4 *Summary of Mechanical Properties of Hexcel F-161/7781 (ECDE-1/0-550) Fiberglass Epoxy (26% Resin)*

Fabrication	Lay-up: Balanced	Vacuum: None	Pressure: 55-65 psi	Bleedout: Vertical and Stepped Edge	Cure: 1 hr/350°F	Postcure: 2 hr/300°F 2.5 hr/400°F	Plies: 8 and 10
Physical Properties	Weight Percent Resin: 26.0 $v_f = 0.59$	Avg. Specific Gravity: 2.01	Avg. Percent Voids: 0.5	Avg. Thickness: 0.008 inch/ply			
Test Methods	Tension: MIL-HDBK-17	Compression: MIL-HDBK-17	Shear: Picture Frame	Flexure: ASTM-D790	Bearing:	Interlaminar Shear: ASTM-D2345	

Temperature		-65°F				75°F				160°F				400°F	
Condition		Dry		Wet		Dry		Wet		Dry		Wet		Dry	
		Avg	SD	Avg	SD	Avg	SD	Avg	SD	Avg	SD	Avg	SD	Avg	SD
Tension															
ultimate stress, ksi	0°	92.4	5.16	80.5	10.87			61.4	3.20	65.7	3.03	50.7	5.72	59.8	3.81
	90°	67.8	10.65	62.3	5.01			50.3	2.61	53.6	5.19	46.2	2.69	35.2	5.16
ultimate strain, %	0°	2.86	2.11	2.37	0.31			1.78	0.13	1.97	0.14	1.58	0.19	1.96	0.08
	90°	2.42	3.14	1.97	0.24			1.65	0.08	1.88	0.12	1.55	0.10	1.38	0.13
proportional limit, ksi	0°														
	90°														
initial modulus, 10^6 psi	0°	4.42		4.49				4.10		3.92		3.72		3.27	
	90°	4.22		4.21				3.76		3.17		3.38		2.86	
secondary modulus, 10^6 psi	0°	3.32		3.14				3.06		3.24		3.07		2.94	
	90°	2.70		2.74				2.62		2.72		2.55		2.46	
Compression															
ultimate stress, ksi	0°	73.2	6.83	74.0	5.02			57.3	4.0	48.9	3.50	44.7	3.25	28.8	3.03
	90°	64.2	3.19	55.8	4.40			37.5	2.28	42.0	2.64	40.1	1.90	18.9	0.69
ultimate strain, %	0°	1.70	0.42	1.65	0.28			1.09	0.17	1.12	0.15	0.84	0.14	0.79	0.03
	90°	1.40	0.14	1.42	0.27			1.26	0.41	1.14	0.23	1.22	0.18	0.71	0.27
proportional limit, ksi	0°	39.0		46.0				42.0		41.0		24.0		15.0	
	90°	28.0		41.0				24.0		36.0		21.0		11.0	
initial modulus, 10^6 psi	0°	4.42		4.47				4.27		4.05		3.94		3.73	
	90°	4.02		4.19				4.12		3.68		3.40		3.07	
Shear															
ultimate stress, ksi	0°-90°	20.1	2.3					16.0	1.64	13.4	1.28				
	±45°														

		-65°F Dry			75°F Dry			160° Dry		
		Avg	Max	Min	Avg	Max	Min	Avg	Max	Min
Flexure										
ultimate stress, ksi	0°				94.10	96.86	89.64			
proportional limit, ksi	0°									
initial modulus, 10^6 psi	0°									
Bearing										
ultimate stress, ksi	0°									
stress at 4% elong., ksi	0°									
Interlaminar Shear										
ultimate stress, ksi	0°				5.56	5.65	5.50			

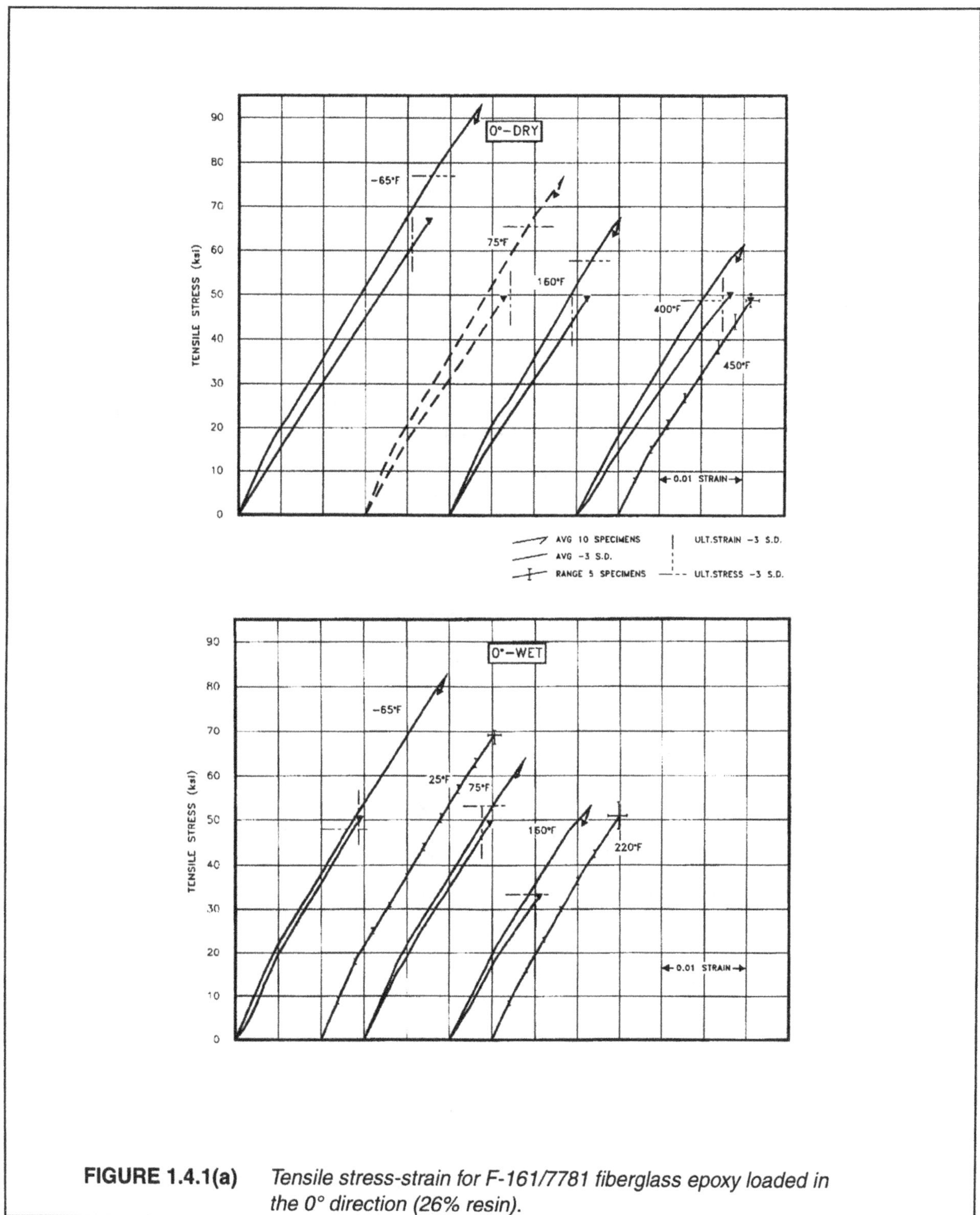

FIGURE 1.4.1(a) *Tensile stress-strain for F-161/7781 fiberglass epoxy loaded in the 0° direction (26% resin).*

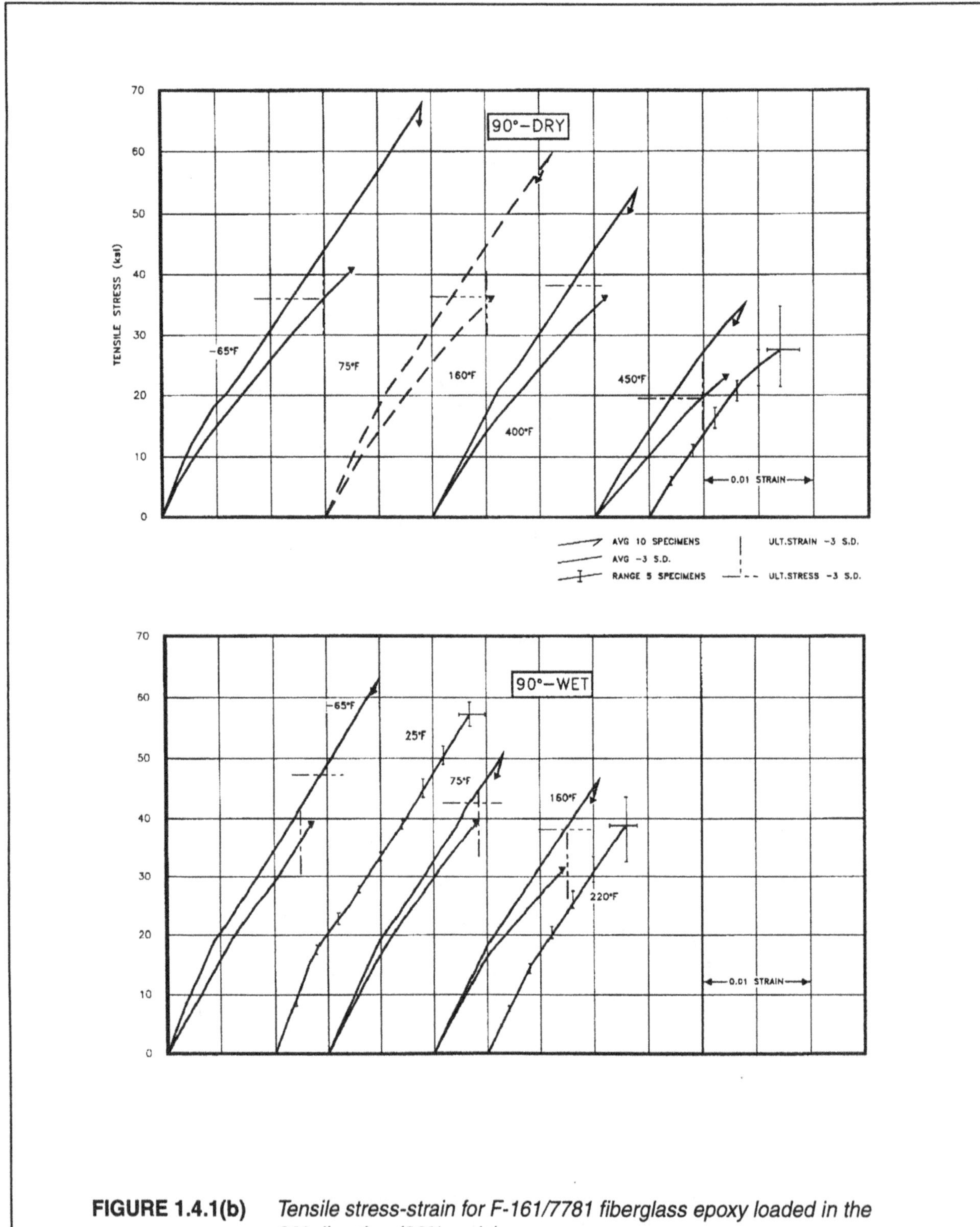

FIGURE 1.4.1(b) *Tensile stress-strain for F-161/7781 fiberglass epoxy loaded in the 90° direction (26% resin).*

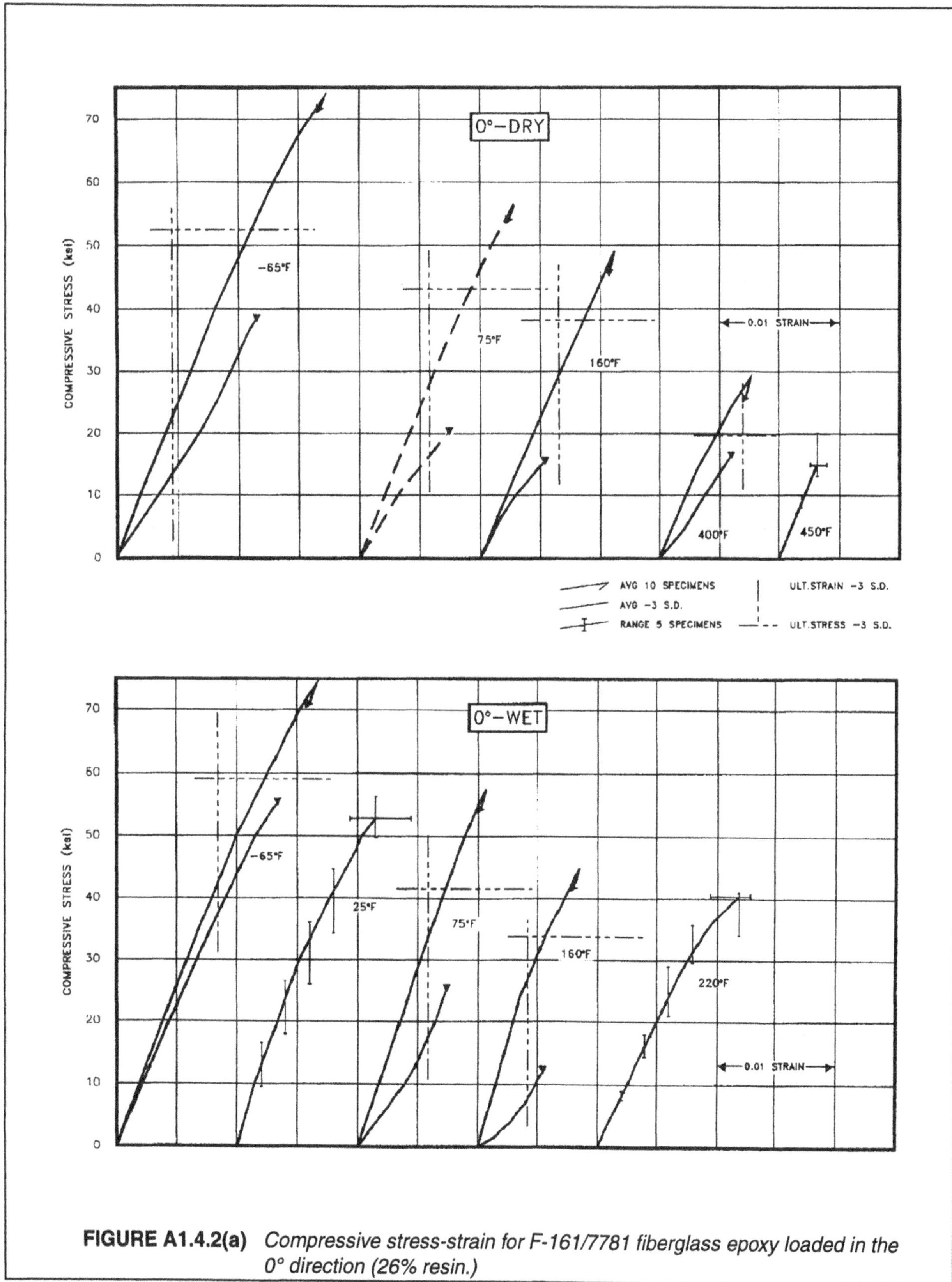

FIGURE A1.4.2(a) *Compressive stress-strain for F-161/7781 fiberglass epoxy loaded in the 0° direction (26% resin.)*

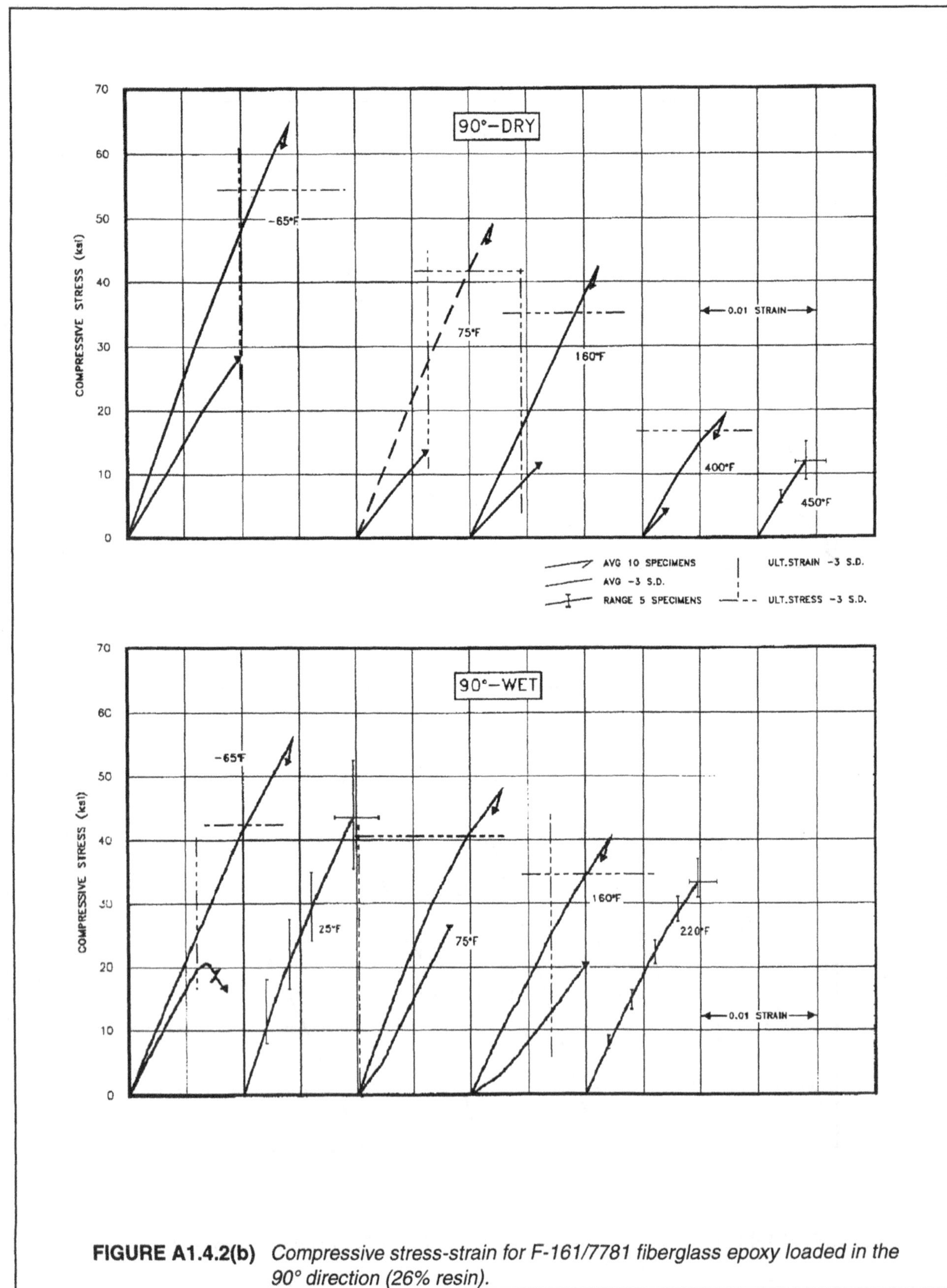

FIGURE A1.4.2(b) *Compressive stress-strain for F-161/7781 fiberglass epoxy loaded in the 90° direction (26% resin).*

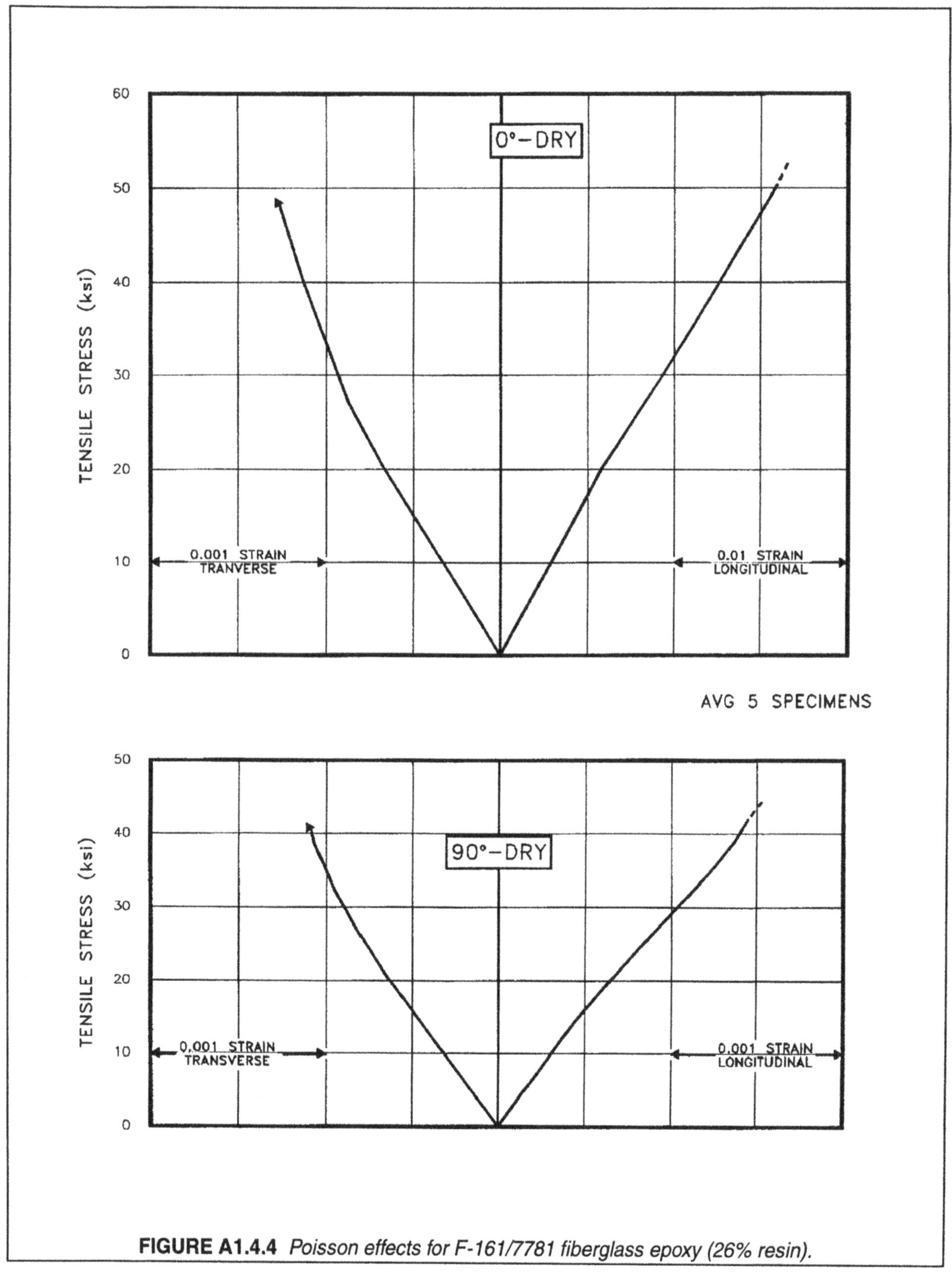

FIGURE A1.4.4 *Poisson effects for F-161/7781 fiberglass epoxy (26% resin).*

APPENDIX A1

TABLE A1.5 *Summary of Mechanical Properties of Hexcel F-161/7781 (ECDE-1/0-550) Fiberglass Epoxy (31% Resin)*

Fabrication	Lay-up: Balanced	Vacuum: None	Pressure: 55-65 psi	Bleedout: Vertical and Stepped Edge	Cure: 1 hr/350°F	Postcure: 2 hr/300°F 2.5 hr/400°F	Plies: 8 and 10
Physical Properties	Weight Percent Resin: 31.0	Avg. Specific Gravity: 1.92	Avg. Percent Voids: 0.6	Avg. Thickness: 0.009 inch/ply			
Test Methods	Tension: MIL-HDBK-17	Compression: MIL-HDBK-17	Shear: Picture Frame	Flexure: ASTM-D790	Bearing:	Interlaminar Shear:	

Temperature		-65°F				75°F				160°F				400°F	
Condition		Dry		Wet		Dry		Wet		Dry		Wet		Dry	
		Avg	SD	Avg	SD	Avg	SD	Avg	SD	Avg	SD	Avg	SD	Avg	SD
Tension															
ultimate stress, ksi	0°	85.2	4.68	82.3	4.97			64.0	2.04	60.1	3.75	51.4	4.23	47.3	4.87
	90°	70.0	5.24	67.9	2.98			53.5	2.91	49.3	0.95	39.8	3.50	31.0	1.95
ultimate strain, %	0°	2.93	0.14	2.53	0.18			2.10	0.06	2.02	0.10	1.66	0.17	1.66	0.18
	90°	2.50	0.21	2.41	0.22			1.90	0.11	1.86	0.06	1.47	0.09	1.25	0.09
proportional limit, ksi	0°														
	90°														
initial modulus, 10^6 psi	0°	4.22		4.30				3.84		3.69	3.72	3.65		3.09	
	90°	3.97		4.15				3.68		3.37	3.34	3.30		2.75	
secondary modulus, 10^6 psi	0°	3.13		3.01				3.03		2.97	0.04	2.88		2.94	
	90°	2.62		2.96				2.62		2.55	0.25	2.46		2.47	
Compression															
ultimate stress, ksi	0°	73.1	5.18	66.0	10.75			54.4	7.04	50.6		45.9	5.39	32.8	6.04
	90°	58.4	3.17	57.5	11.56			47.3	4.73	42.2		38.7	4.19	25.8	8.27
ultimate strain, %	0°	1.86	0.21	1.72	0.32			1.33	0.28	1.52		1.04	0.23	0.95	0.24
	90°	1.61	0.29	1.44	0.36			1.10	0.21	1.30		0.99	0.22	0.87	0.28
proportional limit, ksi	0°	44.0		38.0				33.0		32.0		25.0		16.0	
	90°	33.0		33.0				30.0		--		21.0		15.0	
initial modulus, 10^6 psi	0°	3.90		4.04				4.03		3.42		4.06		3.50	
	90°	3.56		3.84				3.96		3.23		4.01		3.07	
Shear															
ultimate stress, ksi	0°-90°	20.5	2.23					15.9	0.72	13.7	0.82				
	±45°														

		-65°F Dry			75°F Dry			160° Dry		
		Avg	Max	Min	Avg	Max	Min	Avg	Max	Min
Flexure										
ultimate stress, ksi	0°				90.23	93.74	87.29			
proportional limit, ksi	0°									
initial modulus, 10^6 psi	0°									
Bearing										
ultimate stress, ksi	0°									
stress at 4% elong., ksi	0°									
Interlaminar Shear										
ultimate stress, ksi	0°				5.56	5.65	5.50			

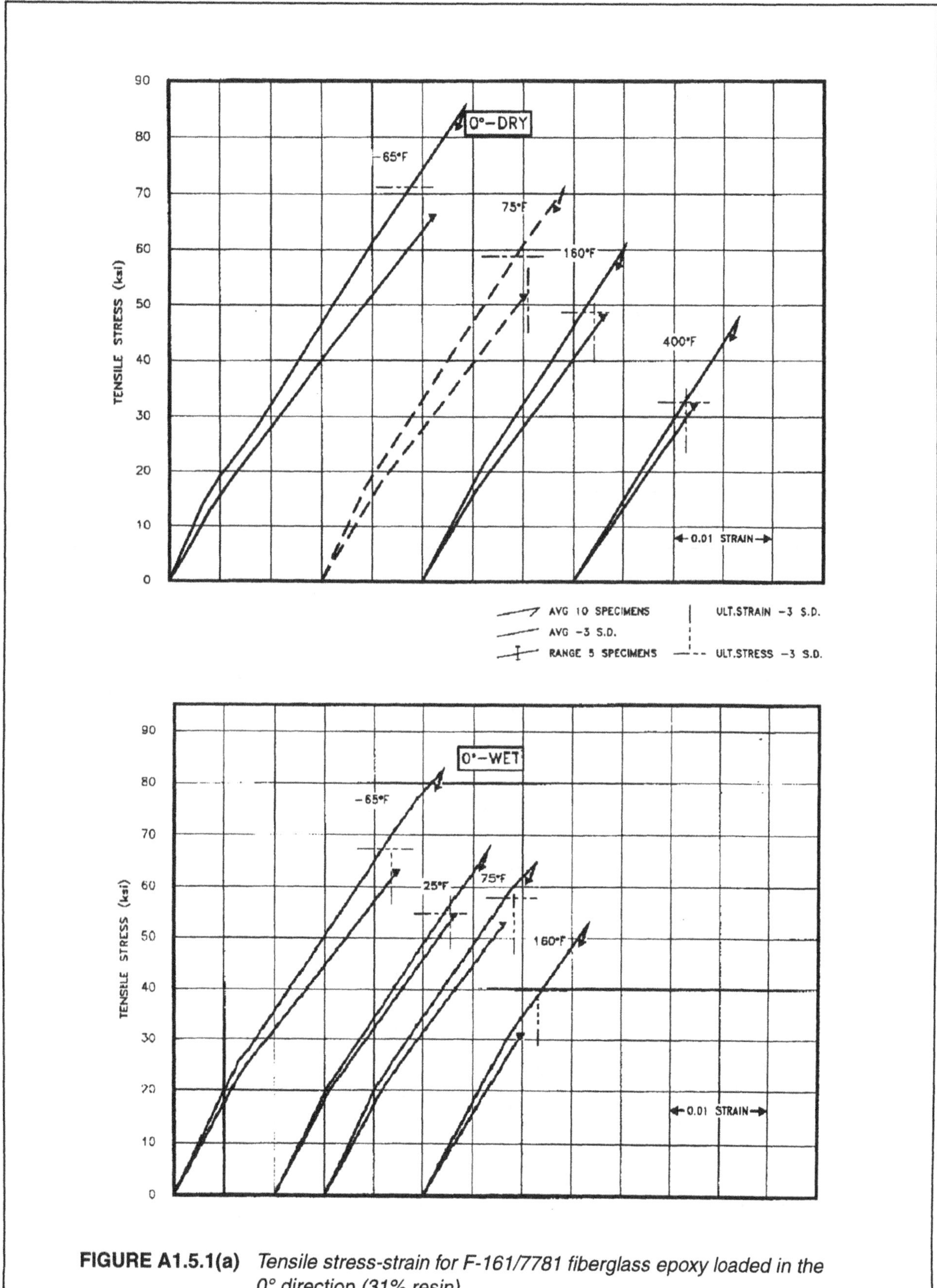

FIGURE A1.5.1(a) *Tensile stress-strain for F-161/7781 fiberglass epoxy loaded in the 0° direction (31% resin).*

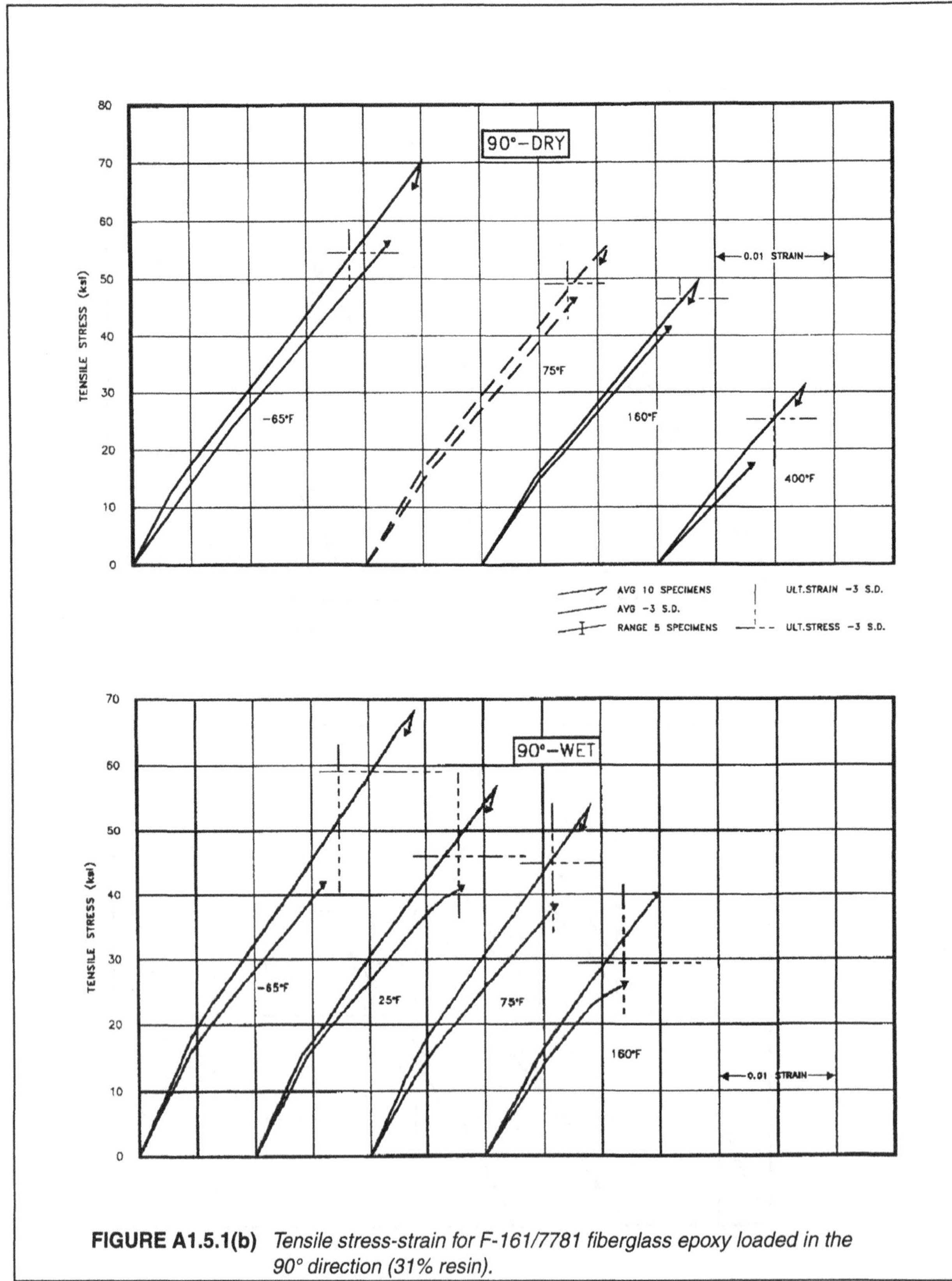

FIGURE A1.5.1(b) *Tensile stress-strain for F-161/7781 fiberglass epoxy loaded in the 90° direction (31% resin).*

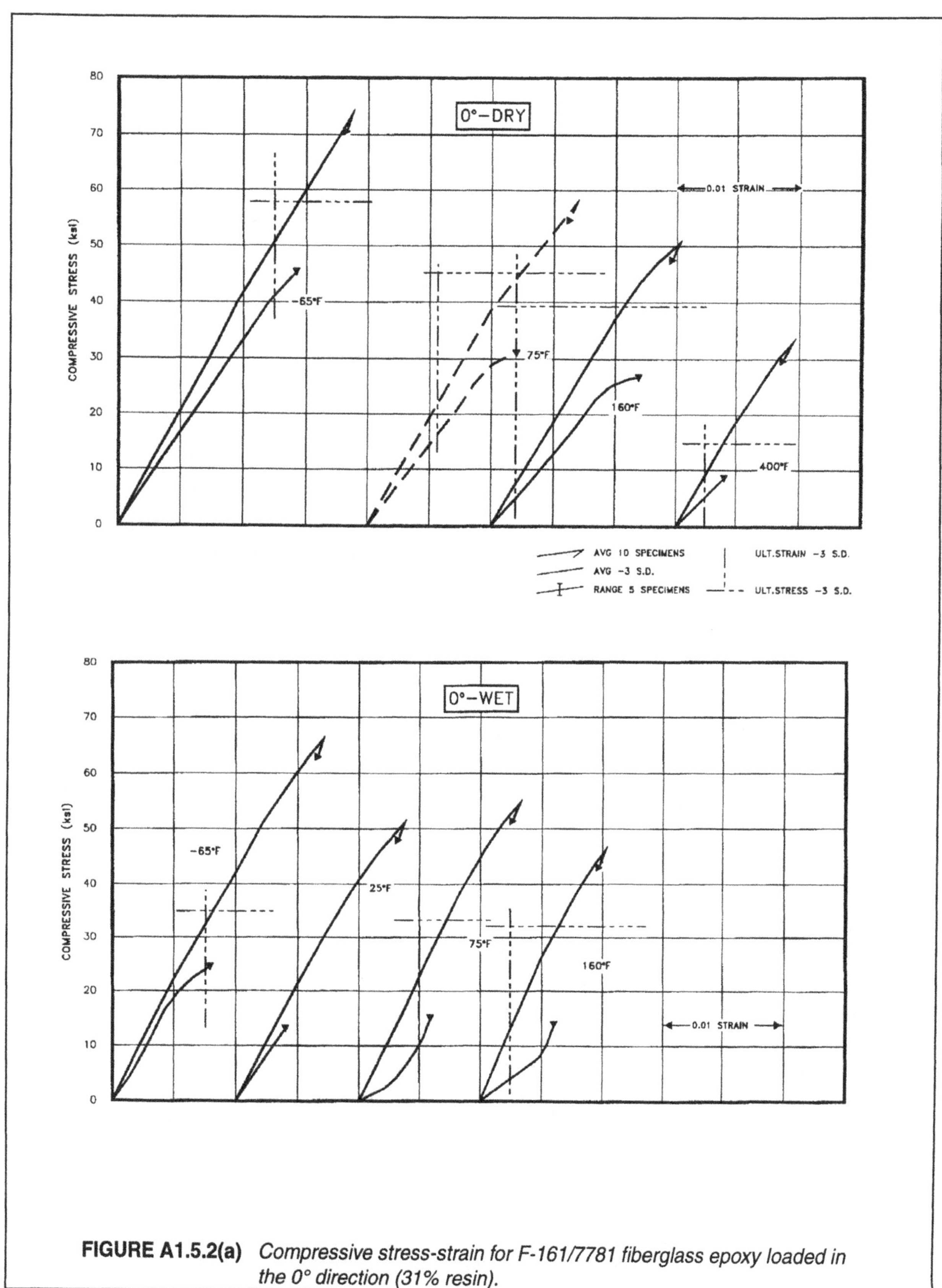

FIGURE A1.5.2(a) *Compressive stress-strain for F-161/7781 fiberglass epoxy loaded in the 0° direction (31% resin).*

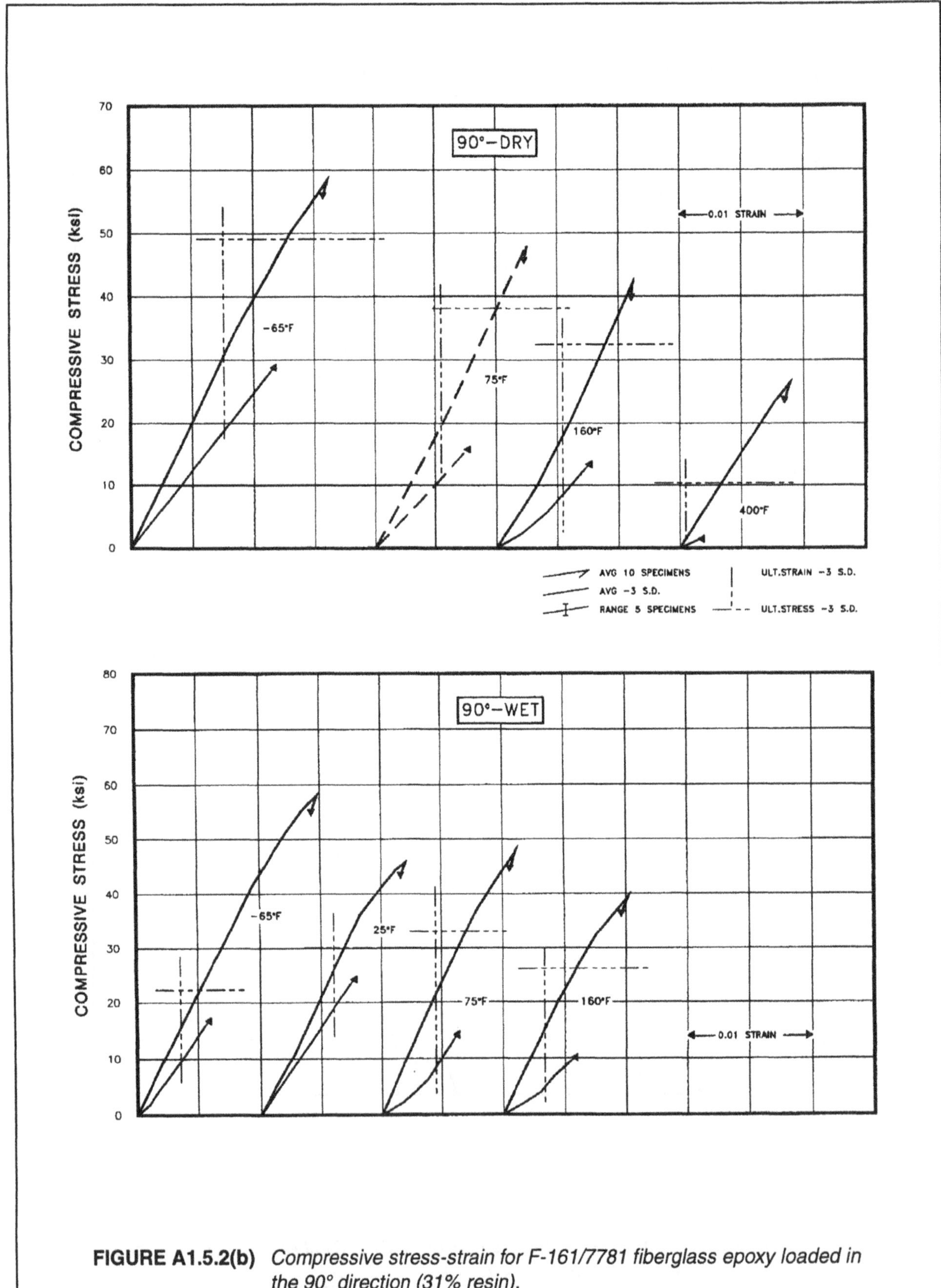

FIGURE A1.5.2(b) *Compressive stress-strain for F-161/7781 fiberglass epoxy loaded in the 90° direction (31% resin).*

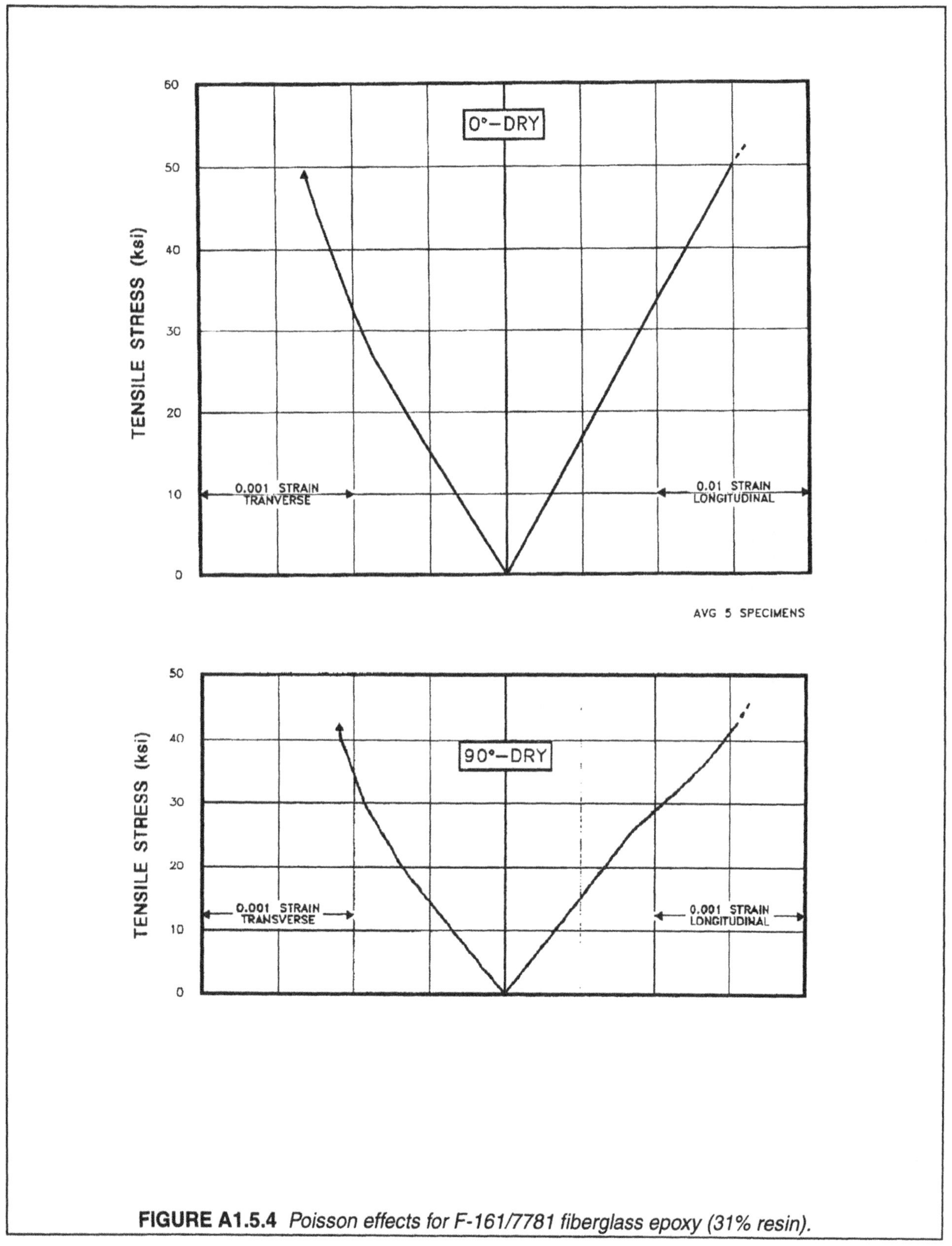

FIGURE A1.5.4 *Poisson effects for F-161/7781 fiberglass epoxy (31% resin).*

TABLE A1.6 *Summary of Mechanical Properties of Hexcel F-161/7781 (ECDE-1/0-550) Fiberglass Epoxy (36% Resin)*

Fabrication	Lay-up: Balanced	Vacuum: None	Pressure: 55-65 psi	Bleedout: Vertical and Stepped Edge	Cure: 1 hr/350°F	Postcure: 2 hr/300°F 2.5 hr/400°F	Plies: 8
Physical Properties	Weight Percent Resin: 35.6	Avg. Specific Gravity: 1.86	Avg. Percent Voids: 0.9	Avg. Thickness: 0.010 inch/ply			
Test Methods	Tension: MIL-HDBK-17	Compression: MIL-HDBK-17	Shear: Picture Frame	Flexure: ASTM-D790	Bearing:	Interlaminar Shear:	

Temperature		-65°F				75°F				160°F				400°F	
Condition		Dry		Wet		Dry		Wet		Dry		Wet		Dry	
		Avg	SD	Avg	SD	Avg	SD	Avg	SD	Avg	SD	Avg	SD	Avg	SD
Tension															
ultimate stress, ksi	0°	83.9	2.85	73.0	2.89			55.5	2.57	61.9	2.24	45.0	1.85	39.2	3.40
	90°	68.7	4.19	63.9	1.61			48.9	2.67	51.9	3.25	37.6	0.99	32.0	1.44
ultimate strain, %	0°	3.30	0.18	2.79	0.02			2.12	0.14	2.61	0.08	1.59	0.07	1.45	0.13
	90°	2.80	0.18	2.41	0.05			1.95	0.09	2.18	0.19	1.50	0.05	1.35	0.08
proportional limit, ksi	0°														
	90°														
initial modulus, 10^6 psi	0°	3.84		3.81				3.58		3.25		3.35		2.96	
	90°	3.67		3.81				3.30		3.13		3.18		2.51	
secondary modulus, 10^6 psi	0°	2.81		2.75				3.04		2.49		3.04		2.74	
	90°	2.65		2.67				2.72		2.39		2.70		2.22	
Compression															
ultimate stress, ksi	0°	76.2	5.88	68.8	4.36			55.1	2.63	54.7	5.49	46.0	5.66	31.0	8.08
	90°	56.0	4.56	52.9	6.32			47.0	6.78	36.9	1.47	35.3	3.30	23.2	3.26
ultimate strain, %	0°	2.13	0.28	1.64	0.23			1.36	0.32	1.90	0.56	1.32	2.41	1.02	0.23
	90°	1.75	0.48	1.58	0.57			2.00	0.89	1.29	0.09	1.27	2.40	0.91	0.14
proportional limit, ksi	0°	28.0		24.0				24.0		32.0		22.0		17.0	
	90°	18.0		17.0				16.0		28.0		17.0		12.0	
initial modulus, 10^6 psi	0°	4.10		4.50				3.87		3.45		3.36		2.87	
	90°	4.00		4.10				3.64		2.87		2.88		2.63	
Shear															
ultimate stress, ksi	0°-90°	19.6	1.04					15.0	0.70	12.7	0.62				
	±45°														

		-65°F Dry			75°F Dry			160° Dry		
		Avg	Max	Min	Avg	Max	Min	Avg	Max	Min
Flexure										
ultimate stress, ksi	0°				86.31	92.16	79.07			
proportional limit, ksi	0°									
initial modulus, 10^6 psi	0°									
Bearing										
ultimate stress, ksi	0°									
stress at 4% elong., ksi	0°									
Interlaminar Shear										
ultimate stress, ksi	0°				5.56	5.65	5.50			

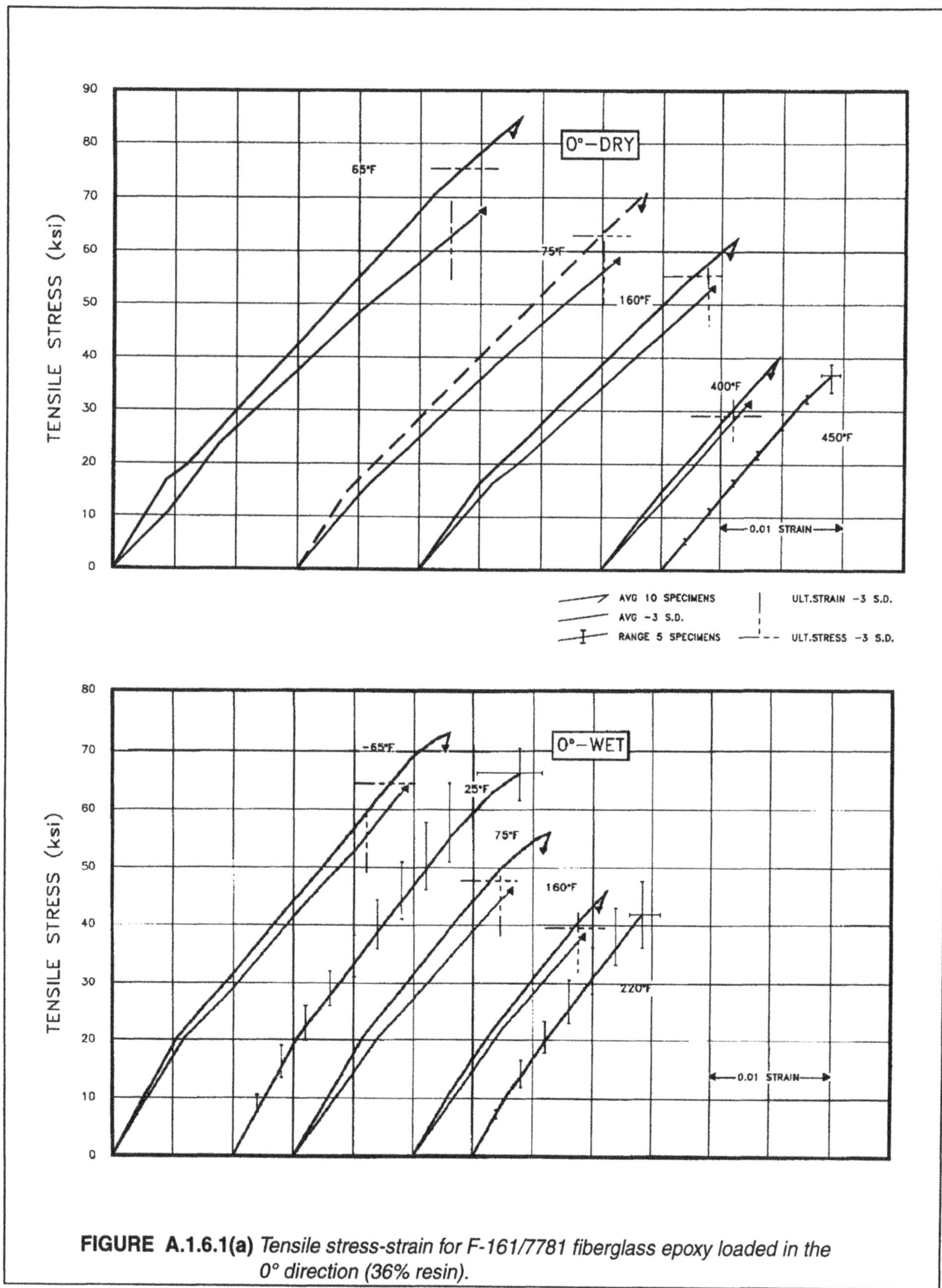

FIGURE A.1.6.1(a) *Tensile stress-strain for F-161/7781 fiberglass epoxy loaded in the 0° direction (36% resin).*

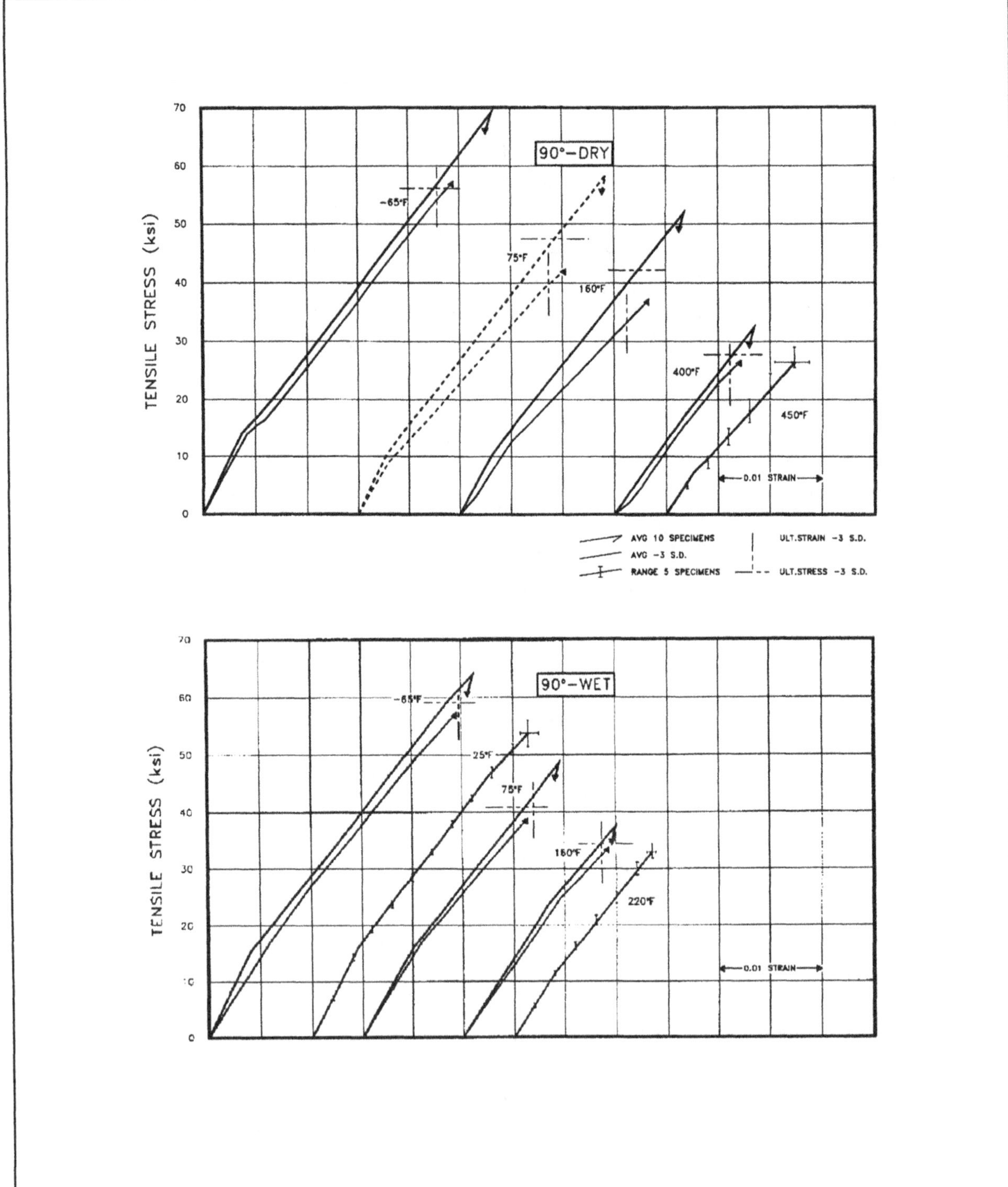

FIGURE A1.6.1(b) *Tensile stress-strain for F-161/7781 fiberglass epoxy loaded in the 90° direction (36% resin).*

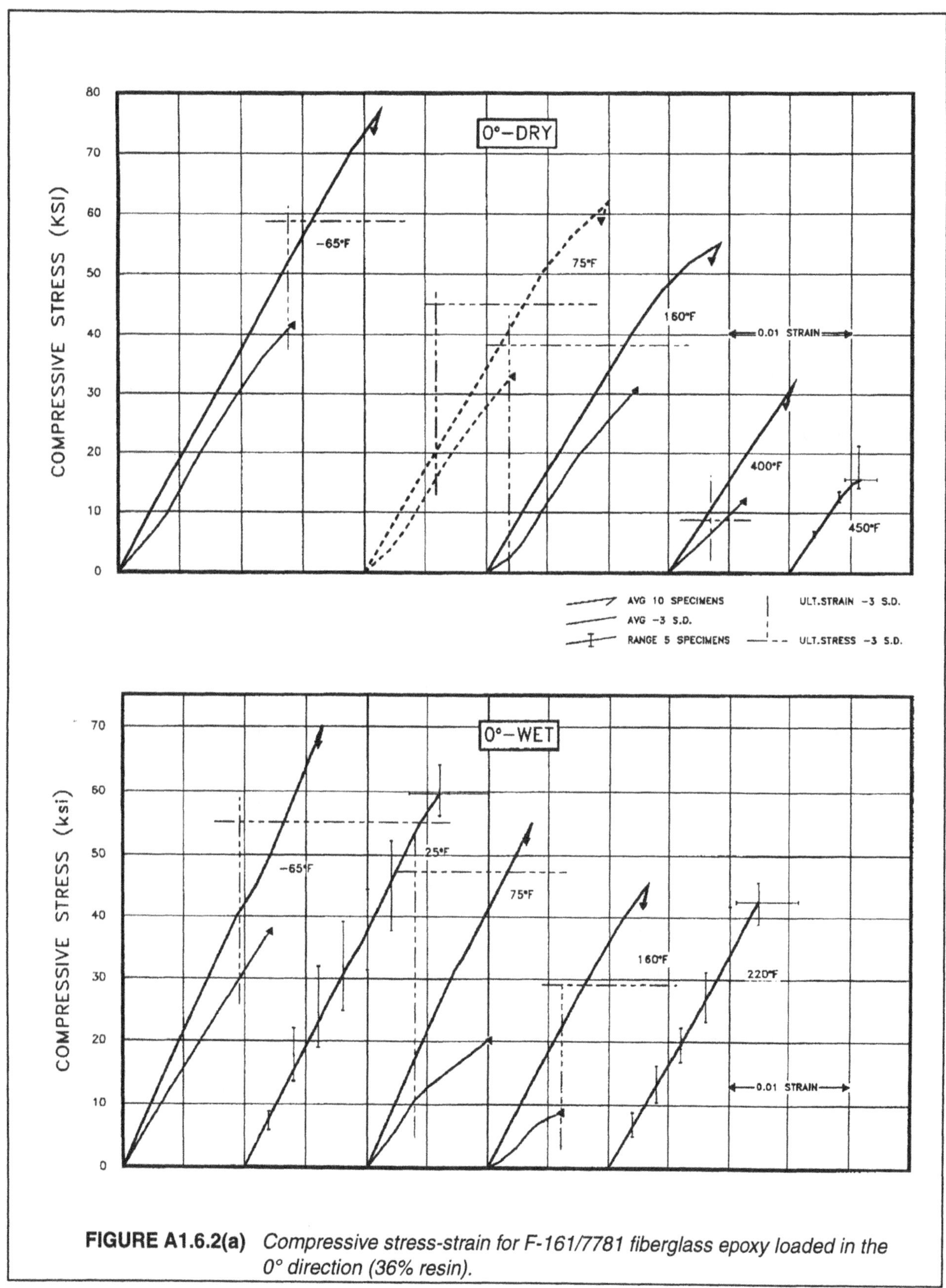

FIGURE A1.6.2(a) *Compressive stress-strain for F-161/7781 fiberglass epoxy loaded in the 0° direction (36% resin).*

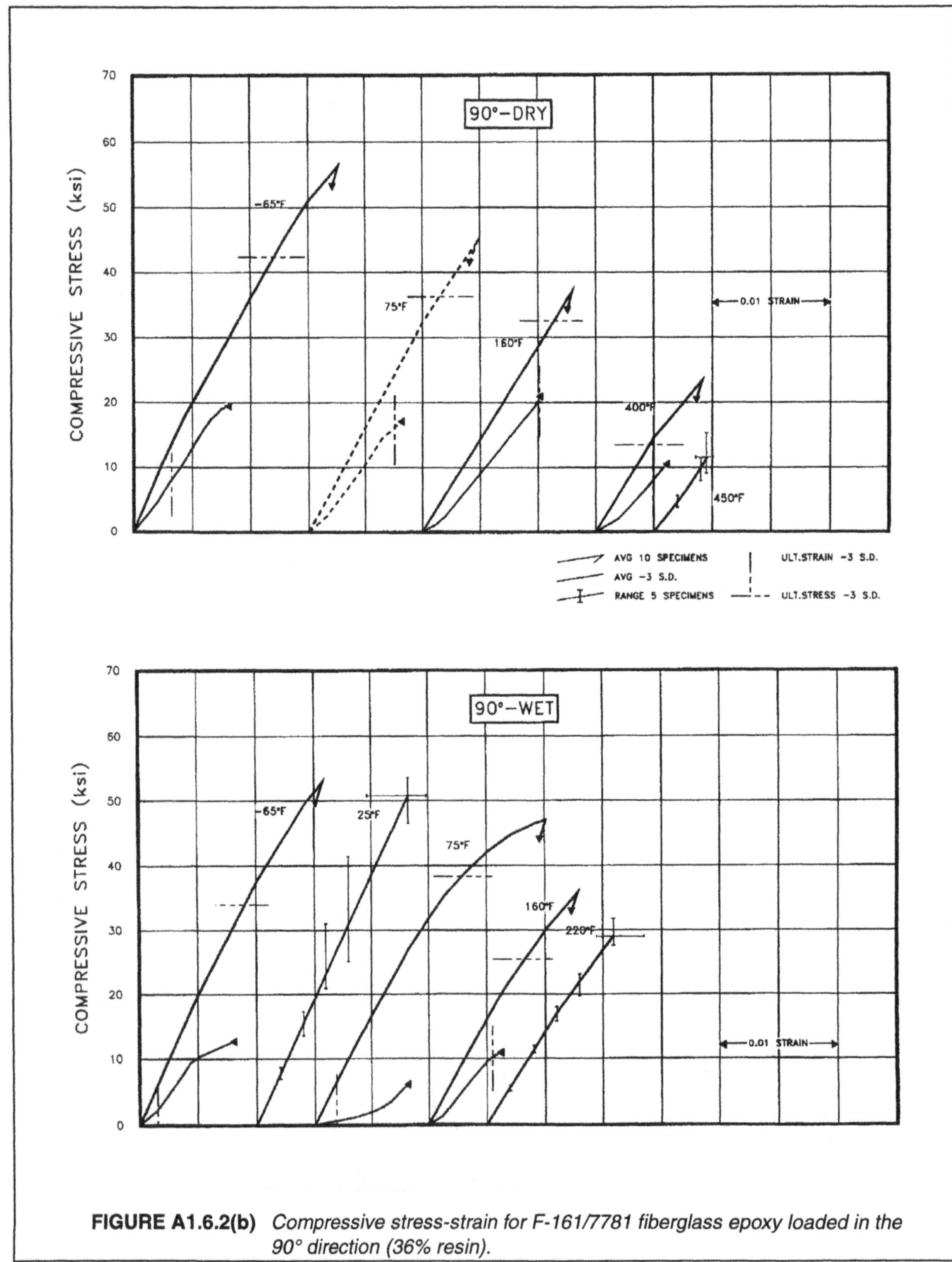

FIGURE A1.6.2(b) *Compressive stress-strain for F-161/7781 fiberglass epoxy loaded in the 90° direction (36% resin).*

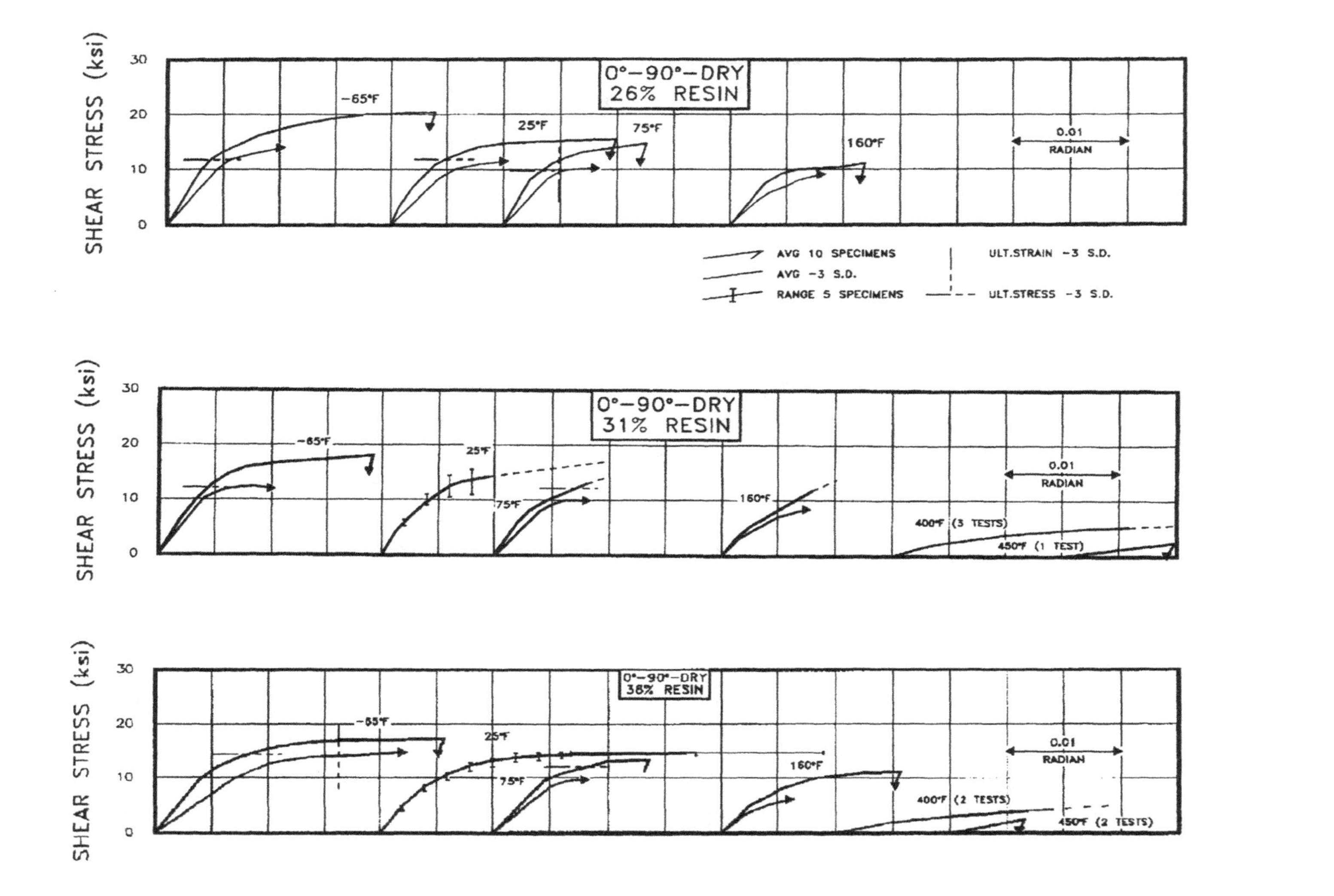

FIGURE A1.6.3 *Picture frame shear for F-161/7781 fiberglass epoxy (26%, 31%, 36% resin).*

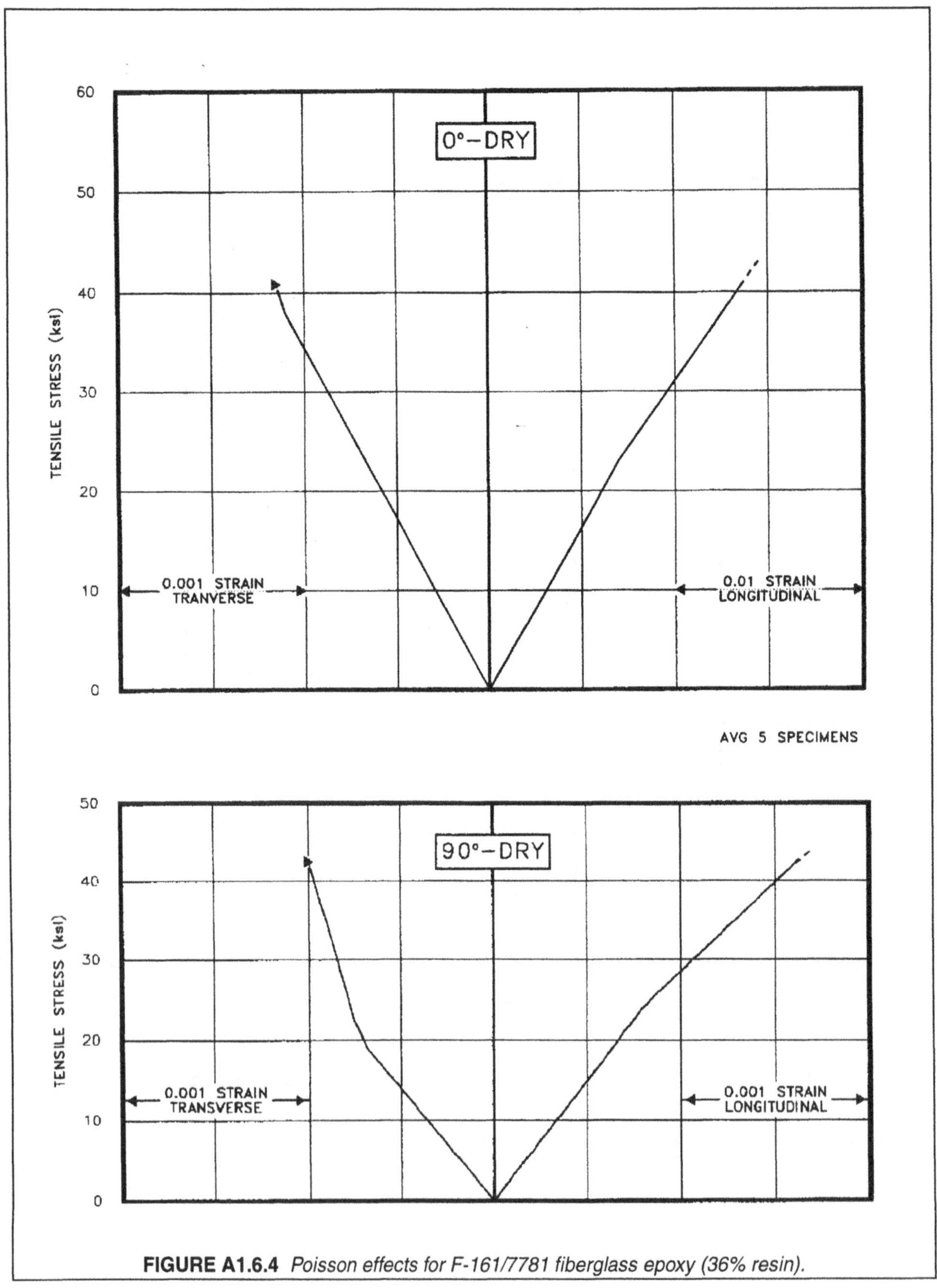

FIGURE A1.6.4 *Poisson effects for F-161/7781 fiberglass epoxy (36% resin).*

APPENDIX A1

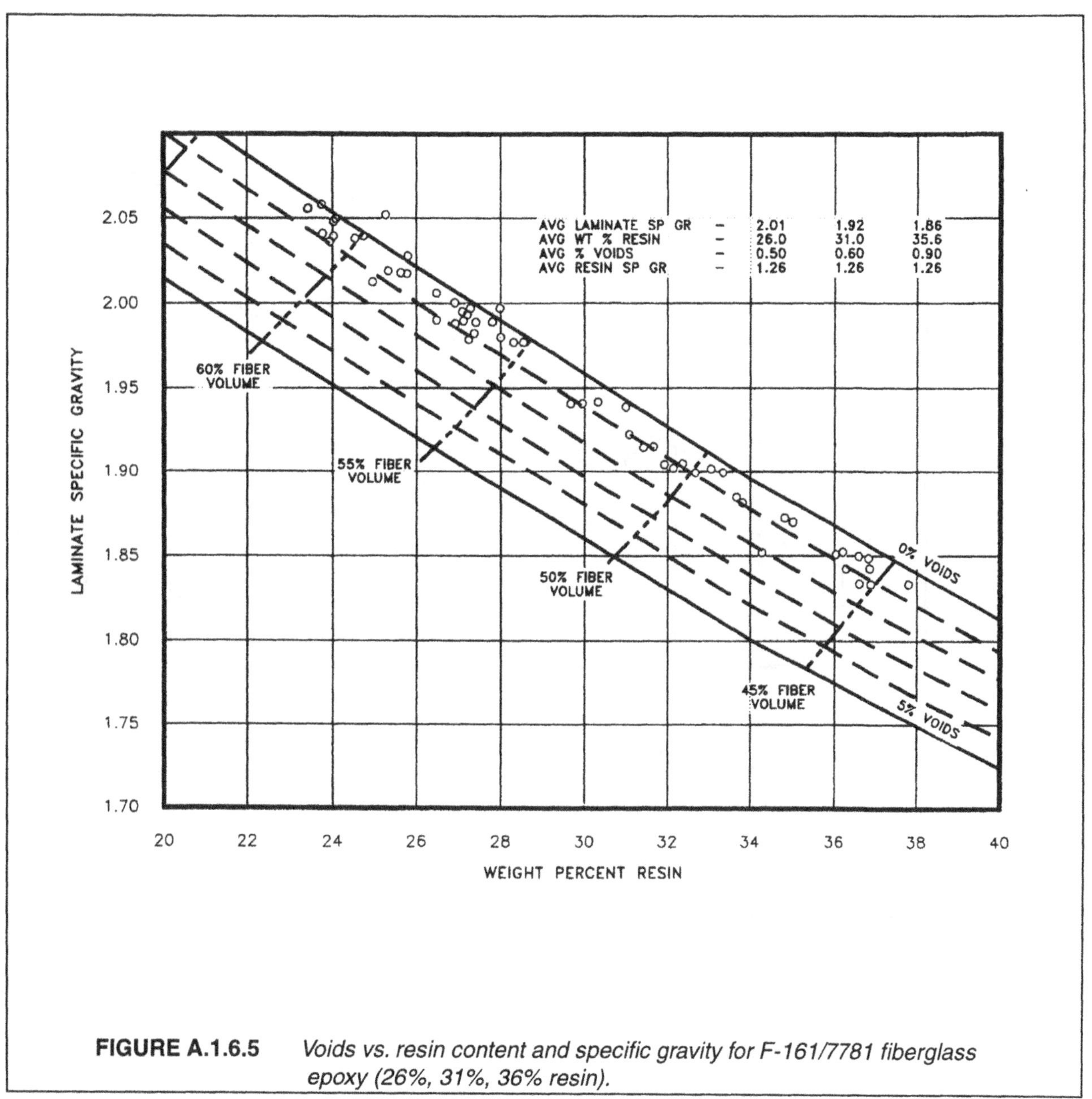

FIGURE A.1.6.5 *Voids vs. resin content and specific gravity for F-161/7781 fiberglass epoxy (26%, 31%, 36% resin).*

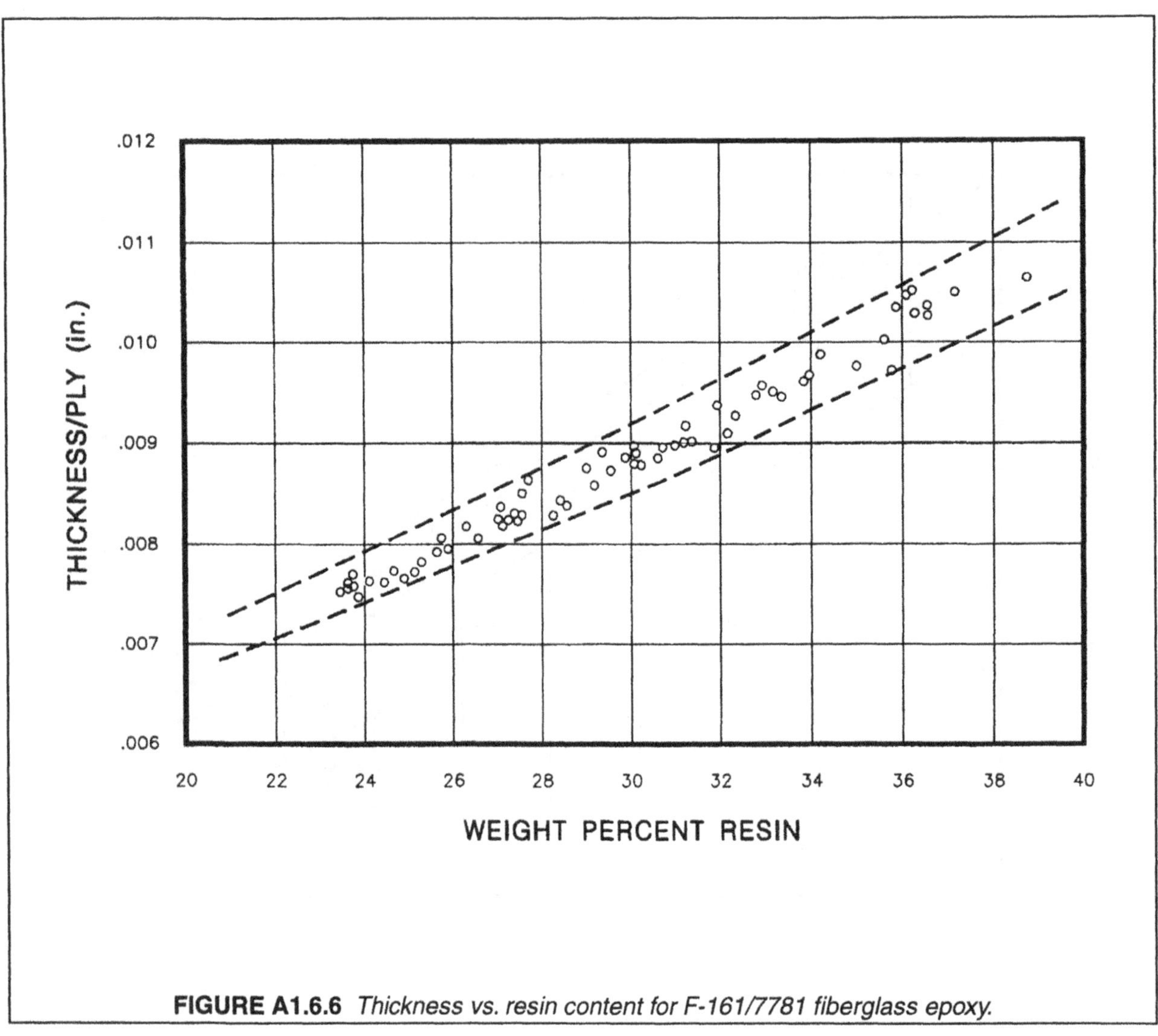

FIGURE A1.6.6 *Thickness vs. resin content for F-161/7781 fiberglass epoxy.*

APPENDIX A1

TABLE A1.8 *Summary of Mechanical Properties of Narmco N588/7781 (ECDE-1/0-550) Fiberglass Epoxy*

Fabrication		Lay-up: Balanced		Vacuum: None		Pressure: 45-55 psi		Bleedout: Vertical		Cure: Stepwise to 350°F; 1hr/350°F		Postcure: None		Plies: 8	
Physical Properties		Weight Percent Resin: 32.8 $v_f = 0.51$			Avg. Specific Gravity: 1.91				Avg. Percent Voids: 1.0			Avg. Thickness: 0.075 inches			
Test Methods		Tension: ASTM-D638 TYPE 1			Compression: MIL-HDBK-17		Shear: Rail		Flexure: ASTM-D790		Bearing: ASTM-D953			Interlaminar Shear: Short Beam	
Temperature		-65°F				75°F				160°F				400°F	
Condition		Dry		Wet		Dry		Wet		Dry		Wet		Dry	
		Avg	SD	Avg	SD	Avg	SD	Avg	SD	Avg	SD	Avg	SD	Avg	SD
Tension															
ultimate stress, ksi	0°	71.4	2.4	63.8	3.3	58.4	2.1	50.0	2.3	48.8	3.0	35.0	2.0	40.4	3.4
	90°	59.3	3.3	50.6	2.4	47.2	3.8	41.1	2.7	41.4	2.0	28.9	2.8	33.3	3.8
ultimate strain, %	0°	2.41	0.09	2.06	0.15	2.05	0.18	1.61	0.12	1.59	0.15	1.13	0.07	1.26	0.07
	90°	2.35	0.17	1.96	0.12	1.81	0.16	1.55	0.16	1.67	0.10	1.17	0.14	1.25	0.12
proportional limit, ksi	0°	26.6	1.7	28.7	2.5	23.3	1.1	25.4	2.8	21.0	1.7	29.9	2.0	24.3	
	90°	19.3	0.8	19.2	1.6	17.6	0.8	18.1	1.4	17.3	2.5	20.9	1.3	14.3	
initial modulus, 10^6 psi	0°	3.64		3.85		3.71		3.57		3.58		3.10		3.13	0.17
	90°	3.41		3.37		3.56		3.23		2.92		2.63		2.80	0.23
secondary modulus, 10^6 psi	0°														
	90°														
Compression															
ultimate stress, ksi	0°	99.2	5.9	87.4	5.8	74.0	3.6	63.5	3.2	59.0	2.4	49.5	1.9		
	90°	83.4	3.5	71.8	4.1	62.9	2.9	53.7	1.7	50.9	1.5	40.7	1.8		
ultimate strain, %	0°	2.52	0.26	2.30	0.25	1.89	0.15	1.65	0.19	1.60	0.12	1.38	0.06		
	90°	2.30	0.27	2.06	0.20	1.87	0.14	1.58	0.15	1.63	0.16	1.29	0.08		
proportional limit, ksi	0°	42.7	2.6	46.2	2.5	44.5	3.2	39.8	3.6	37.6	2.7	30.7	2.7		
	90°	40.8	3.8	42.4	2.7	35.3	3.7	34.4	2.3	31.2	2.4	24.4	1.6		
initial modulus, 10^6 psi	0°	4.32		4.15		4.18		4.11		3.88		3.70			
	90°	4.08		3.83		3.68		3.72		3.41		3.41			
Shear															
ultimate stress, ksi	0°-90°	22.6				16.0	1.05			13.8					
	±45°														

		-65°F Dry			75°F Dry			160° Dry		
		Avg	Max	Min	Avg	Max	Min	Avg	Max	Min
Flexure										
ultimate stress, ksi	0°	105.0	115.6	95.6	90.4	102.6	84.5	79.3	87.8	74.0
proportional limit, ksi	0°	69.6	75.9	59.0	68.9	72.4	64.6	64.8	72.2	57.2
initial modulus, 10^6 psi	0°	3.48	3.62	3.42	3.36	3.60	3.20	3.19	3.27	3.09
Bearing										
ultimate stress, ksi	0°	84.6	92.5	77.9	68.4	71.3	66.0	48.4	53.6	44.2
stress at 4% elong., ksi	0°	29.3	30.9	26.5	26.2	27.4	25.3	21.8	22.8	20.6
Interlaminar Shear										
ultimate stress, ksi	0°	8.84	9.16	8.56	8.35	8.56	8.05	7.39	7.72	6.47

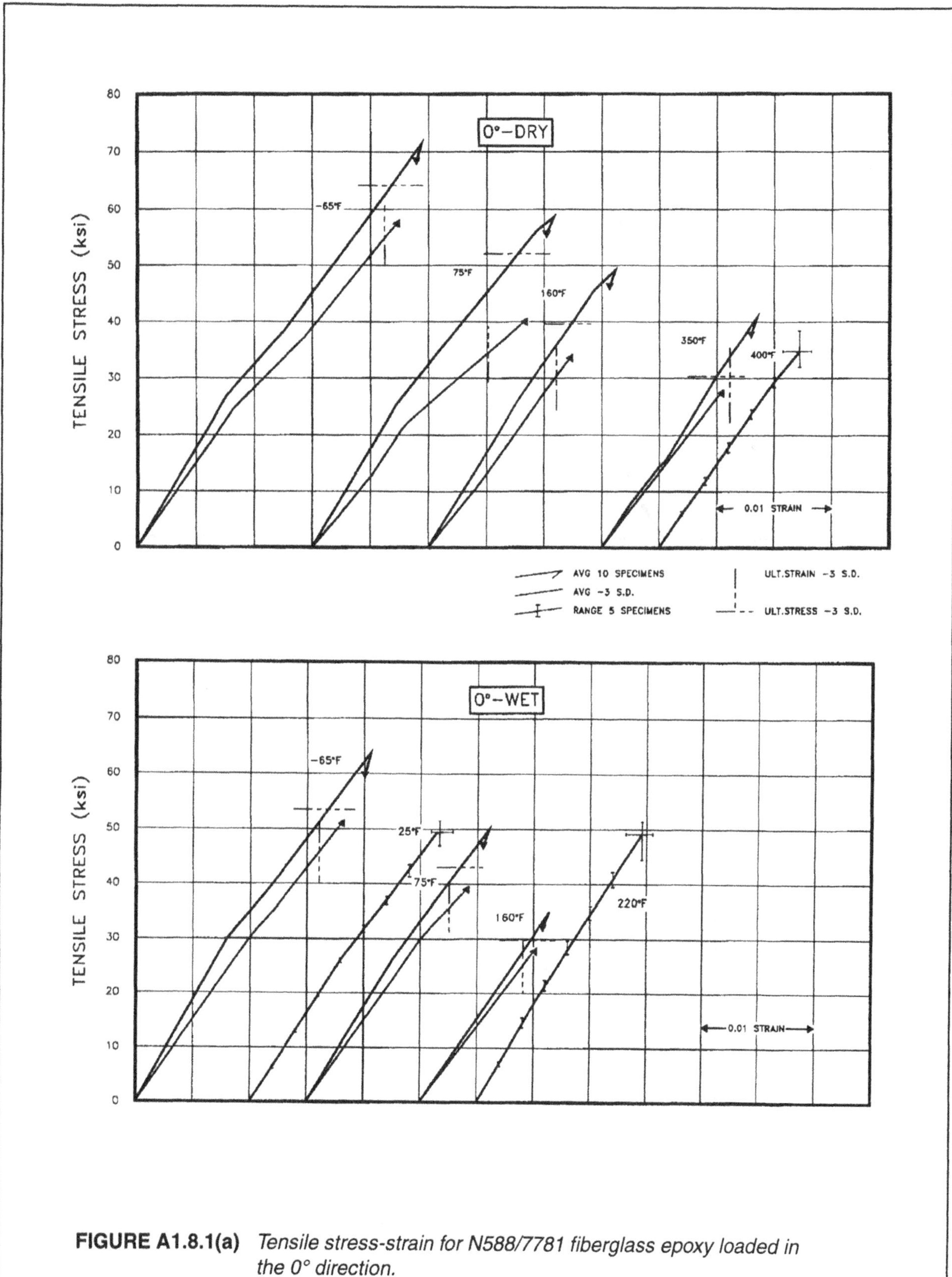

FIGURE A1.8.1(a) *Tensile stress-strain for N588/7781 fiberglass epoxy loaded in the 0° direction.*

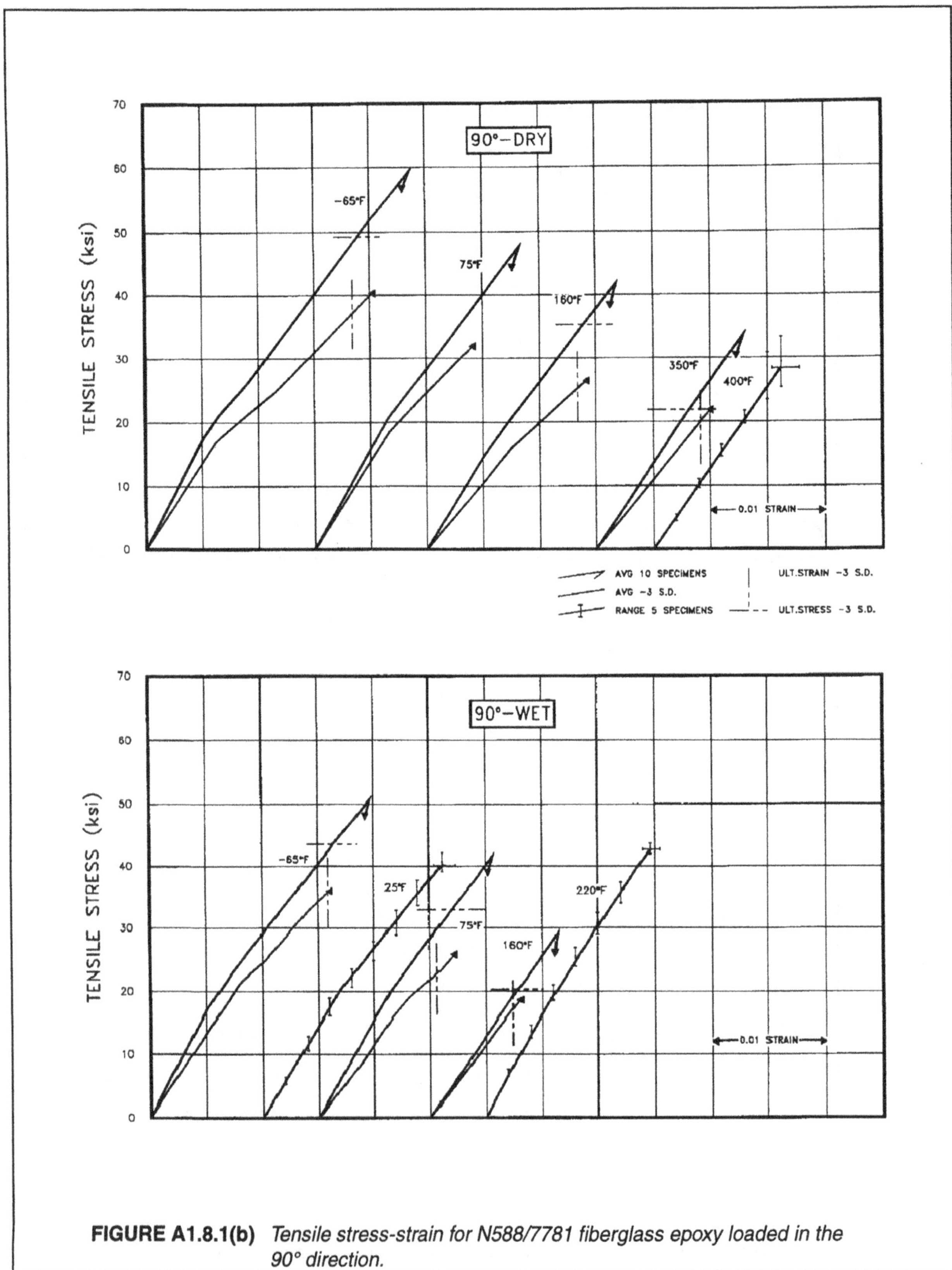

FIGURE A1.8.1(b) *Tensile stress-strain for N588/7781 fiberglass epoxy loaded in the 90° direction.*

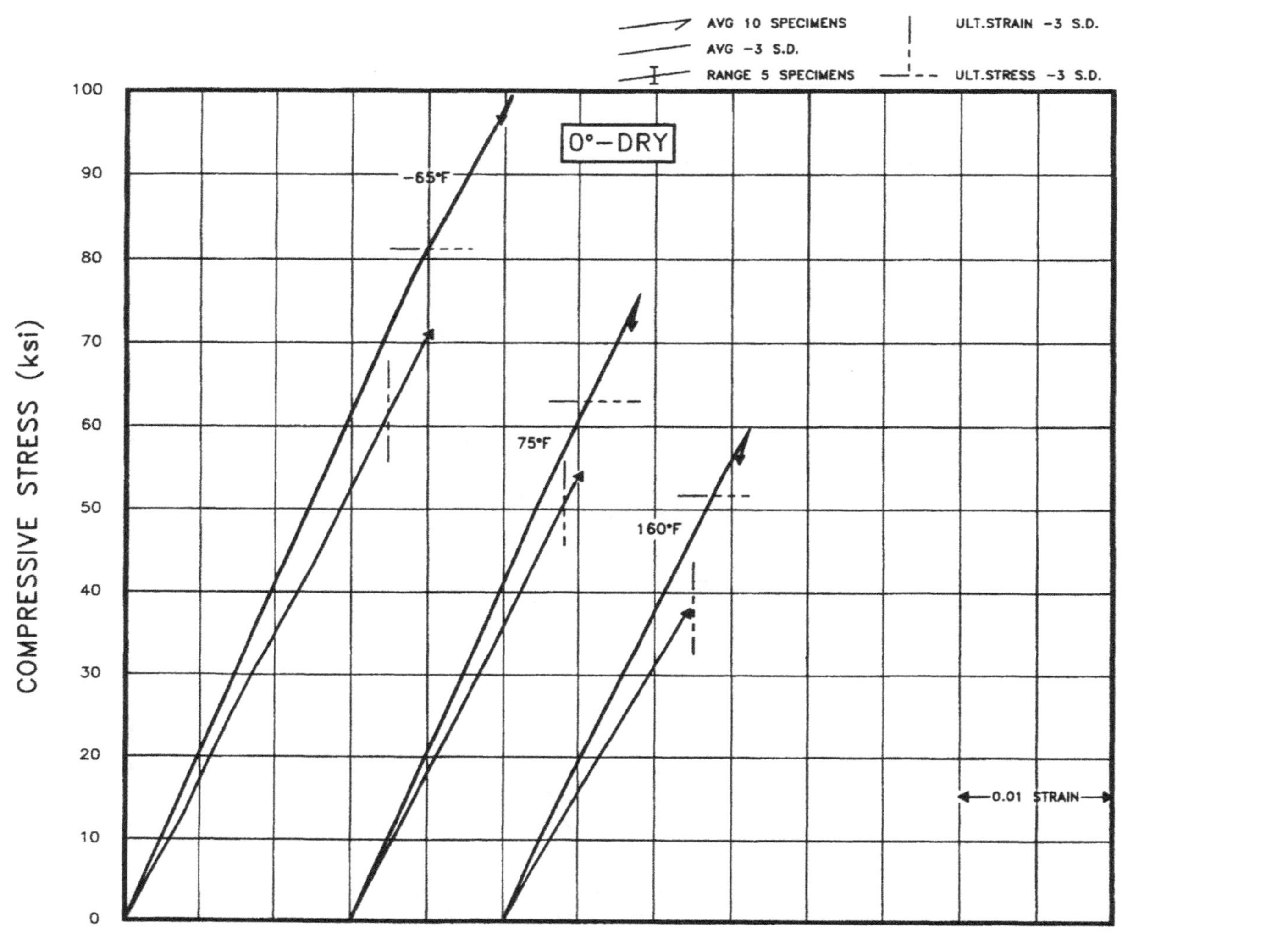

FIGURE A1.8.2(a) *Compressive stress-strain for N588/7781 fiberglass epoxy loaded in the 0° direction*, continued on next page.

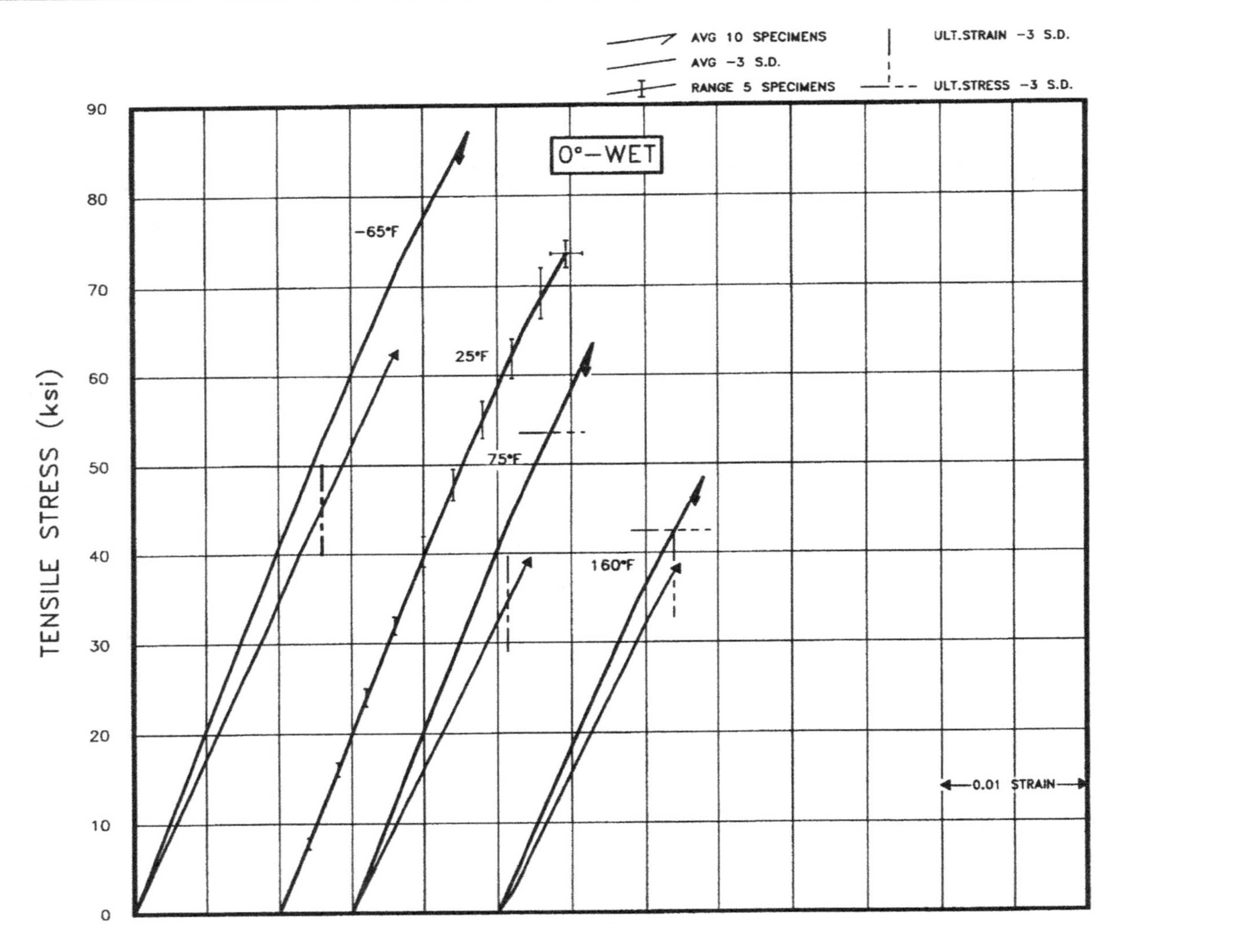

FIGURE A1.8.2(a) *Compressive stress-strain for N588/7781 fiberglass epoxy loaded in the 0° direction*, concluded.

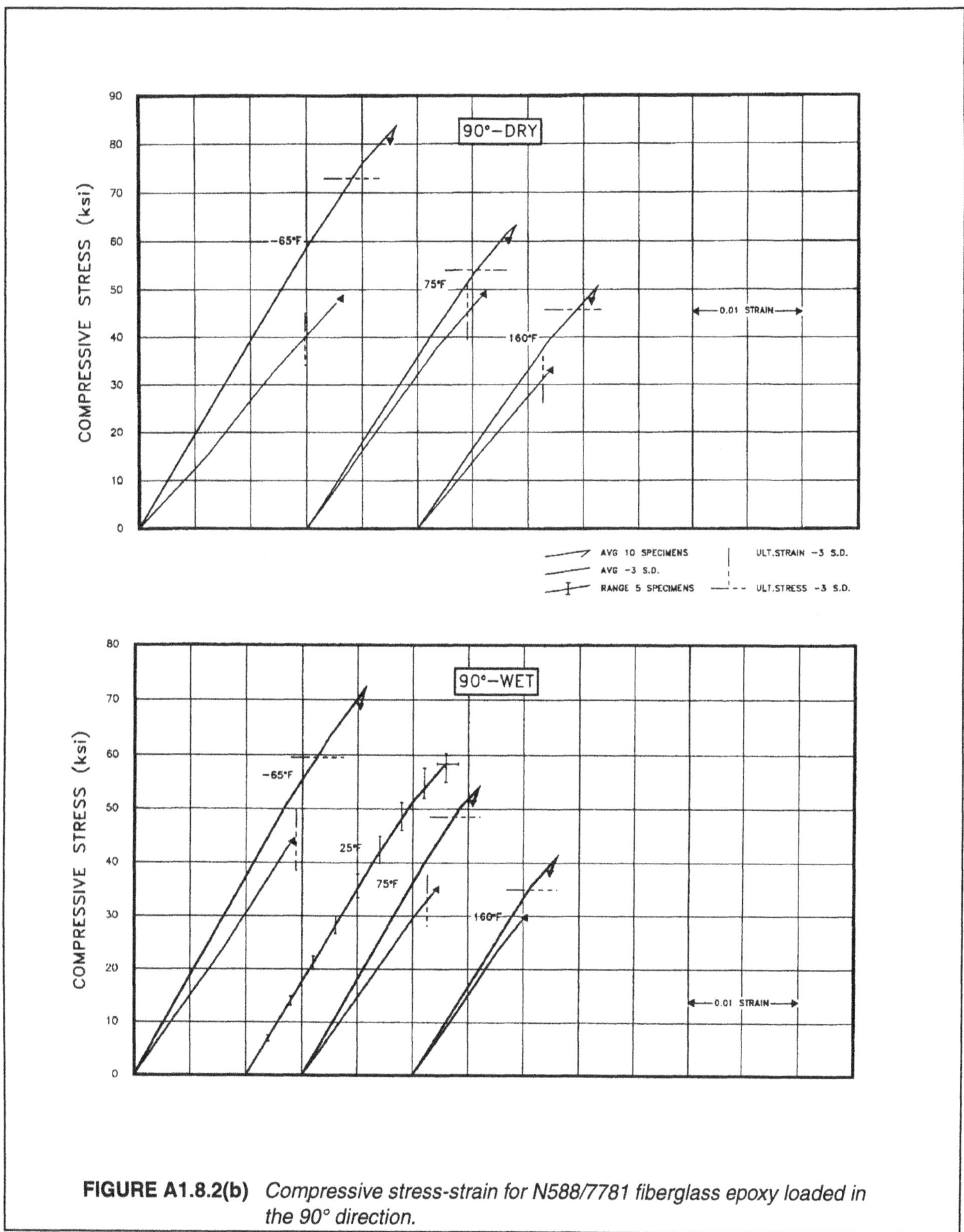

FIGURE A1.8.2(b) *Compressive stress-strain for N588/7781 fiberglass epoxy loaded in the 90° direction.*

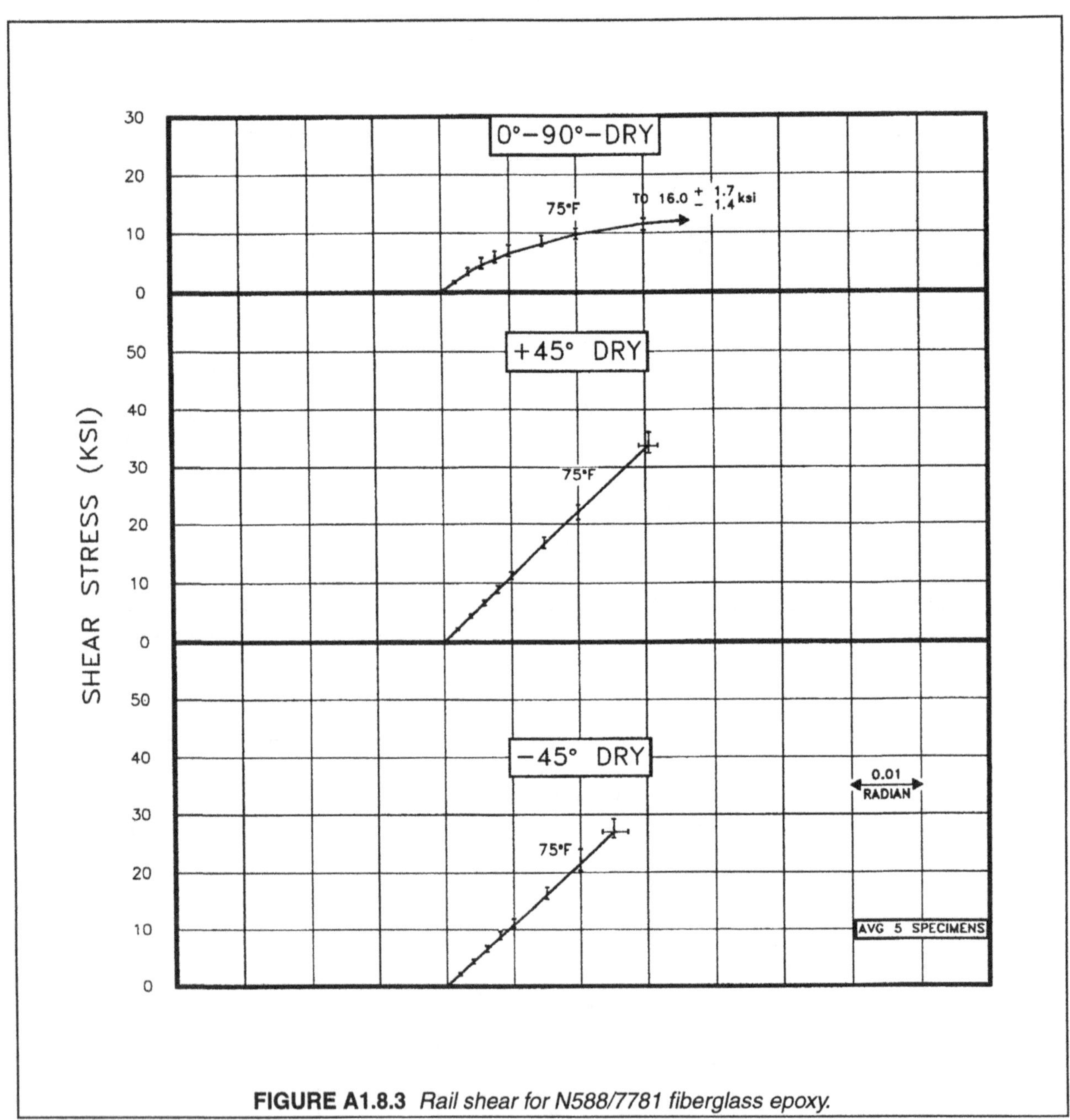

FIGURE A1.8.3 *Rail shear for N588/7781 fiberglass epoxy.*

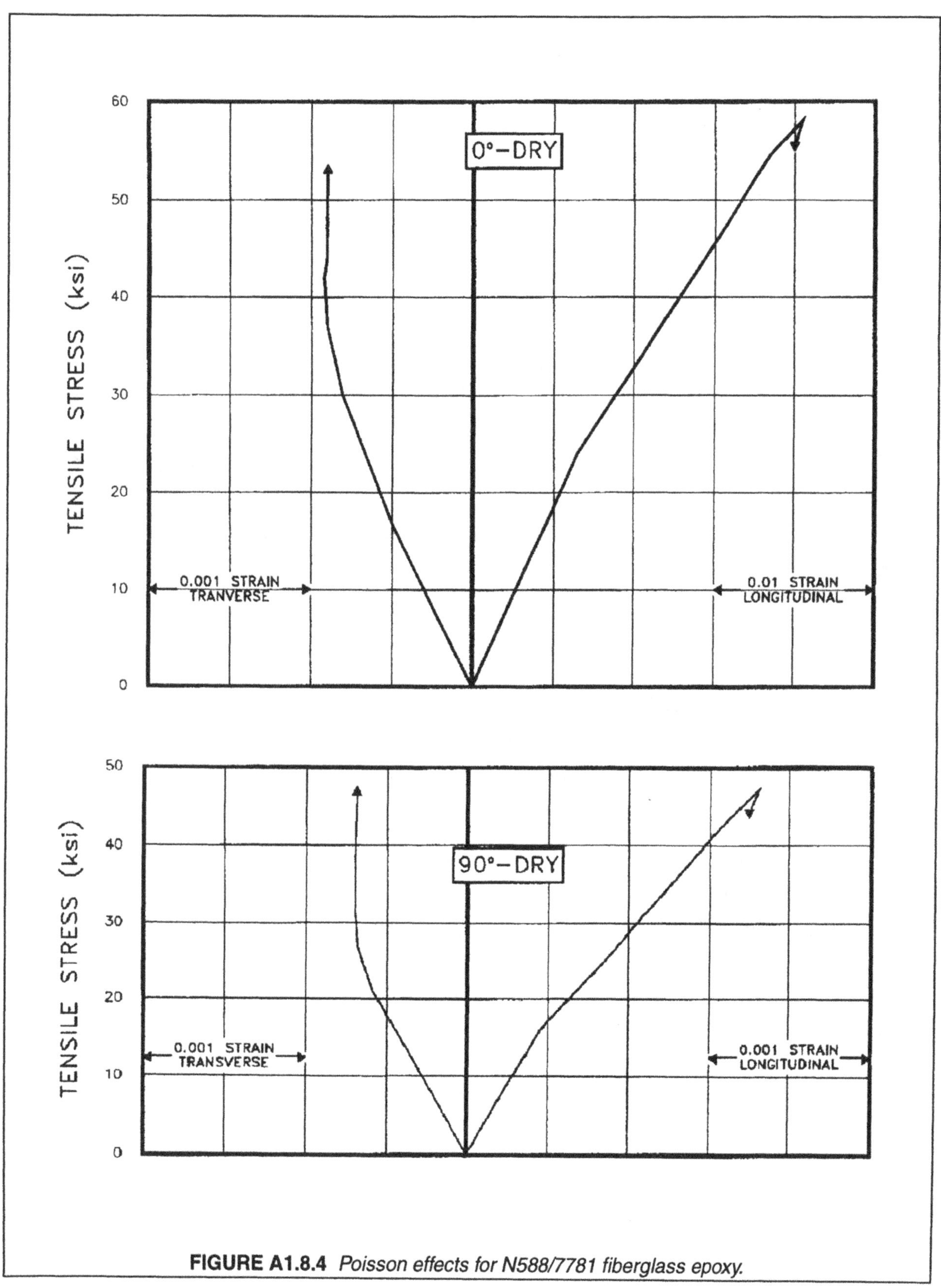

FIGURE A1.8.4 *Poisson effects for N588/7781 fiberglass epoxy.*

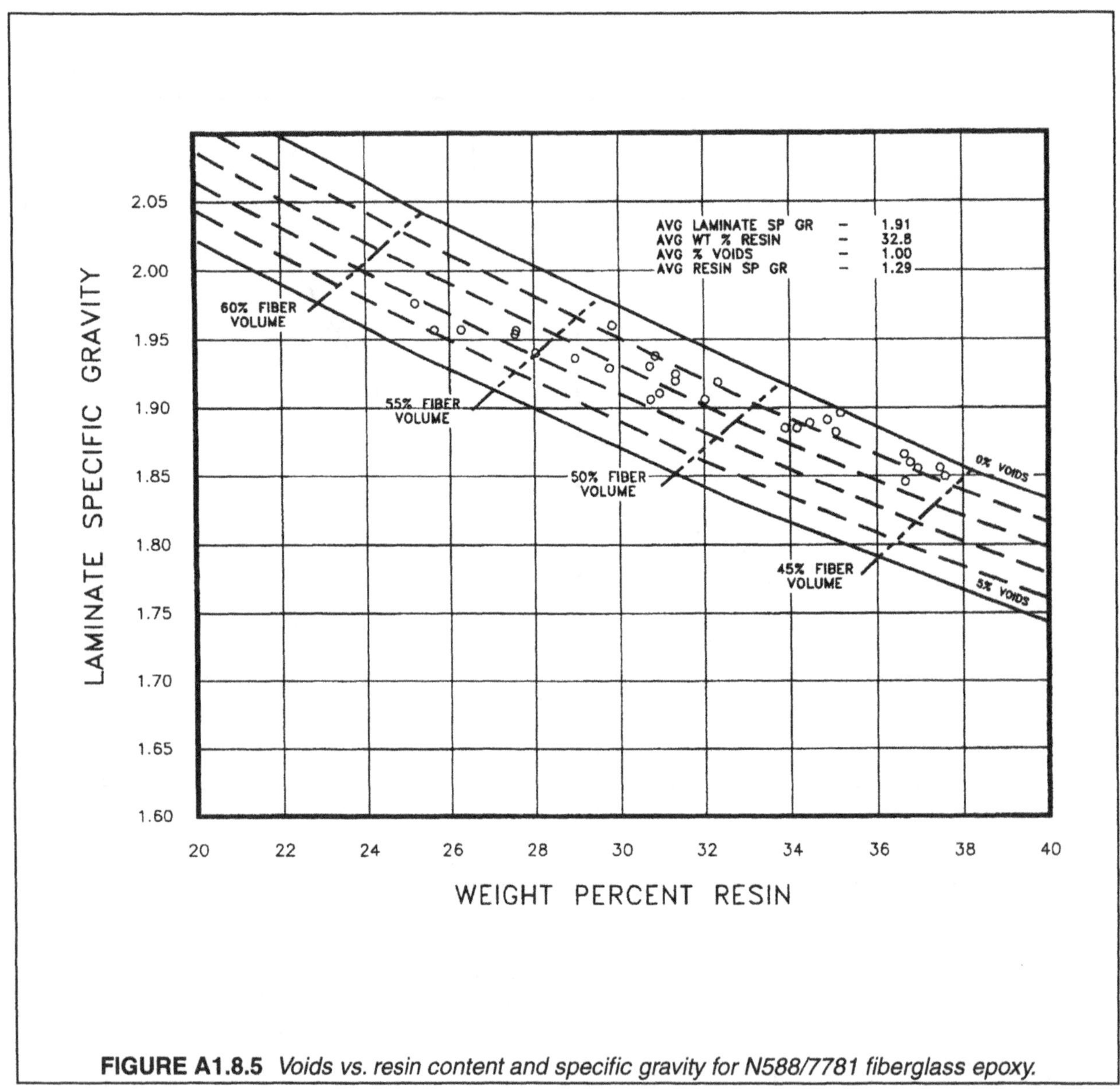

FIGURE A1.8.5 *Voids vs. resin content and specific gravity for N588/7781 fiberglass epoxy.*

APPENDIX A1

TABLE A1.40 *Summary of Mechanical Properties of Narmco N506/7781 (ECDE-1/0-A1100) Fiberglass Phenolic.*

Fabrication	Lay-up: Balanced	Vacuum:	Pressure:	Bleedout: Vertical	Cure:	Postcure:	Plies: 8
Physical Properties	Weight Percent Resin: 25.3 - 32.3	Avg. Specific Gravity: 1.72 - 1.85	Avg. Percent Voids: Figure 4.40.5	Avg. Thickness: 0.071 - 0.095 inches			
Test Methods	Tension: ASTM-D638 TYPE 1	Compression: MIL-HDBK-17	Shear: Rail	Flexure: ASTM-D790	Bearing: ASTM-D953	Interlaminar Shear: Short Beam	

Temperature		-65°F				75°F				160°F				400°F	
Condition		Dry		Wet		Dry		Wet		Dry		Wet		Dry	
		Avg	SD	Avg	SD	Avg	SD	Avg	SD	Avg	SD	Avg	SD	Avg	SD
Tension															
ultimate stress, ksi	0°	48.1	2.4	49.8	3.3	38.9	1.5	37.2	1.8	35.3	1.4	30.6	3.0	21.6	1.6
	90°	37.9	1.8	40.0	2.7	31.5	1.5	32.1	1.4	27.9	1.7	26.2	2.2	21.6	1.7
ultimate strain, %	0°	1.76	0.07	1.76	0.13	1.33	0.14	1.34	0.13	1.19	0.10	1.15	0.14	0.69	0.05
	90°	1.63	0.08	1.65	0.13	1.26	0.15	1.32	0.07	1.11	0.07	1.11	0.14	0.78	0.06
proportional limit, ksi	0°	13.6	0.9	18.1	1.2	13.5	0.6	17.0	1.0	13.9	1.0	14.9	0.70	9.7	1.1
	90°	9.9	0.4	12.5	0.9	9.2	0.8	12.8	0.7	10.3	0.8	11.6	0.70	8.6	0.5
initial modulus, 10^6 psi	0°	3.40	0.21	3.35	0.20	3.94	0.69	3.14	0.26	3.74	0.41	3.01	0.19	3.57	0.24
	90°	3.08	0.29	3.04	0.22	3.54	0.41	2.81	0.24	3.33	0.37	2.78	0.21	3.18	0.30
secondary modulus, 10^6 psi	0°														
	90°														
Compression															
ultimate stress, ksi	0°	66.7	6.2	65.9	5.0	59.7	4.7	54.5	7.1	50.6	2.3	49.2	4.2		
	90°	57.7	5.8	56.2	5.8	49.0	4.6	48.7	4.0	43.0	4.3	42.9	3.7		
ultimate strain, %	0°	1.85	0.09	1.69	0.18	1.58	0.14	1.49	0.12	1.45	0.06	1.40	0.12		
	90°	1.70	0.21	1.63	0.13	1.40	0.09	1.43	0.07	1.37	0.12	1.31	0.15		
proportional limit, ksi	0°	45.8	3.8	38.5	7.9	39.0	2.4	41.2	4.6	39.9	2.4	35.0	1.7		
	90°	35.2	3.8	34.4	5.0	32.6	4.4	35.5	3.0	32.4	3.1	31.1	3.3		
initial modulus, 10^6 psi	0°	3.90	0.19	4.17	0.29	3.95	0.28	3.89	0.26	3.68	0.21	3.67	0.12		
	90°	3.69	0.25	3.68	0.17	3.70	0.20	3.57	0.20	3.30	0.23	3.45	0.21		
Shear															
ultimate stress, ksi	0°-90°	13.8				12.3	0.97			11.4					
	±45°														

		-65°F Dry			75°F Dry			160° Dry		
		Avg	Max	Min	Avg	Max	Min	Avg	Max	Min
Flexure										
ultimate stress, ksi	0°	68.2	72.8	65.2	58.4	64.0	52.1	52.7	56.3	47.4
proportional limit, ksi	0°	59.3	66.1	54.6	48.9	56.8	42.5	42.4	46.2	38.8
initial modulus, 10^6 psi	0°	2.97	3.04	2.88	2.89	2.99	2.78	2.97	3.06	2.82
Bearing										
ultimate stress, ksi	0°	65.7	73.2	57.0	58.9	64.0	46.8	49.5	55.8	44.5
stress at 4% elong., ksi	0°	25.1	26.0	23.7	24.5	24.9	23.8	21.6	22.6	20.7
Interlaminar Shear										
ultimate stress, ksi	0°	4.83	5.10	4.29	4.64	4.92	3.94	4.62	4.88	4.08

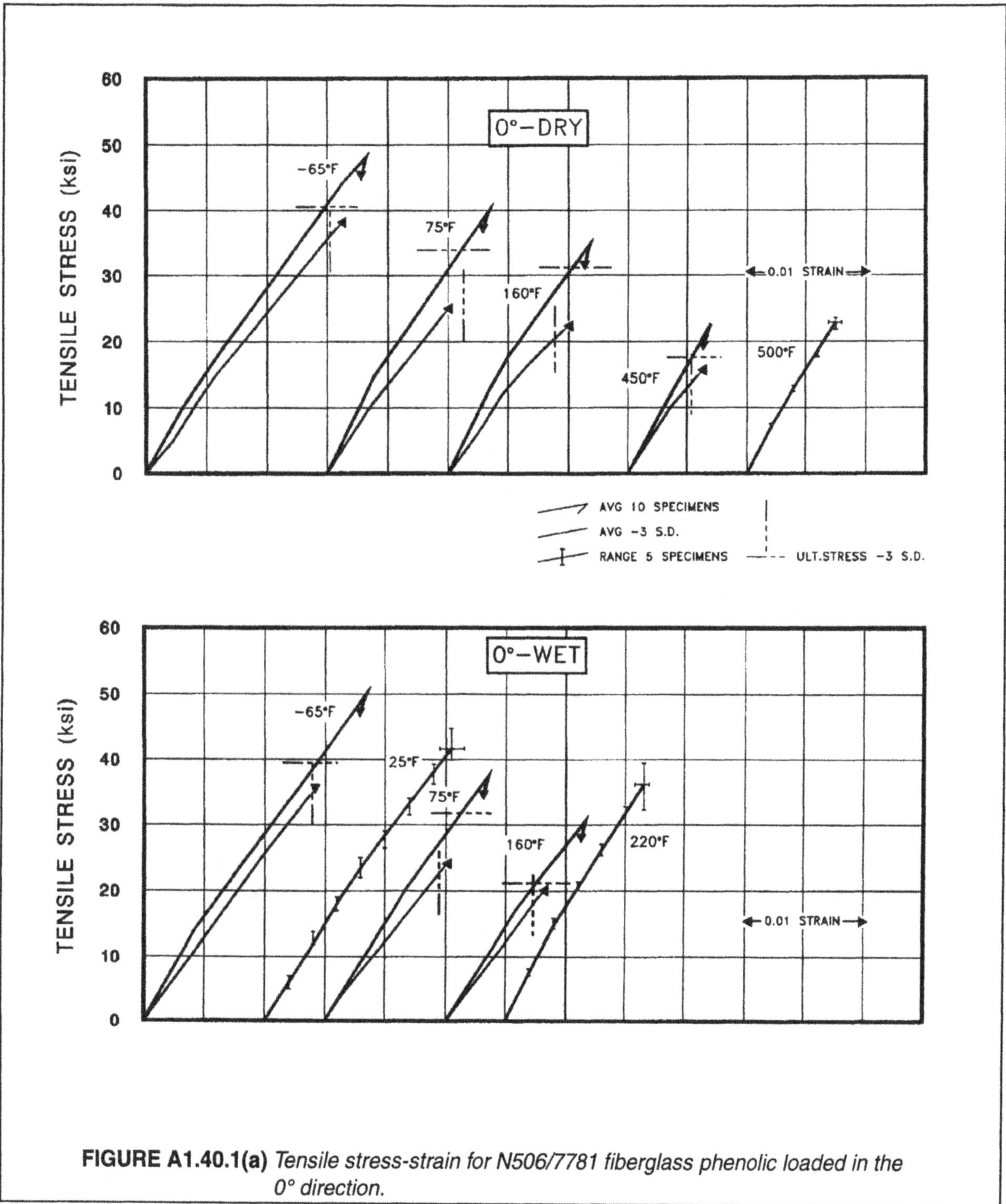

FIGURE A1.40.1(a) *Tensile stress-strain for N506/7781 fiberglass phenolic loaded in the 0° direction.*

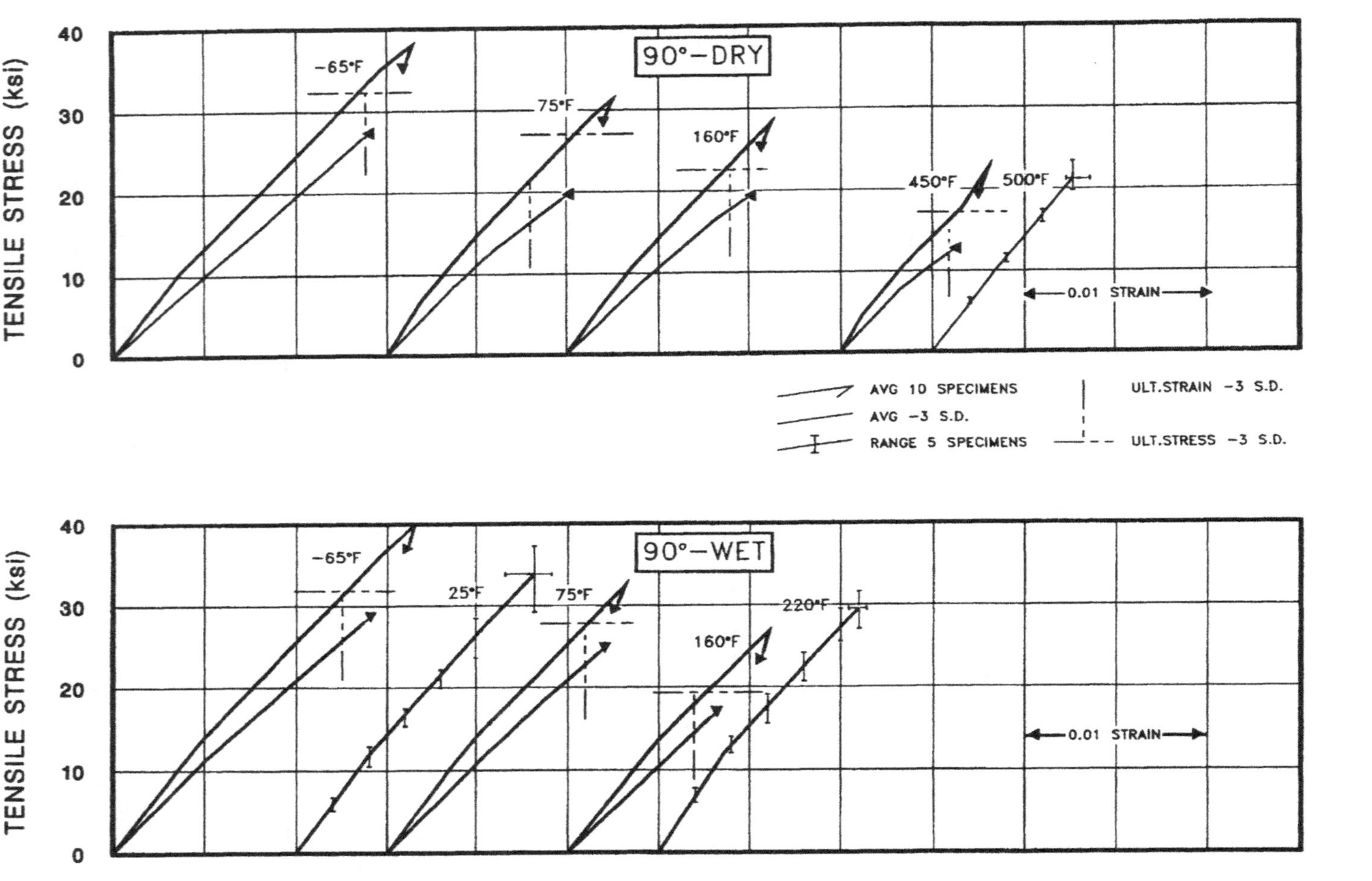

FIGURE A1.40.1(b) *Tensile stress-strain for N506/7781 fiberglass phenolic loaded in the 90° direction.*

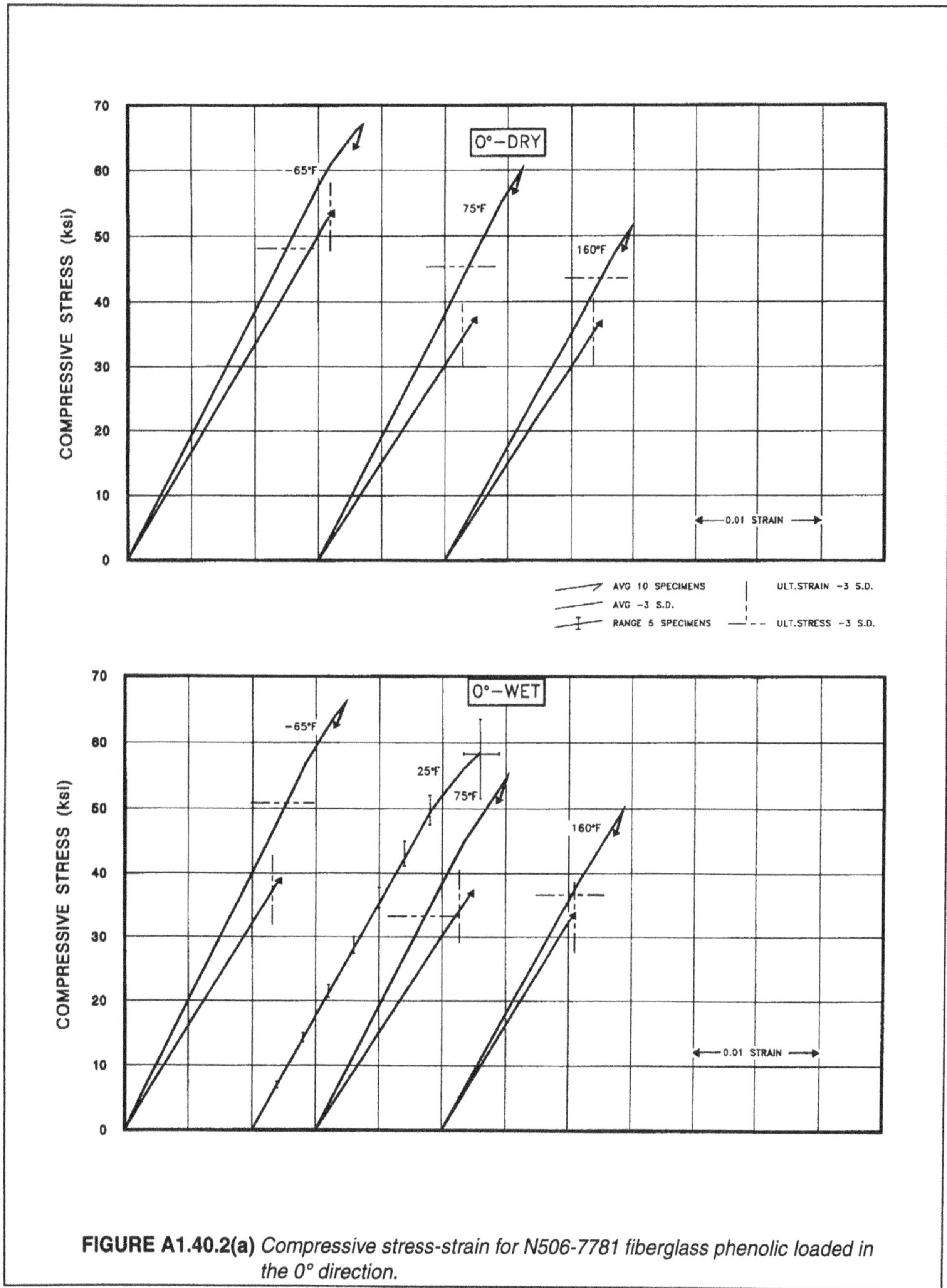

FIGURE A1.40.2(a) *Compressive stress-strain for N506-7781 fiberglass phenolic loaded in the 0° direction.*

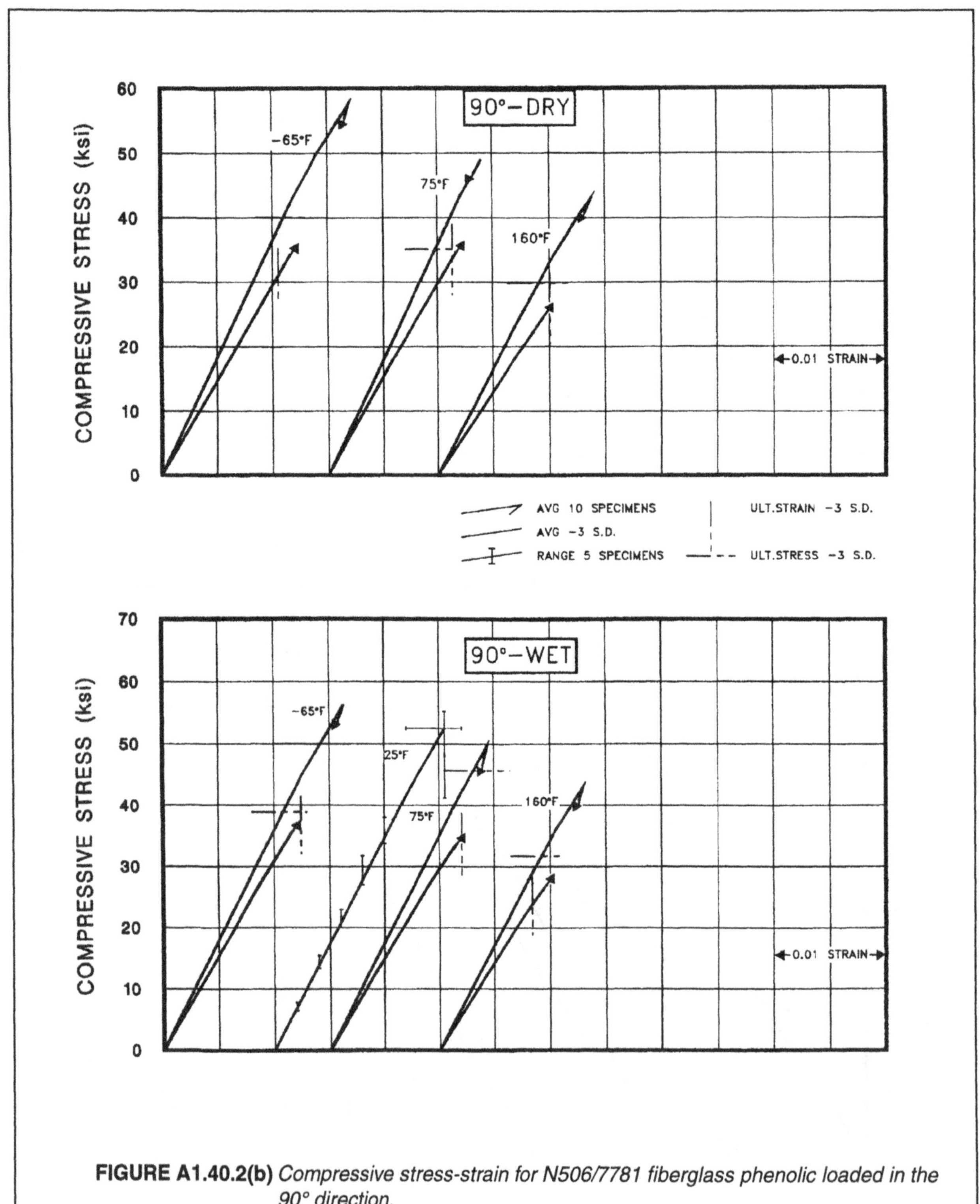

FIGURE A1.40.2(b) *Compressive stress-strain for N506/7781 fiberglass phenolic loaded in the 90° direction.*

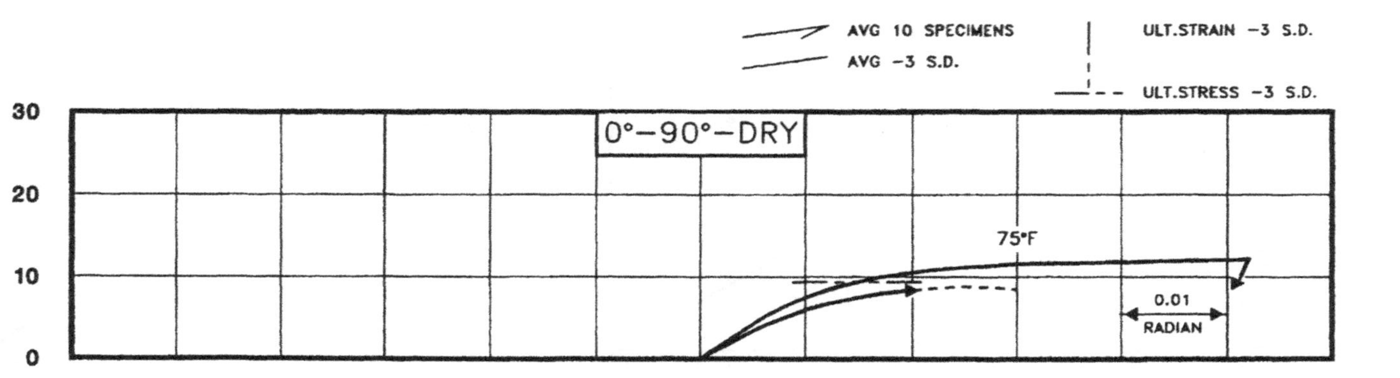

FIGURE A1.40.3 *0° - 90° rail shear for N506/7781 fiberglass phenolic.*

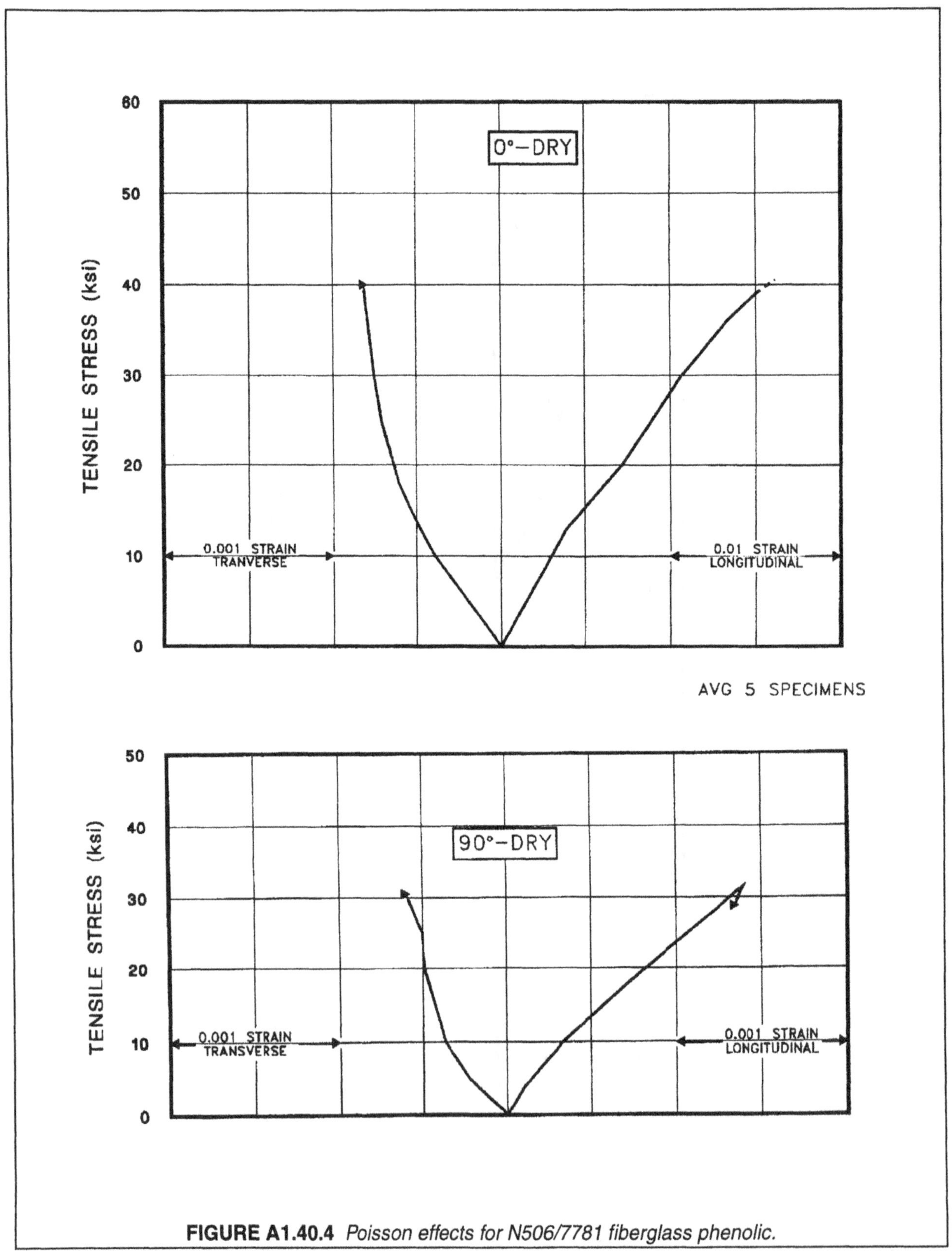

FIGURE A1.40.4 *Poisson effects for N506/7781 fiberglass phenolic.*

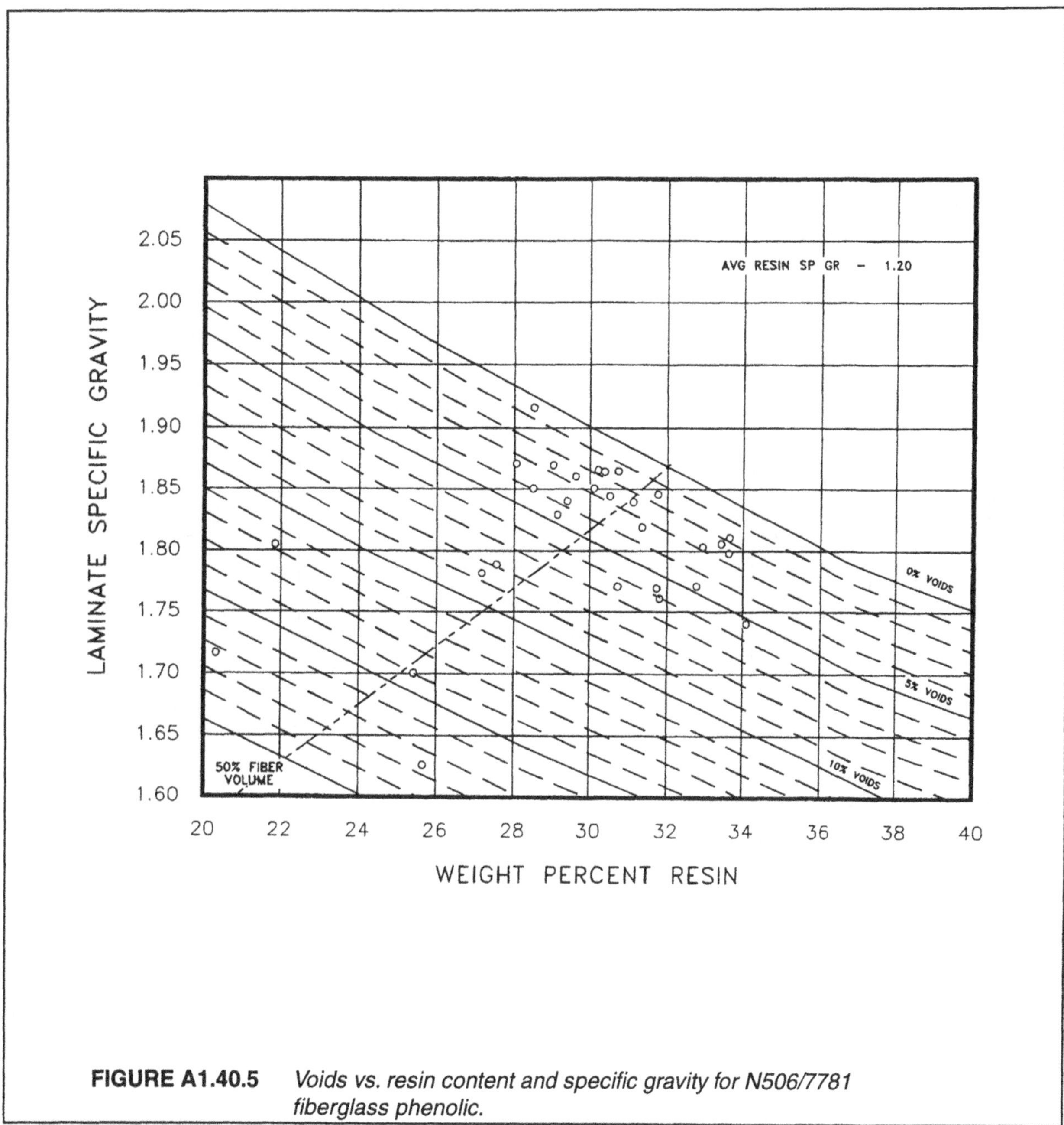

FIGURE A1.40.5 *Voids vs. resin content and specific gravity for N506/7781 fiberglass phenolic.*

APPENDIX A1

TABLE A1.110 *Summary of Mechanical Properties of Narmco 5505 Boron-Epoxy (100%-0° Direction) (Tentative).*

Fabrication	Lay-up: Parallel	Vacuum: 2 ins	Pressure: 50 ± 5 psi	Bleedout:	Cure: 1.5hr/ 350°F ± 10°F	Postcure: 2hr/350°F	Plies: 6
Physical Properties	Weight Percent Resin:	Avg. Specific Gravity:	Avg. Percent Voids:	Avg. Thickness: 0.005 in/ply			
Test Methods	Tension: Tab-ended	Compression: Sandwich Beam	Shear:	Flexure: 4 Point Loading	Bearing:	Interlaminar Shear: Short Beam	

Temperature		-67°F				75°F				260°F				375°F	
Condition		Dry		Wet		Dry		Wet		Dry		Wet		Dry	
		Avg	SD	Avg	SD	Avg	SD	Avg	SD	Avg	SD	Avg	SD	Avg	SD
Tension															
ultimate stress, ksi	0°	201.1				208.3				191.6				167.3	
	90°	10.5				8.7				6.5				3.3	
ultimate strain, %	0°	6390				6930				6660				6150	
	90°	3250				3710				4970				6920	
proportional limit, ksi	0°	141.8				175.5				140.0				79.5	
	90°														
initial modulus, 10^6 psi	0°	32.0				30.9				29.6				28.6	
	90°														
secondary modulus, 10^6 psi	0°														
	90°														
Compression															
ultimate stress, ksi	0°	482.3				378.0				303.3				143.9	
	90°														
ultimate strain, %	0°	13670				10830				8920				4466	
	90°														
proportional limit, ksi	0°	333.5													
	90°														
initial modulus, 10^6 psi	0°	35.7				34.8				34.6				35.8	
	90°														
Shear															
ultimate stress, ksi	0°-90°														
	±45°														

		-65°F Dry			75°F Dry			160° Dry		
		Avg	Max	Min	Avg	Max	Min	Avg	Max	Min
Flexure										
ultimate stress, ksi	0°									
proportional limit, ksi	0°									
initial modulus, 10^6 psi	0°									
Bearing										
ultimate stress, ksi	0°									
stress at 4% elong., ksi	0°									
Interlaminar Shear										
ultimate stress, ksi	0°									

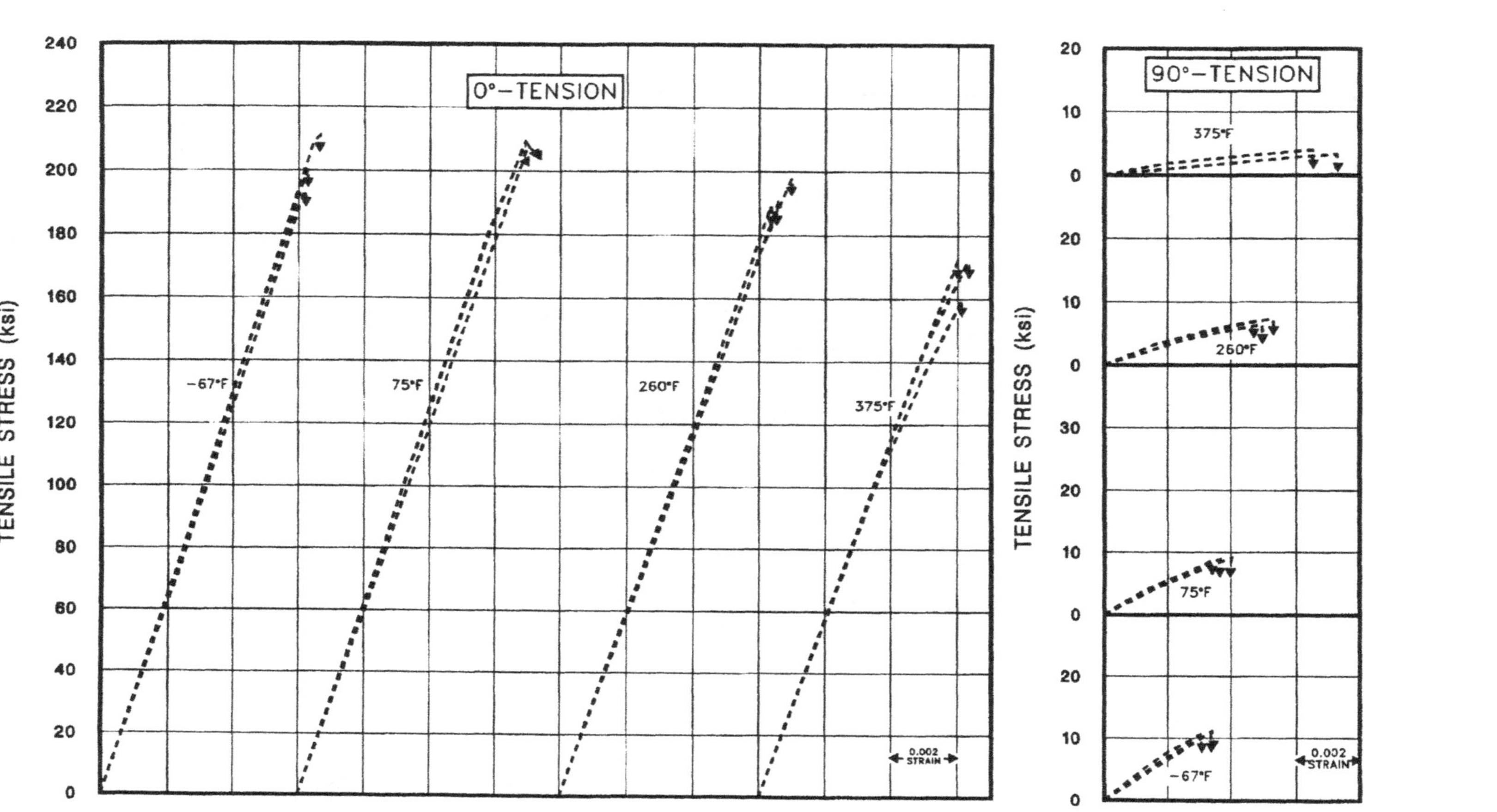

FIGURE A1.110.1 *Tensile stress-strain for AVCO 5505 boron/epoxy (100% - 0° orientation/50.3% to 35% fiber volume) loaded in the 0° and 90° directions. Individual tests shown.*

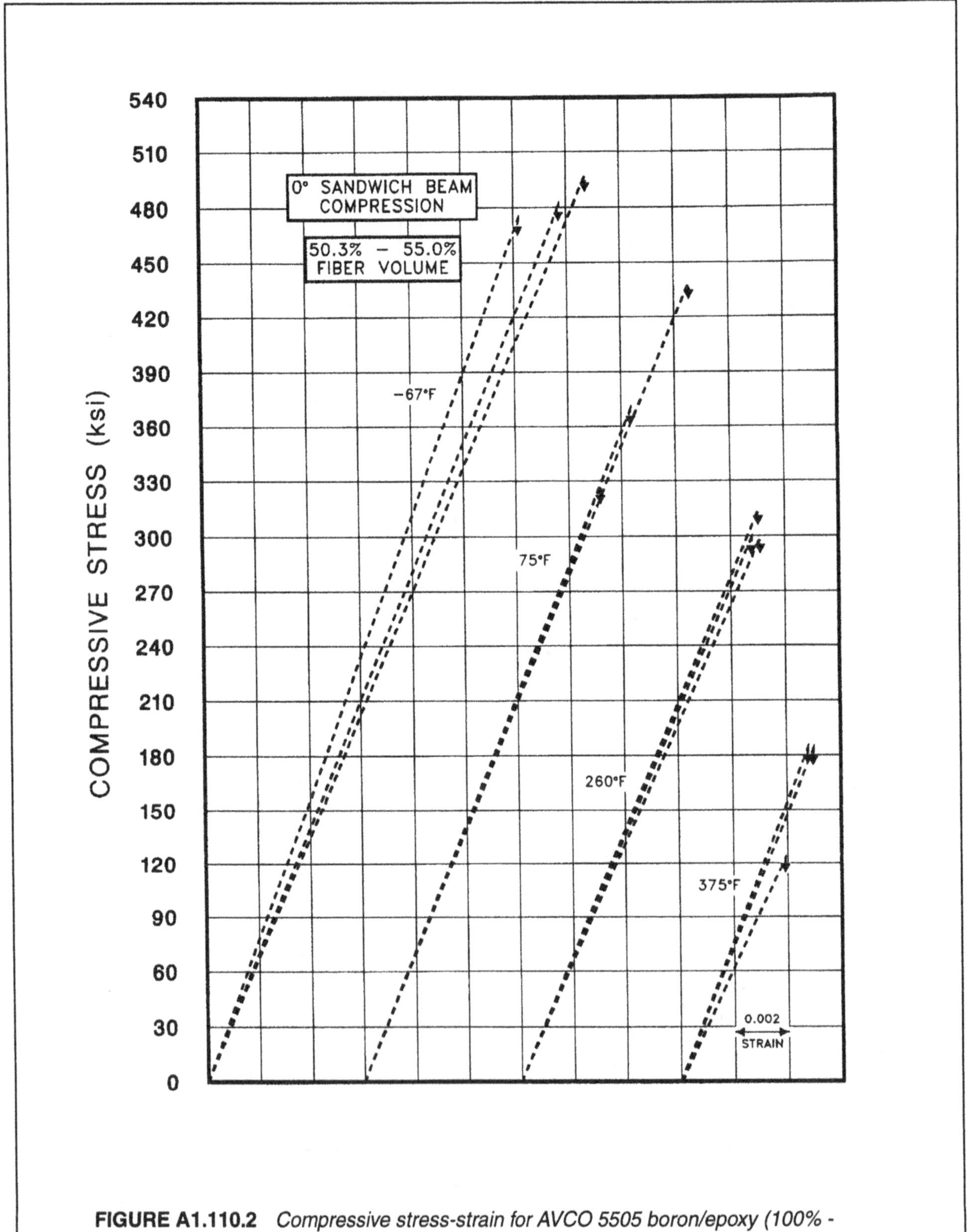

FIGURE A1.110.2 *Compressive stress-strain for AVCO 5505 boron/epoxy (100% - 0° orientation loaded in the 0° direction. Individual tests shown.*

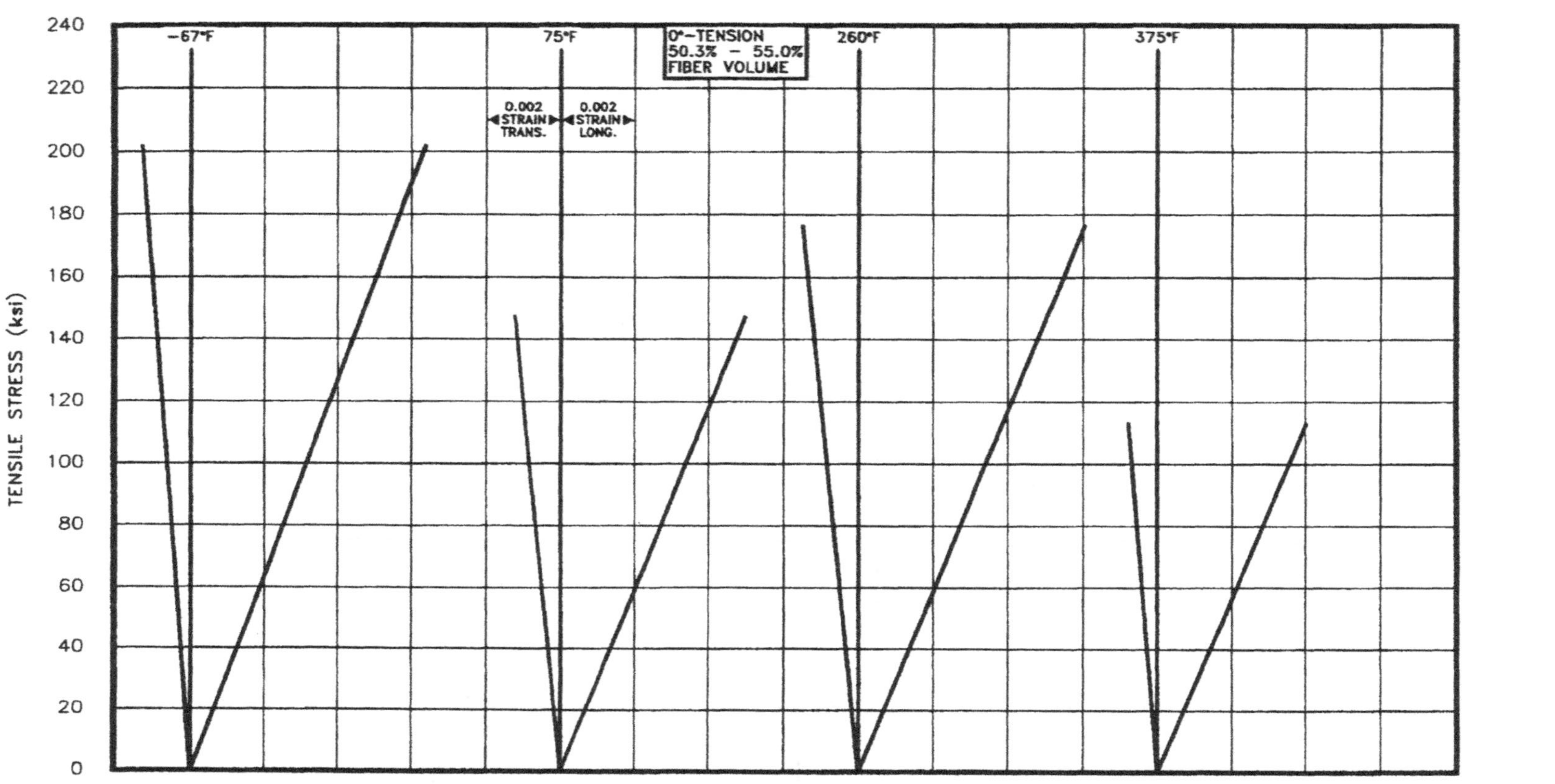

FIGURE A1.110.3 *Poisson effects for AVCO 5505 boron/epoxy (100% - 0° direction).*

APPENDIX A1

TABLE A1.111 *Summary of Mechanical Properties of Narmco 5505 Boron-Epoxy (0°-90° Crossply) (Tentative)*

Fabrication	Lay-up: [2(0/90)]S	Vacuum: 2 ins	Pressure: 50 ± 5 psi	Bleedout:	Cure: 1.5hr/ 350°F ± 10°F	Postcure: 2hr/380°F	Plies: 6
Physical Properties	Weight Percent Resin:	Avg. Specific Gravity:	Avg. Percent Voids:	Avg. Thickness: 0.005 in/ply			
Test Methods	Tension: Tab-ended	Compression:	Shear: Picture Frame	Flexure:	Bearing:	Interlaminar Shear:	

Temperature		-67°F				75°F				260°F				375°F	
Condition		Dry		Wet		Dry		Wet		Dry		Wet		Dry	
		Avg	SD	Avg	SD	Avg	SD	Avg	SD	Avg	SD	Avg	SD	Avg	SD
Tension															
ultimate stress, ksi	0°	99.9				103.9				98.5				91.9	
	90°	23.6				17.8				11.4				8.1	
ultimate strain, %	0°	5400				5710				5830				5780	
	90°	15850				24470									
proportional limit, ksi	0°	53.0				77.7				48.6				48.6	
	90°														
initial modulus, 10^6 psi	0°	18.9				18.0				17.5				16.5	
	90°														
secondary modulus, 10^6 psi	0°														
	90°														
Compression															
ultimate stress, ksi	0°														
	90°														
ultimate strain, %	0°														
	90°														
proportional limit, ksi	0°														
	90°														
initial modulus, 10^6 psi	0°														
	90°														
Shear															
ultimate stress, ksi	0°-90°	19.5				17.3									5.4
	±45°	65.7				63.7									33.3

		-65°F Dry			75°F Dry			160° Dry		
		Avg	Max	Min	Avg	Max	Min	Avg	Max	Min
Flexure										
ultimate stress, ksi	0°									
proportional limit, ksi	0°									
initial modulus, 10^6 psi	0°									
Bearing										
ultimate stress, ksi	0°									
stress at 4% elong., ksi	0°									
Interlaminar Shear										
ultimate stress, ksi	0°									

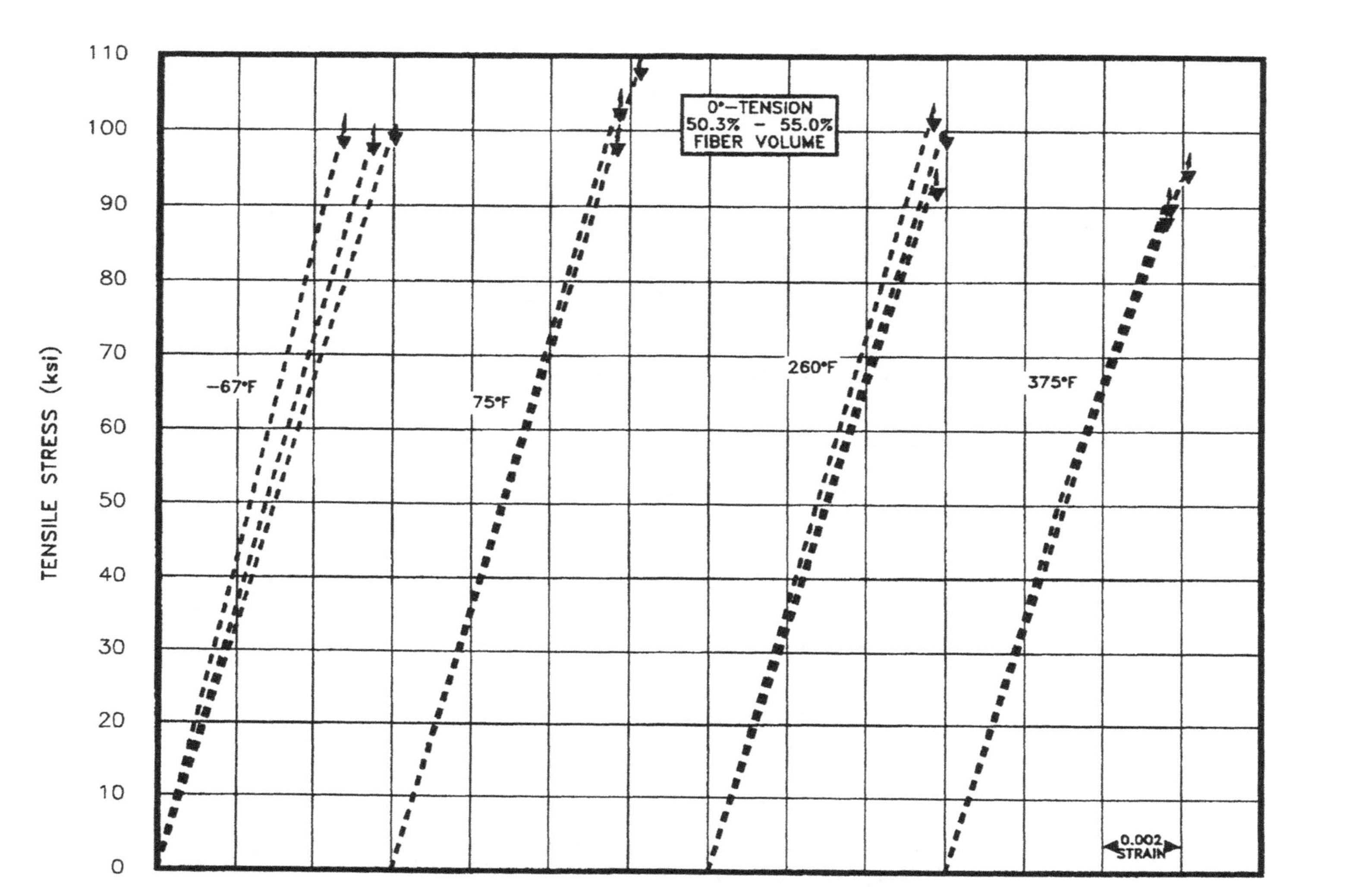

FIGURE A1.111.1(a) *Tensile stress-strain for AVCO 5505 boron/epoxy (0° - 90° crossply) loaded in the 0° direction. Individual tests shown.*

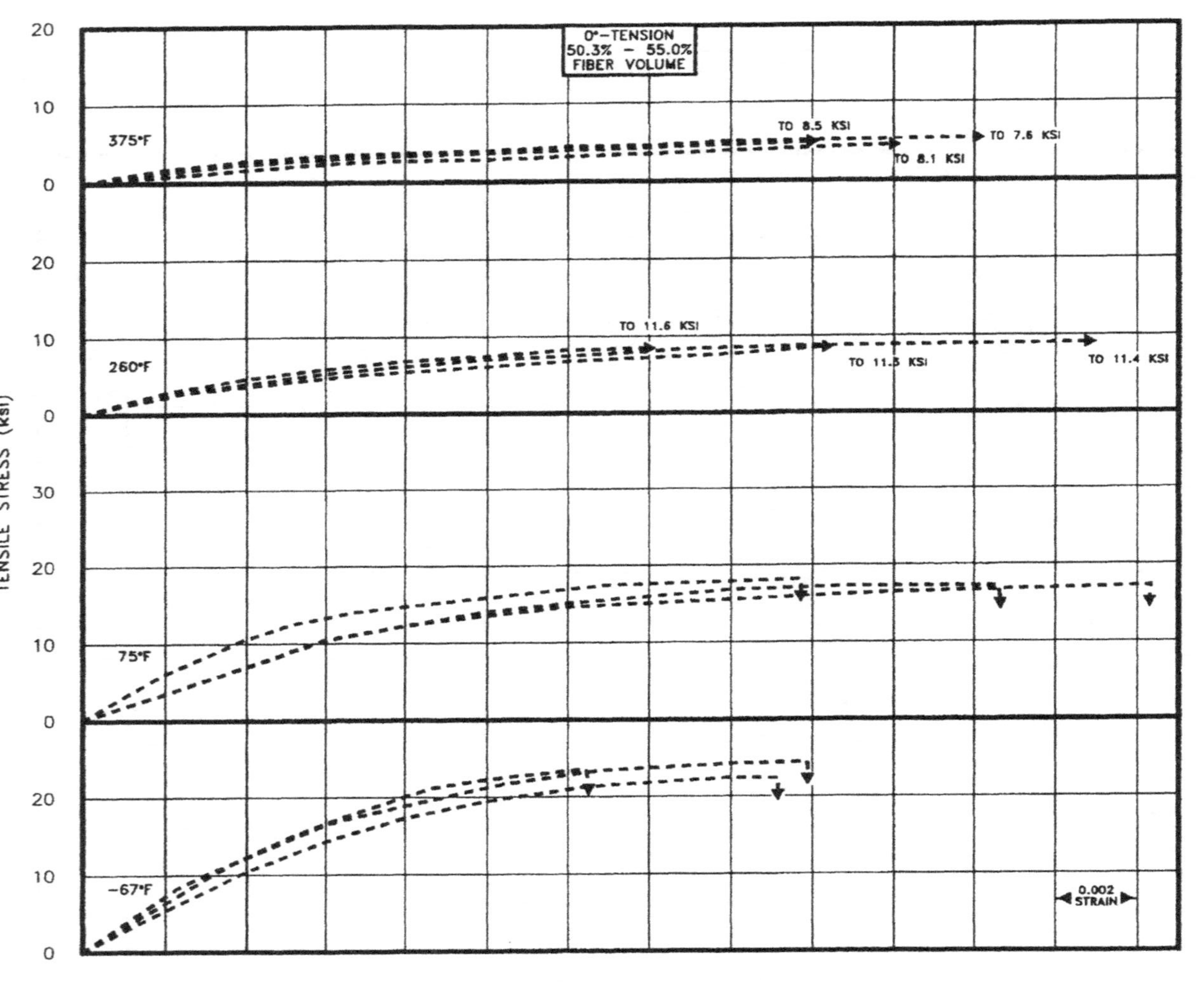

FIGURE A1.111.1(b) *Tensile stress-strain for AVCO 5505 boron/epoxy (0° - 90° crossply) loaded in the 45° direction. Individual tests shown.*

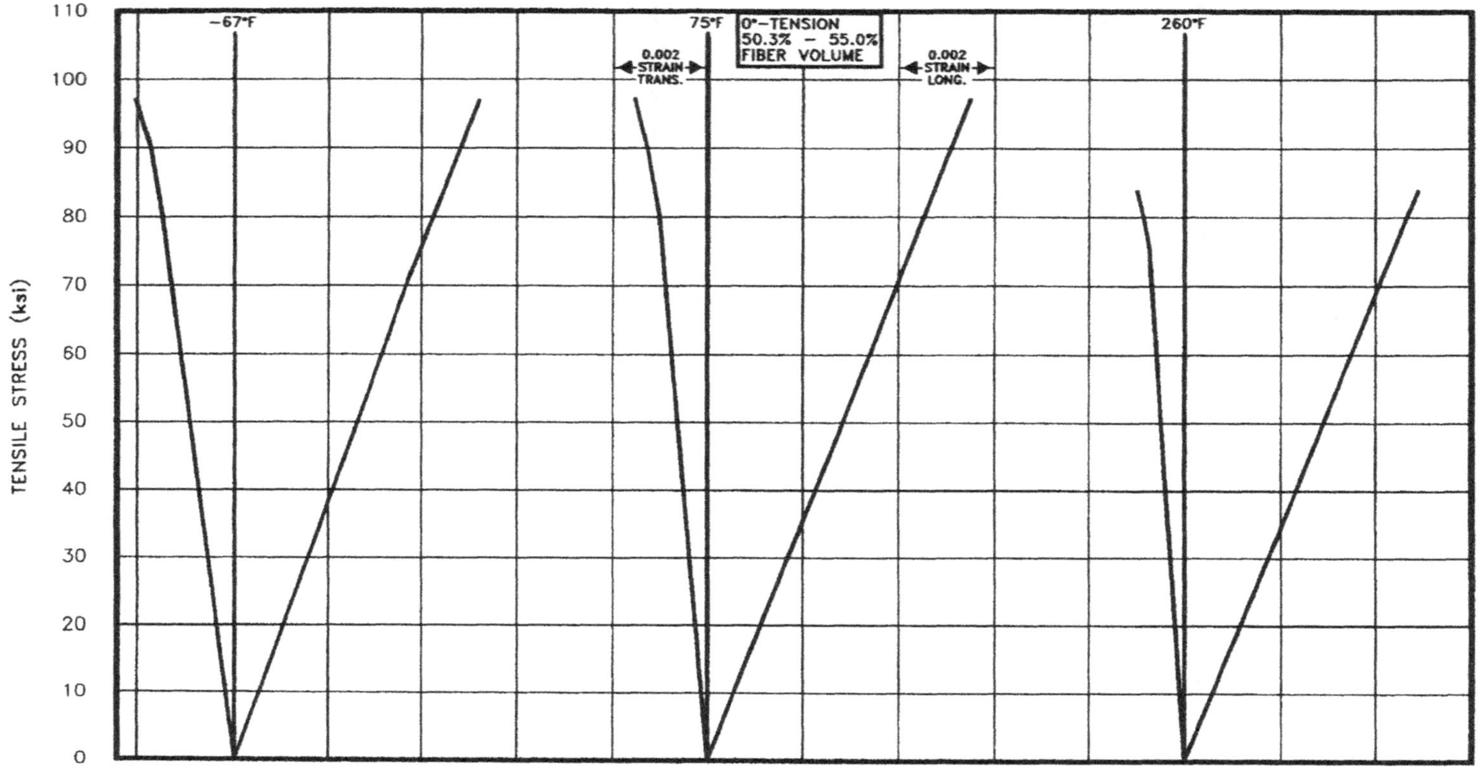

FIGURE A1.111.3 *Poisson effects for AVCO 5505 boron/epoxy (0° - 90° crossply).*

REFERENCES

A1.2 S. J. Dastin and others, *Determination of Principal Properties of "E" Fiberglass High Temperature Epoxy Laminates for Aircraft,* Grumman Aircraft Engineering Corporation, DAA21-68-C-0404, August 1969.

A1.3.4.1(a) ASTM D638, "Tensile Properties of Plastics," *Annual Book of ASTM Standards*, ASTM, Philadelphia, PA.

A1.3.4.1(b) P. D. Shockey and others, *Structural Airframe Application of Advanced Composite Materials*, General Dynamics, IIT Research Institute, Texaco Experiment, AFML-TR-69-01, **IV**, AF 33(615)-5257, October 1969.

A1.3.4.2 ASTM D695, "Compressive Properties of Rigid Plastics," *Annual Book of ASTM Standards*, ASTM, Philadelphia, PA.

A1.3.4.3 K. H. Boller, *A Method to Measure Intralaminar Shear Properties of Composite Laminates*, Forest Products Laboratory, AFML-TR-69-311, March 1970.

A1.3.4.4 ASTM D2733-68T, "Interlaminar Shear Strength of Structural Reinforced Plastics at Elevated Temperatures," *Annual Book of ASTM Standards*, ASTM, Philadelphia, PA (canceled January 15, 1986 and replaced by ASTM D3846).

A1.3.4.5 ASTM D790-70, "Flexural Properties of Unreinforced and Reinforced Plastics and Electrical Insulating Materials," *1971 Annual Book of ASTM Standards*, ASTM, Philadelphia, PA, 1971.

A1.3.4.6 ASTM D953, "Bearing Strength of Plastics," *Annual Book of ASTM Standards*, ASTM, Philadelphia, PA.

A1.4.5 G. C. Grimes and G. J. Overby, *Boron Fiber Reinforced/Polymer Matrix Composites - Material Properties*, Southwest Research Institute, January 1970.

INDEX

VOLUME 2

PAGE

PAGE

Material Property Data

CONCLUDING MATERIAL

Custodians:
Army - MR
Navy - AS
Air Force - 11

Review activities:
Army - AR, AT, AV, IE, MI, TE
Navy - SH
Air Force - 13
DLA - DH(DCMC-OF)

Preparing activity:
Army - MR

(Project CMPS-0163)